BLUE DARKER THAN BLACK

BLUE DARKER THAN BLACK

A THRILLER

MIKE JENNE

YUCCA

Yucca Publishing books may be purchased in bulk at special discounts for sales promotion, corporate gifts, fund-raising, or educational purposes. Special editions can also be created to specifications. For details, contact the Special Sales Department, Yucca Publishing, 307 West 36th Street, 11th Floor, New York, NY 10018 or yucca@skyhorsepublishing.com.

Yucca Publishing® is an imprint of Skyhorse Publishing, Inc.®, a Delaware corporation.

Visit our website at www.yuccapub.com.

10 9 8 7 6 5 4 3 2 1

Library of Congress Cataloging-in-Publication Data is available on file.

Jacket design by Haresh R. Makwana
Jacket photo credit: Gemini Mission courtesy of NASA

Print ISBN: 978-1-63158-066-6
Ebook ISBN: 978-1-63158-072-7

Printed in the United States of America

For Andrew Grant Hindman,
A young man who appreciates the treasures
found in words and numbers:
Dream big, as big as the sky, and follow fast after those dreams.

Author's Note

The Cold War is raging—a highly classified AF space program—in modified Gemini capsules—astronauts fly missions to intercept and destroy Soviet satellites suspected of carrying nuclear weapons. At Wright-Patterson Air Force Base in Ohio, project Blue Gemini is led by Air Force Major General Mark Tew (whose health is progressively failing) and his civilian deputy, retired General Virgil Wolcott.

Scott Ourecky joined the Air Force as a brilliant engineer and mathematician, but repeatedly failed the aptitude test to become a pilot. His recognized ability sees him take right seat in a flight to space with the most proficient pilot assigned to the Project, Major Drew Carson (who yearns to fly in combat in Vietnam).

A Delta Airlines stewardess, Bea Harper, daughter of an Air Force pilot killed in the Korean War, agreed to marry Ourecky only if he promises not to become a pilot.

Air Force Airman Matthew Henson is trained as a covert operative to support Blue Gemini's global search and rescue operation and is later dispatched to Africa to establish a contingency recovery site.

Two astronauts are killed in a launch accident during Blue Gemini's maiden flight.

An Air Force sergeant, Eric Yost, a would-be spy, monitors the Project's hangar convinced they are hiding UFO remains. Deep in debt from gambling losses, he plans to contact several magazines.

Soviet Lieutenant General Rustam Abdirov is selected by the Soviet General Staff to develop Krepost, as a nuclear orbital bombardment system. Seeking a powerful, compact computer for the Krepost, Abdirov gets a protégé, Major General Gregor Yohzin to arrange a GRU (Soviet military intelligence) operative, Colonel Felix Federov to steal a Gemini computer from a Smithsonian Institution warehouse in Maryland and for Major Anatoly Morozov to go to Ohio to investigate rumors that the Air Force is storing captured UFOs at Wright-Patterson Air Force Base.

Before Apollo 11 astronauts land on the moon, Ourecky and Carson are secretly launched into orbit to intercept and destroy a suspect Soviet satellite. Despite a major power failure on their spacecraft, they disobey General Tew's orders to return to Earth, and instead continue their mission. In so doing, they discover a previously unknown class of Soviet reconnaissance satellites. Upon their return to Earth, Tew grounds the pair.

1

OVER THE MOON

Flight Crew Office
Aerospace Support Project, Wright-Patterson Air Force Base, Ohio
2:35 p.m., Thursday, July 17, 1969

Pondering the most expedient solution to a complex equation, Major Scott Ourecky tapped his pencil eraser on the desktop, scratched his ear, and then reached for his favorite slide rule. He eased it from its cordovan leather sheath and admired it before placing it into action. The trusty Deitzgen slip-stick had served him well, from his college days in Nebraska to the ordnance research labs at Eglin, and now it had accompanied him all the way to orbit and back.

Since arriving at Blue Gemini over a year ago, he had come full circle on his tangential journey; he was now back to compiling calculation "cheat sheets" for future intercept missions and preparing formulas to be processed into computer programs. The only significant change was his environs; instead of laboring in the secluded depths of the Project's basement, crammed into a musty space only slightly larger than a broom closet, he was afforded access to the spacious Flight Crew office on the second floor. It was the inner sanctum of the inner sanctum, and by virtue of having flown into space, he now possessed a permanent passkey to the pilots' hallowed bastion. Perhaps only Bruce Wayne had a cooler asylum, but even that could be debated. Yes, Batman had his Bat Cave and a really groovy rocket-propelled car, but Ourecky's office came complete with a ticket to orbit, even though his all-access pass was temporarily revoked.

His attention was distracted by a news alert faintly emanating from a radio on his desktop. Anxious, he twisted the volume knob on the small AM receiver and twiddled with its flimsy telescoping antenna to optimize the signal. Apollo 11 had blasted off yesterday morning; its three astronauts were well on their way to the *moon,* and he diligently tried to stay abreast of the mission. Half-expecting bad news, he breathed a sigh of relief as he listened to an announcer state that the lunar crew had fired the engine of their Service Module to successfully execute a mid-course correction maneuver. They were due to arrive in lunar orbit on Saturday, and if all went well, Neil Armstrong and Buzz Aldrin would touch down on the moon on Sunday. Reflecting on the moment, Ourecky closed his eyes and leaned back in his chair. Men landing on the *moon.* It was almost too much to believe.

He opened his eyes to focus on a large color world map taped to the wall. A gift from Gunter Heydrich, autographed by all of the controllers in the Blue Gemini mission control facility, the chart was a mission tracking map for their flight in June. It depicted the graceful undulating parabolas of orbital paths tracing over the Earth's surface, as well as the far-flung contingency recovery zones that would have been their safe harbor if the flight had ended early. Ourecky set aside his slide rule and looked at the heaping backlog of worksheets yet undone. Here, he was immersed in his natural element, applying arcane mathematics and physics to define the paths of objects in space, but he reminded himself that it was not long ago that he had actually followed those parabolas as he orbited the Earth.

A plaintive coo disrupted his thoughts; he looked up to see a mottled gray dove perched on the red brick windowsill. He smiled at the bird through the soot-smudged windowpane. The skittish dove wagged its tail feathers and jerkily nodded its head in reply, as if acknowledging a secret shared between the two, and then quickly flittered away.

Ourecky longed to fly. As much as he didn't relish the thought of being jammed back in the Box for pre-mission simulations, he desperately yearned to go upstairs again. It didn't look to be in the cards, at least within the foreseeable future, since General Tew was holding fast to his vow that he and Carson would remain indefinitely—if not permanently—grounded. Despite Tew's reluctance, there was a faint glimmer of hope; Virgil Wolcott had quietly confided that the pair would fly within a year, perhaps even sooner, once Tew had sufficient time to calm down. In the meantime, he and Carson were tasked with ensuring that Crew Three—Parch Jackson and Mike Sigler—were adequately prepared to go up on Blue Gemini's next mission, which was currently scheduled for December.

Whether by circumstance or Tew's design, he and Carson rarely saw each other during working hours. Ourecky normally divided his time between the Flight Crew Office and working with the computer programmers on the fourth floor. While he toiled at his equations and paperwork, Carson spent most of his days in the simulator hangar, relegated to the unenviable position of perpetual CAPCOM. Tew apparently believed that Carson had instigated the impromptu mutiny on their flight—which was not entirely true—so the general seemed intent on doling the pilot's punishment from a heavier ladle. On the other hand, maybe Tew thought that Carson and Ourecky had grown *too* close and their disparate working conditions would eventually drive a wedge between the two.

Ourecky and Carson still met to run and work out at the base gymnasium long before dawn. Occasionally, they met for dinner or drinks after duty, but that was becoming progressively more rare, since Carson was typically worn out by the end of his twelve-hour shifts covering the CAPCOM desk.

Although Ourecky was notionally being punished, the arrangement lent him a huge degree of freedom. So long as he stayed ahead of the programmers, which wasn't a particularly difficult feat, he was essentially free to set his own routine, so he normally worked long days—twelve to fourteen hours at a stretch—from Monday morning until Thursday afternoon. Consequently, every Thursday, he was able to zoom home before the afternoon rush and be there when Bea returned from her weekly five-day circuit of flying back and forth from Atlanta. Every weekend was effectively a three-day vacation, and they took full advantage, savoring every minute they could spend together.

Bea. He sighed as he looked at a framed picture of Bea and him, taken last Christmas at his parents' house in Nebraska. In the month's time since he had returned from orbit, they were finally able to live almost like a normal married couple. With his more predictable schedule, they had drawn considerably closer, finally sharing the kind of emotional intimacy that had been severely lacking in the past few hectic months. So, as much as he wanted to fly again, he wasn't anxious to sacrifice the progress he had made at home, and he hoped that when the time came, he would be able to strike an effective balance.

Simulator Facility, Aerospace Support Project
4:30 p.m.

Major Drew Carson adjusted his painfully tight headset, hoping to alleviate some of the aggravating pressure on his ears, and then kneaded his throbbing temples. It felt like his head was clamped in a cast iron bench vise that was being gradually torqued down on his ears. Just eight hours into another twelve-hour stint at the CAPCOM desk, he was miserable. The rest of the nation was over the moon, fixated on Apollo 11's historic flight, and yet here they were, cooped up in this hangar, preparing for a secret space mission that was still months away.

Struggling to remain alert, he cracked his knuckles, took a sip of lukewarm coffee, and then glanced at the mission clock on his console:

the GET—Ground Elapsed Time—was 38:12:18. The two men in the Box had been conscious for over thirty-eight grueling hours; by now, they were long past frazzled. Having endured countless hours in the Box, Carson knew well their agony. Time was their unmerciful enemy. Without sleep, their brains were turning to mush. What were once logical thoughts were now mired in a muck of exhaustion and distraction. The stiff seat backs were unrelenting, but by now the aching cramps in their spines had been replaced by numbness punctuated by sporadic sharp spasms. Thirty-eight hours, and still four more yet to go.

At this point, they were likely suffering from at least mild hallucinations. Carson knew what they were experiencing; at one point or another, he had seen and felt it all. They would reach to throw switches, only to see them suddenly vanish. The rest of their instruments would also refuse to stay put; it was almost impossible to maintain a disciplined scan as their dials and indicator lights swam around on the gray face of their control panels. The "eight-ball" attitude indicator would spin and dance like a dervish possessed. A phantom buzz of static would plague their earphones, periodically interrupted by faint garbled voices demanding an immediate response. Leering gremlins would flagrantly lurk in the cabin, taunting them as they tugged and yanked at the critical wiring behind the breaker panels.

Carson pitied Jackson and Sigler. Outwardly, he hoped for their success, but secretly he longed for their failure. The plain truth was that they weren't ready, and they weren't going to *be* ready, not next week, nor in a month, nor six months from now when the next mission was scheduled to fly. It was futile to believe otherwise.

It was their own fault that they hadn't grabbed any rest. Unlike him and Ourecky, they hadn't developed a working rhythm to allow each other to doze for a few minutes at a time. Even brief catnaps would fend off the onset of hallucinations, but the sleep deprivation they endured was reflective of a much more significant deficiency. Taken as individuals, they were tremendously proficient test pilots, but despite their personal competence, the two men just couldn't work effectively as a cohesive team. In Wolcott's homespun Oklahoma argot, Jackson and Sigler just didn't geehaw.

Their incompatibility was no closely guarded secret; everyone here knew it and talked openly about it. Even Jackson and Sigler lamented their shortcomings and candidly expressed doubts about whether they could ever hope to accomplish that which had seemingly been so easy for Carson and Ourecky.

The root cause of their problems was that they simply didn't trust one another. They were civil enough outside the Box, but once they were locked inside the simulator and a mission profile was in progress, they bickered and second-guessed each other almost constantly. Jackson was almost always skeptical of Sigler's maneuver solutions and regularly insisted that the right-seater rework his calculations. When they failed to hit their marks on executing the maneuvers, Sigler berated Jackson for not correctly flying his fixes. As a result, they just slipped further and further behind. If that was not enough, the two perpetually feuded over the cabin environmental controls, so the uncomfortable cockpit was always too warm or too cold for at least one of them.

As he spent his days eavesdropping on their grumbling and angry rants via a hot VOX mike, Carson felt less like a liaison, and more like a marriage counselor for a hopelessly doomed union. When he was able to talk to them during contact windows, he tried his best to calm them down and nudge them in the right direction. Although he was supposed to be a strictly impartial intermediary, he found himself often gently prodding and coaxing the quarrelsome pair towards the solutions that they should be developing entirely on their own.

To make matters even worse, Tew had decreed that they adhere to the excruciatingly strict rules for a "hard" lock-in—no stretch breaks or latrine calls—just as Carson and Ourecky had endured back in January. Glancing over his shoulder towards the back row of consoles, Carson observed that both Tew and Wolcott were present. It wasn't uncommon for Wolcott to frequently linger in the hangar, but Tew almost never ventured out of the main building. Carson was curious why he seemed so concerned now, since he rarely showed any more than a passing interest in previous simulated missions.

Watching the clock, he knew that a decisive moment was near. According to the simulated mission's profile, the crew had been out of contact for the past seventy-two minutes. During this interval, they should have executed a significant phase shift maneuver.

Carson watched the clock as he listened intently. Finally, nineteen seconds past the designated start of the contact window, he heard Jackson's hoarse voice over the intercom: "CAPCOM, this is . . . Scepter Three. We, uh, executed the phase shift burn as scheduled. Ready for . . . data download?"

"Scepter, this is CAPCOM," replied Carson. "Go for download." A few seconds later, he watched a small green light blink on his control console, indicating that telemetry data was being "received" from the simulated spacecraft notionally passing overhead in orbit. In reality, Jackson had misjudged the timing at a crucial juncture, approximately an hour prior, and it was unlikely the crew could recover from the blunder in sufficient time to make their intercept.

As the download continued, Carson asked, "Scepter, are you ready to copy reentry guidance?"

Barely coherent, Sigler answered, "We are . . . ready to copy."

Knowing that their hands were painfully cramped, Carson slowly read the current instructions for primary and contingency reentry, concluding with, "Scepter, you are still go for reentry to PRZ One-Two on your twenty-eighth rev. How copy?"

"CAPCOM, this is Scepter. We copy PRZ One-Two on Rev Two-Eight," stated Sigler. He read back the contingency reentry data in an agonizingly slow monologue, occasionally slurring his words.

"Roger, Scepter. Good copy. Do you have visual or radar acquisition on the target?" asked Carson. Even though he already knew the answer, he watched the clock carefully; the contact window was due to slam shut in just a few seconds and they wouldn't talk again for almost another hour.

Jackson's exasperated tone conveyed far more than his words. "Uh . . . negative," he muttered. "Uh . . . we have *not* acquired the target . . . we . . ."

Replicating the spacecraft's passage out of radio range, there was a faint warbling noise before the intercom abruptly clicked off. Switching off the voice loop, Carson scowled as he opened a black leatherette binder. Although it was the responsibility of Heydrich and his controllers to analyze the telemetry data and issue a formal verdict on whether the crew had been successful with their last burn, Carson could refer to a collection of graphs—formulated by Ourecky—to quickly assess the mission's status.

Balancing the binder in his lap, he flipped through a series of predicted progress diagrams until he found the one that corresponded to the last maneuver. Although he had long since memorized the numbers and their relevance, he traced his finger along a red-penciled curve on the graph and double-checked the parameters on the X and Y axes. He scratched a faint pencil mark where the two graph lines converged, and saw that the crew was woefully behind the curve. Even if the hapless pair somehow possessed an inexhaustible stock of fuel and consumables, they could not *possibly* compensate for their lapses and still salvage the mission. Try as they might, the deed was undone and would remain so.

Carson quietly cursed as he considered the consequences; at this point, save for abandoning the profile outright, the only option was to call a restart. A restart meant that this ordeal would drag on through the weekend, so Carson had little else to look forward to except three more days with his sore butt glued to this uncomfortable chair. Frowning, he looked towards the rear of the room and slowly shook his head at Virgil Wolcott.

4:42 p.m.

Leaning over a console, Wolcott nodded solemnly at Carson as he resisted the urge to smile. He gazed towards Tew, who was standing alongside him. In his haste to ground Carson and Ourecky, Tew seemed absolutely intent on proving that Jackson and Sigler were sufficiently competent to fly the next mission. In the meantime, the last mission's serendipitous success was whispered along the elite circuit of high-ranking Air Force officials who were aware of Blue Gemini. Congratulatory calls and accolades—all couched in very vague and nonspecific language—continued to pour in.

And now, there was yet another wrinkle. It was highly likely that they would launch another mission before December. Only days after the Project's triumphant flight in June, Admiral Tarbox had apparently persuaded some extremely powerful people of a pressing need to employ the Gemini-I to destroy a new Soviet satellite. While Wolcott and Tew were privy to only sketchy details, the proposed mission was supposedly a hypersensitive requirement that would mandate a drastic shift to the flight schedule, perhaps even requiring a launch in a matter of weeks. They should learn more tomorrow morning, when the Ancient Mariner and his retinue arrived for a meeting.

Unless he changed his mind in the interim, Tew intended to go into tomorrow's assembly with his second string in tow. As the prime—and theoretically, the *only*—flight crew available for the mission, Jackson and Sigler were slated to attend, but Tew was being so stubbornly inflexible that he was fencing off the briefing from Carson and Ourecky.

Although he had known Tew for decades, his friend's obstinate behavior baffled Wolcott; the only logical explanation was that he intended to use the Crew Three's failures as a foil to fend off Tarbox's emergent mission.

Now, Tew was clearly chagrined, obviously grappling with the inevitable reality that Jackson and Sigler weren't going to be ready to fly anytime in the foreseeable future. It didn't matter whether the mission was launched in December or next week.

Opening a foil-lined packet of Red Man chewing tobacco, Wolcott nudged Tew's elbow. "They ain't gonna make it on this run, Mark," he said bluntly. "Sorry to disappoint you, but it ain't happenin'. Do you want me to terminate early and reset them, or just shut it down altogether?"

Audibly gnashing his teeth, angrily glowering at the Box as he came to grips with grim reality, Tew did not reply.

"Mark? Did you not hear me, buddy?"

"I *heard* you, Virgil," replied Tew. His muted voice hardly masked his frustration. "We're still scheduled to meet with Leon Tarbox tomorrow?"

"Yup," replied Wolcott. "The Ancient Mariner and his minions are s'posed to be here at zero nine. Maybe then the great mystery will be revealed."

As if struggling to swallow something particularly distasteful, Tew grimaced and then muttered, "Well, then shut down this damned fiasco, Virgil. We're just wasting our time right now. Yank those two boys out of the Box, and . . ."

"And *what*, boss?" implored Wolcott, stuffing a thick wad of damp tobacco into his lower lip.

"You know *what*, Virg. Inform Carson and Ourecky that I want them to sit in tomorrow morning. As the back-up crew, but *no* more than that."

"Will do, Mark."

As Tew stormed out of the hangar, Wolcott picked up the phone and called Carson in the front row. "Pour the coffee on the coals and fetch the mules, Carson. We're *done* for the day."

Holding the receiver to his ear, Carson faced Wolcott and nodded.

"That ain't all, buster," added Wolcott. "General Tew wants you and your fellow cowpoke to attend a high-level meeting tomorrow morning. Be advised that you are now out of Purgatory, but just barely. Congrats, Carson: you've been elevated to back-up crew status."

Carson smiled broadly, replied consent, and hung up the receiver.

Wolcott grinned slyly, turned towards Heydrich, and quietly said, "I think you owe me a sawbuck, Gunter."

"You're a very shrewd judge of character, Virgil," growled the German engineer, shaking his head as he drew out his wallet. He fished out a five-dollar bill and palmed it to Wolcott. "It's almost uncanny."

"Not really," drawled Wolcott, folding the note before slipping it into the pocket of his denim cowboy shirt. "I just know my horses, pard."

**Aerospace Support Project, Wright-Patterson Air Force Base, Ohio
8:35 a.m., Friday, July 18, 1969**

Wearing a suit and tie, as instructed, Ourecky sat patiently as he waited
for the proceedings to begin. Although he was curious about the meeting and what might eventually come of it, he wasn't especially happy
about missing out on precious time with Bea. All he knew was that
Carson had told him that they were back on the flight roster—at least
on a conditional basis—as the back-up crew for Jackson and Sigler.
He wasn't sure what that entailed or whether they would resume their
normal training regimen, but the development almost certainly implied
that his weekly routine would soon change. Nervously tapping his fingertips on the table, he glanced towards Carson; seated to his left, as
always, the handsome pilot grinned like he had just won the fattest
jackpot in Vegas.

Ourecky glanced up as the door creaked open. Resembling haggard
survivors of the Bataan death march, Jackson and Sigler slowly staggered into the conference room. Since Tom "Big Head" Howard had
perished in February's launch catastrophe, Jackson was the tallest of
the pilots; standing erect, he would scrape close to five-eleven, but now
he was hunched over as if in abject pain. Built like a sprinter, he was
thin—almost painfully so—with narrow shoulders and hips. His dark
brown hair was cut in a flattop, which looked a week past due for a trim.
Mike Sigler was two inches shorter than his command pilot, with a pug
nose, closely shorn receding blonde hair, and the solid physical build of
a collegiate wrestler.

Sighing in relief, the two men slumped into their chairs. Obviously
slow to recover from the Box's stresses, they were jittery and their eyes
were bloodshot. To complement their horrendous physical appearances,
both men seemed steeped in shame and humiliation, like a pervasive
stench of body odor that could not be showered or scrubbed away.
Like Ourecky, they obviously weren't sure of the meeting's purpose and
probably suspected that they might be bumped from December's flight
as a result of their failure yesterday. Even if they weren't immediately

benched, Tew might be calling them all together to issue Crew Three a stern warning of what might happen if they didn't get their act together.

Sigler gestured at a pitcher of ice water at the center of the table; in a thin, raspy voice, he murmured, "Please . . ."

Barely able to lift the vessel, Jackson poured a glass and slid it to his counterpart. Grasping the tumbler between two trembling hands, the right-seater quickly slurped it down and then quietly thanked the pilot. Sigler wore a thick pad of blood-tinged gauze taped to the inside of his left wrist and forearm. It was the telltale wear mark where the metal glove cuff ring persistently chafed the skin. As a pre-Box prophylactic measure, Ourecky had long ago learned to protect that sensitive flesh with a generous wrapping of white adhesive bandage tape; he was very surprised that Sigler had not arrived at the same solution.

"I guess we really blew it yesterday," mumbled Jackson, looking towards Carson.

"Not *we*," asserted Sigler apologetically. "It was all my fault, Parch. I let you down."

Ourecky was surprised. Their conduct was in marked contrast to what Carson had related to him earlier. Soft-spoken and contrite, the pair showed absolutely no sign of the surly behavior and incessant grousing Carson had described.

"Just another bad day in the Box, guys," noted Carson, twisting an end of his neatly trimmed moustache. "We've all had them. No need to dwell on it."

After several minutes of uncomfortable silence, they were joined by Tew, Wolcott and Heydrich. Only the two generals seemed to have any clue about what was to ensue, and they remained reticent.

After the three men sat down, Jackson quietly spoke: "I'm sorry we let you down, General. . . . We're ready to accept any consequences . . ." The buzz of a desktop intercom interrupted his apology.

Cupping his ear as he leaned in the direction of his desk, Tew shook his head to cut off Jackson's apology. Through the intercom, his aide announced, "The admiral's plane has arrived, sir. They'll be up shortly."

"We're waitin' with bated breath," snapped Wolcott.

"Gentlemen, let's set aside what happened yesterday and focus on matters at hand," instructed Tew. "Let's just see what Tarbox and his people have to say. Regardless of what it is, I want everyone to remain calm and keep their opinions to themselves. General Wolcott and I will do all the talking."

"Yup," added Wolcott. "Unless we specifically call on you, we don't need anyone pipin' in."

A few minutes later, led by Tarbox, the Navy contingent arrived and took their places at the table. Obviously following the admiral's fashion sense, the four members of his entourage were attired in almost identical snug-fitting off-the-rack Brooks Brothers suits in either black or dark gray. With matching white shirts and solid-color ties, the staidly dressed monochromatic clique could be readily mistaken for a gaggle of accountants or a squad of FBI agents.

Ourecky was mildly surprised to see that Ed Russo accompanied Tarbox's group. He was aware that after the tragic launch accident in February, Russo—now a full-fledged lieutenant colonel—had returned to the Manned Orbiting Lab project in California. According to rumor, when the MOL effort was summarily cancelled in June, Russo was shifted to a temporary assignment within a classified Navy effort overseen by Tarbox. Ourecky looked towards Carson; seething, the pilot wasn't very adept at concealing his festering scorn towards Russo.

Seated at the admiral's right hand like a favored son, Russo had obviously ingratiated himself to the Ancient Mariner. Like Admiral Rickover, the autocratic overseer of the Navy's nuclear program, Tarbox was granted immense latitude in handpicking officers for critical assignments. He subjected each prospective candidate to a lengthy series of excruciating interviews to assess their technical knowledge, judgment, and personal reliability. Consequently, it was a virtual certainty that Russo—even though an Air Force officer—had endured the same gauntlet as the other men in Tarbox's inner circle of trusted advisors.

Ourecky studied Tarbox. He knew him from his days in El Segundo but mostly just by reputation; he could tally their actual encounters on one hand and still have fingers left over. The acerbic admiral always reminded him of a malicious elf from a childhood fairy tale; he seemed like he would be far more comfortable in a lofty room atop a castle's tower, gleefully spinning straw into gold in exchange for some desperate damsel's firstborn.

After exchanging cursory greetings with Tew and Wolcott, Tarbox cleared his throat and curtly nodded at Russo.

"This will be a joint Navy-Air Force venture," declared Russo, speaking on cue. He paused to solemnly hand neatly bound briefing books to Tew, Wolcott, and Heydrich. "The objective is to interdict a Soviet maritime radar surveillance platform—Object 4201—launched in May."

Scrutinizing a diagram that depicted the new satellite alongside the second stage of a Titan II booster, Wolcott whistled through his chipped teeth. "Whew . . . that's a mighty big critter."

"*When?*" demanded Tew tersely. With one hand, he perched black-framed reading glasses on his florid nose as he quickly leafed through the briefing book's pages.

"No later than mid-September, General," replied Russo.

September? thought Ourecky. *Surely this has to be some sort of joke.*

"*Out* of the question," replied Tew, slamming his book shut as he turned his attention to Tarbox. "I thought you were coming here with something substantive, not some convoluted pipe dream. I'm not going to rush our boys into harm's way, particularly when this damned thing may be nothing more than a discarded booster."

"This platform is very real, Mark, and *very* dangerous to our national security," growled Tarbox. "And it's crucial that we scuttle it as quickly as possible. You need to set aside your qualms because I can *assure* you that this mission will fly."

"That's all well and good, Leon, but you're assuming that we will have hardware available to execute your mission. That's not necessarily the case."

As if on cue, Russo pulled a document out of a folder and slid it across the table to Tew. "General, this is an inventory of what you currently have on hand at the HAF in San Diego," he asserted arrogantly.

"Including spare parts and back-up flight computers. You currently have two complete mission-ready stacks. One is ready for encapsulation and transit to the PDF at Johnston Island."

Ourecky resisted the urge to shake his head. He despised Russo almost as much as Carson did and just could not comprehend how such a slimy character could have evolved into such a disruptive force. In his brief time as a liaison officer here, he had obviously gleaned a tremendous amount of inside information and was now exploiting it to the Project's detriment. A wooden steed jammed with Greek warriors probably couldn't wreak nearly as much havoc as this erstwhile emissary.

As the men discussed what equipment was available and what was not, Ourecky read the pages of Wolcott's briefing book as the retired general slowly flipped through them. Gathering all the pertinent facts about orbital inclinations and timing, he felt confident that he and Carson could execute the mission with minimal preparation. In fact, the profile was *so* similar to June's mission, they could probably launch as soon as the hardware was ready, tomorrow if necessary. Despite this, Blue Gemini was commanded by General Tew, and if he was reluctant, he obviously had good reason.

Glaring at Russo, Tew pointed at Jackson and Sigler. "There's another flaw with your plan," he said. "*If* we elect to undertake your mission, these two gentlemen will be the crew to fly it. As it stands, they have been training for a mission scheduled to launch in December, and they are not yet certified for *any* flight, much less one that will be executed in less than two months. So even though we may possess the hardware, we don't have a ready crew to scramble."

"If that's your prime crew," croaked Tarbox, looking towards Carson and Ourecky. "Then who are these other two?"

"They're our, uh, *back-up* crew," replied Tew. "They haven't even started training for the mission in December, so they're not certified either."

"Carson and Ourecky flew last month," noted Russo smugly, as if he had actually contributed something momentous to that effort.

"*You're* Carson and Ourecky?" growled Tarbox, extending a hand to shake theirs. "Congratulations! That was a fine piece of flying, you two. *Superb* work."

"We're mighty proud of them," observed Wolcott. "Top-notch hands, they are."

Russo gestured towards Ourecky and added, "Admiral, Major Ourecky took *the* picture." As if Tarbox needed reminding, Russo produced the now infamous photograph depicting the brass data plate on the Soviet reconnaissance satellite previously known as Object 2368-B.

"Amazing," blurted Tarbox, covetously examining the glossy print. He seemed to be on the verge of drooling. "*Absolutely* amazing."

Carson glared at Russo with unbridled malice, as if he were summoning daggers and hatchets to fly from his eyes and into the shiny forehead of his nemesis.

Sensing Carson's barely latent hostility, Wolcott chuckled and then drawled, "Yup, Ourecky snapped *the* picture, but let's not forget that our man Carson here played a *significant* role in making that happen."

Nodding, Tarbox handed the photo back to Russo. "And they're your *back-up* crew for this mission, Mark?" he asked, looking askance at Tew. "Since time is so short and they already have operational experience, why don't you assign them to fly my mission?"

"I'll take that under advisement," answered Tew. "But unless there's some compelling reason to convince me otherwise, *Jackson* and *Sigler* will fly the next mission. If they're ready to go in September, fine, but otherwise . . ."

"We'll *just* see about that," scoffed Tarbox.

"So what's with the danged ants in your pants, Leon?" Wolcott sneered. "Shucks, it ain't like that big ol' satellite is going anywhere anytime soon. Why are you so riled up to whang it?"

"Good point," added Tew. "Why don't you enlighten us?"

"There's a major fleet exercise called 'Operation Peacekeeper' scheduled for mid-September in the North Atlantic," divulged Tarbox, examining his watch. "It involves an evaluation of some new anti-submarine warfare technology and procedures, and we would prefer that the Soviets weren't watching over our shoulders. We want this monster knocked down as expeditiously as possible."

"That's fine," replied Tew. "But I don't see how this supposed time-sensitive issue merits doing business in such a haphazard manner. Sure, we might be able to do this, but at what potential cost? If you're so intent on safeguarding your new ASW technology, why don't you delay the tests or at least conduct them somewhere that it's less likely you'll be monitored?"

Not responding, the volatile admiral fumed for a few seconds before vowing, "Mark my words, gentlemen, this *will* be done. The faster you accept that notion, the more time that you will have to prepare." He abruptly stood up, and his entourage quickly gathered their materials and followed suit. Before leaving the room, he added, "My staff will draft a memorandum of understanding to spell out our specific requirements and expectations. It will be delivered here on Monday morning. Good day, gentlemen."

Tew stood up as the last of Tarbox's protégés departed. Grasping his stomach, he ambled slowly to his desk, extracted a blue flask containing milk of magnesia, swigged straight from the bottle, and replaced it. His face was pale, almost entirely without color, and beads of sweat dotted his forehead.

"This is *not* good," he declared, regaining his seat at the conference table. "I'm sure you're all aware that the admiral wields a lot of power and can exert a lot of influence, and I'm confident that he will relentlessly pursue this fiasco until he either gets his way or he's slapped down. At this juncture, we have to assume that he'll be successful, but I'm also confident that I can present a strong argument to adhere to our current flight schedule."

"That's fine, pard, but what if that sumbitch pushes hard enough to force our hand?" asked Wolcott. "Whether we cotton to it or not, we may end up firing earlier than later."

Tew wiped his glistening forehead with a handkerchief before replying, "If that happens, then we'll react accordingly. In the meantime, we will make sure that we have the time-sensitive pieces in place. With that said, here's my plan. Virgil, call the HAF in San Diego and make sure that they're ready to complete encapsulation of the stack and load it on the LST on extremely short notice."

"Consider it done, boss."

"And Virgil, since we're really not sure when this thing will fly, I suppose it goes without saying that we'll need to *drastically* accelerate training."

"Gunter, can you restart this morning?" asked Wolcott reluctantly, looking at his watch.

Swallowing, Heydrich answered, "*Ja*, but . . ."

"Wonderful. Gunter, after we clear out of this corral, ring up your boys and tell them to fire up the Box. Also let them know that they'll be working this weekend, and probably the next several weekends to come. Jackson, I want you and Sigler to skedaddle straight over to the hangar and jump right back into the saddle. You'll knuckle down until you get it right."

The weary pair looked as if they had been sentenced to a firing squad. Ourecky was sure that he saw tears welling in Sigler's bloodshot eyes.

Wolcott slid one briefing book towards Jackson and the other towards Carson. "Before any of you walk out of here, take some time to bone up on what the Navy is hankering to do."

Looking towards Carson and Ourecky, Tew emphatically declared, "For you two gentlemen, the only change is semantic. Yes, you are now elevated to back-up crew status, but I'm making that change strictly to ensure redundancy. Don't delude yourselves: Jackson and Sigler *will* fly this mission. I'm only showing you some temporary leniency because I want you two training and working *together* again. Do you understand?"

Carson and Ourecky nodded glumly.

Wolcott looked towards a freshly printed TELEX on Tew's desk and noted, "You know, Mark, maybe you should accept that invitation. A trip outside the office might do you a world of good, and you would be right smack in the middle of history as it's being made."

"As if we're not already?" replied Tew.

"You know what I mean, boss. If nothing else, since Leon seems so anxious to sling his face cards around, it would be an excellent opportunity to show him that you can also play at the high stakes table. Tarnations, Mark, that invite couldn't have dropped in our lap at a more opportune time. How often will you have the president's undivided attention?"

"Very true," answered Tew. "I guess it wouldn't hurt for me to make a trip out to the PDF. But before we get too far ahead of ourselves, we're not even sure that this mission is viable. For all we know, this shot may be outside the range of our capabilities. Tarbox may already know that, and may have come here just to rattle us."

Loosening his back tie, Heydrich looked up from his briefing book and interjected, "Mark, we'll run these numbers through our computers. I should have a solid answer for you by the middle of next week."

"It *can* be done, General," averred Ourecky confidently. "In fact, this profile is very similar to our last mission. I can't speak for Crew Three, but Major Carson and I could probably fly it right now, with minimal preparation."

"Ourecky, are you *that* sure?" asked Tew.

"I am, sir."

"Well, Carson, before we arbitrarily pack you two boys into a rocket and shoot you into space tomorrow, how do *you* feel about it?" asked Wolcott.

"I would have to look at it more closely," answered Carson. "And although I do tentatively agree with Scott's assessment, I'm definitely not going to decline any opportunity to train."

"What's your take, Gunter?" asked Tew. "Is there any chance that Ourecky is right?"

Heydrich sat up straight and answered, "I'll certainly have to study it more closely, but I suspect that he's correct. But even then, there are a lot of pieces to the puzzle, some of them completely outside our control, like the tracking network."

"Correct. Assuming that we're compelled to go sooner than later, how difficult will it be to divert the ARIAs and other tracking assets?" asked Tew, examining a desk calendar. "When is the next lunar flight scheduled?"

Heydrich looked towards the ceiling. "If nothing goes wrong this weekend that would cause a delay to the program, the next Apollo mission should go in November."

"You don't think they will make it on Sunday?" asked Tew, raising his eyebrows.

"I would be *very* surprised," replied Heydrich. "I think that they might try for the landing, but I strongly doubt that they will make it. The parameters are just *too* close."

"Okay. Assuming that we can lock down the tracking assets, how about the recovery network?"

"Isaac needs a minimum of three weeks' notice to mobilize and deploy his troops," answered Wolcott.

"Well, give him a heads up that he might have to deploy on short notice," stated Tew. "Tell him he is authorized to immediately start sneaking his LSO teams into their assigned countries once we have a better handle on the orbital tracks. I know that's an expensive gamble, but we'll figure out the budget issues later."

"Done, boss."

"I'm going to do my damnedest to head off this mission or at least delay it, but we have to be ready regardless. Now, unless anyone has any questions, I need to call the White House to make sure their invitation is still open."

9:25 a.m.

As they entered the sanctuary of the Flight Crew Office, Carson looked to verify that they were alone, closed the door, whirled around at Ourecky and demanded, "What the *hell* was that about, Scott?"

"What do you mean?"

"I've been watching Jackson and Sigler screwing up for the past month. It's like witnessing the same damned train wreck over and over. Did you really have to rub their noses in it?"

"What do you mean?"

"*Gee golly, General Tew, if Crew Three can't do it, then Carson and me can fly that mission without even straining ourselves,*" chided Carson in a mocking voice like a fifth-grader volunteering for the toughest word at a spelling bee. "And moreover, you shouldn't be in such a damned rush to volunteer us for a mission without consulting with me first. Considering that we're a team, or we're supposed to be a team, I think you owe me that."

"I thought . . ."

"Scott, while I will always bow to your theoretical knowledge, you should remember that I have a lot more flying time than you'll ever have, and you should at least respect my position on whether we're adequately trained or not. We've flown exactly *one* mission. Just because you've made it up and back in one piece doesn't give you any leeway to get cocky."

Flustered, Ourecky slouched into his chair and shook his head. "But I thought you wanted to go upstairs again."

"I *do,* but this business isn't like driving down to the local airport and renting a Piper Cub for a hamburger run or a beach trip," snapped Carson, unwrapping a stick of Juicy Fruit. "Yeah, I want to go up, but to be honest, I'm behind Tew on this one. I think it's a lot more practical to stick with our schedule and shoot in December. I don't think that I need to remind you about this project's record: one rocket and crew vaporized on the first mission, and a major battery failure on the second. It's foolhardy for anyone to believe that all the bugs have been worked out."

"But what if Tew changes his mind and we end up with the mission anyway?"

"I don't see that happening, but if it does, we need to spend time in the Box to tune up."

Ourecky all but cringed at the mention of the Box. "You don't think we're ready *now?*" he asked.

"*Ready?*" Carson shook his head vigorously and asked, "Hey, do you know what you call a boxer who jumps right into a championship bout without training?"

"No. What?"

"You call him an ambulance, if he's lucky, a hearse if he's not." Carson sat down at his desk and added, "Stick to what you're good at, Scott, and let me handle the training. *I'll* decide when we're ready."

Delta Airlines 651, 23 miles south of Lexington, Kentucky
4 p.m., Sunday, July 20, 1969

Tightly gripping her stomach, Bea retreated to the sanctuary of the mid-deck galley. Several of her neighbors back in Dayton had been incapacitated with some sort of nasty stomach bug, and it appeared that she had fallen victim as well. She hoped that Scott hadn't also caught it.

"Bea, are you all right?" asked Sally, one of the other stewardesses. Bottle-blonde and petite, Sally was originally from Dallas but now lived in Atlanta. Bea stayed with her occasionally during the week but wasn't overly fond of the raucous parties that seemed to be a nightly occurrence at Sally's apartment complex.

Woefully woozy, struggling to maintain her equilibrium, Bea sagged against a stainless steel shelf in the galley. "Oh, Sally, I'm just *dying* right now. I swear I would have called in sick if I knew I was going to feel this awful."

Adjusting her blouse, Sally shook her head and observed, "Honey, you look terrible. Why don't you get off your feet? You can take the aft jump seat. If you need to stretch out some more, there are open seats in first class. I don't even know why we're bothering to fly today."

"Yeah, I know. Scott's at home, glued to the TV set, just like everyone else in the country. He wanted me to stay home." Bea opened the first aid cabinet and found a bottle of aspirin. She gulped down two tablets and chased them with lukewarm 7-Up.

"You should have listened to him. Look, Bea, the plane's less than a third full. Let me cover for you. Besides, there's Trudy and Joan, if I can ever drag them away from flirting with that rich guy up front."

Bea weakly shook her head as she scooped ice cubes into a plastic cup and then grabbed a couple of miniature bottles from the liquor box. "I'm going to take you up on that. Let me deliver these up to 14-B, and then I'll go back and cozy up in the jump seat for a while."

Walking forward, she regretted that she didn't immediately latch on to Sally's offer. Buffeted by mild turbulence, the plane shuddered slightly. Grasping the cup of ice and bottles in one hand, she used the other to balance herself on seat headrests. By the time she made it to the fourteenth row, her knees were wobbling and her head spun like a carnival ride gone awry.

"Your Johnny Walker Red, sir," she mumbled to the passenger, a neatly dressed regular who flew from Dayton at least once a month. She knew him only by appearance and not by name, but he seemed pleasant enough. At least he wasn't grabby like some of the other Sunday afternoon regulars.

"Thanks, darling," he replied and looked at his watch. "It should be about time to celebrate."

Just then, the captain's jubilant voice blared over the PA system: "Ladies and gentlemen, I have an important announcement! We have just received word that Neil Armstrong and Buzz Aldrin have landed safely on the moon. *Apollo 11 is on the moon!*"

As the handful of passengers cheered and clapped, Bea was suddenly stricken by an overwhelming wave of nausea. Her lunch of chicken salad and crackers was coming up, and there was nothing that she could do to prevent it.

Her panic was momentarily dispelled by years of emergency training; she instantly reviewed her available options, weighing a mad dash back to the mid-deck galley against a frenzied sprint to the first class lavatory, but decided that neither alternative was viable. Abandoning any pretense of ladylike decorum, she snatched an airsickness bag from a seat pocket, ripped it open, and then unceremoniously filled it.

"Sorry," she said sheepishly, sealing the bag closed with its wire tie.

"Are you okay, honey?" asked the passenger sympathetically, offering his napkin.

"I think so," she answered, wiping her lips. "I just wish that you hadn't seen that."

"So, little lady, I take it that you're not a big fan of the space program, huh?" he replied, tipping up one of the miniature bottles.

2

DOWN TO EARTH

Forest Park Apartments, Dayton, Ohio
3:10 p.m., Monday, July 21, 1969

Returning from his weekly trip to empty his post office box, Eric Yost clumsily negotiated his way through the door and into the apartment. He dumped the week's accumulation on the kitchen table, hobbled into the living room, and plopped down in a chair. Grunting, he heaved his cast-encased foot into another chair. This was his fourth cast since December; the tiny bones in his foot and ankle were taking forever to heal. The foot throbbed with pain that never completely subsided, regardless of what he did to abate it. On a positive note, the plaster shackle was due to come off next week. If the follow-up X-rays revealed sufficient healing, Yost would not be fitted with a new cast, but his orthopedic doctor had warned him that he might have to endure the lingering pains for the rest of his life.

He carefully parted the thick curtains, opening them just enough to peer out, and watched the parking lot for several minutes to see if he had been followed. He was absolutely certain that the loan shark was still dispatching his goons to prowl for him. Thankfully, they apparently hadn't discovered he was staying at Kroll's apartment and was driving Kroll's Mustang. With his van safely stashed at a remote parking lot on base, he might as well have vanished from the face of the earth. He hadn't been back to his house on Elm Street since he was ambushed in December, especially since a friend from his old neighborhood informed him that suspicious-looking men occasionally parked on the street and watched the vacant dwelling.

Confident that he hadn't been tailed, Yost pulled the curtains tight and focused his attention on the mail. The burgeoning stack contained little else but bills, junk mail, and bad news. Several pieces concerned the house on Elm Street. Accompanying a long overdue electric bill, a terse letter from Dayton Power & Light threatened to disconnect his power if he didn't pay up in a timely manner. He hadn't bothered to pay his bill in several weeks, so the warning was certainly not a surprise.

The city had shut off his water back in March, but not until after he had received a massive water bill for February. It wasn't difficult to surmise what happened. He hadn't winterized his plumbing before abandoning the house, so without electricity, the pipes surely froze solid in the first cold snap and then later burst. He could only imagine the extent of the damage and was sure that the structure's underpinnings had long since been reduced to a sopping mass of rotted wood and stinking mold. He wondered if the house would ever be worth returning to, or whether he could ever hope to sell it and recoup the money he had sunk into the dump. It was another bottomless money pit courtesy of his ex-wife; Gretchen just couldn't be content living in the perfectly good quarters available on base.

Winnowing through the batch of mail, he found an airmail letter from his former supervisor—Dan Kroll—all the way from Thailand. Kroll's temporary duty assignment had been extended until the end of the year, but he stated that his wife had experienced a falling out with her mother in Oregon and probably would be coming back to Dayton soon. Kroll didn't set an exact date, but made it clear that if Anna returned to Ohio before he came off TDY, Yost would need to make himself scarce. That was not welcome news; although he had planned to be here at least until August, it was highly likely that he would lose both the apartment and use of the Mustang in short order.

He considered his options. Moving back to his old house obviously wasn't practical, since it probably wasn't even habitable. He had already looked into relocating into the transient billets on base, but the downside was that his behavior would be subject to much more intense scrutiny. He also considered volunteering for temporary overseas assignments, even Vietnam if no other gigs were available, but his records had been administratively flagged in such a manner that he could not leave Wright-Patterson. Besides that, even without the administrative flag, he couldn't go TDY until the cast came off for good.

Finally, at the bottom of the stack, he glimpsed a letter that offered at least a glimmering prospect of good news. Yost anxiously ripped open an envelope from *Argosy* magazine. He had held out high hope that the glossy "men's adventure" monthly would be interested in his information about UFOs stored and studied at Wright-Patt, but his heart sank as he recognized that it was just another rejection letter. As he angrily ripped the paper to shreds, he realized that it was not *just* another rejection letter, but it was the *last* rejection letter. With this missive, he had been snubbed by every single publication—eighteen in all—that he had contacted back in February.

Cursing, he shoved aside the heap of mail and contemplated what he had to accomplish before Anna came back. Long before it had disappeared, the cat had succeeded in ruining the carpet and virtually every piece of furniture in the apartment. Kroll would surely be outraged when he found out, but it certainly wasn't Yost's fault, any more than when the feline escaped through the door that he had briefly left open to ventilate the smoke from a pot of charred pinto beans. Yost was fairly certain of the cat's ultimate fate; there was plenty of physical evidence in the parking lot to indicate that it had been brutally mauled by a roaming pack of stray dogs.

Looking around, he seriously thought about tidying up the clutter and disarray. The trash can overflowed with garbage. Topped by a cast iron frying pan dripping with rancid grease, several weeks' worth of dirty dishes were piled in the filthy sink. A foul stench emanated from the bathroom; the toilet regularly overflowed, and Yost had finally gotten tired of trying to repair it. Even though the place was in shambles, he dreaded the thought of vacating Kroll's apartment. Yost's hands shuddered as he realized that it would merely be a matter of time before the loan shark's goons got the drop on him. With no other practical options to elude them, he would probably have to resort to residing in his van again.

Gnashing his teeth and grimacing, he recognized that rather than wallow in misery, it was high time to take decisive action. He reached for a large plastic medicine bottle on the kitchen counter. His pain management needs allowed him to legally tap into a nearly endless supply of government issue codeine tablets. While he wasn't particularly fond of the chronic constipation that came with the narcotic, he had grown accustomed to its soothing blanket of numbness. He popped open the bottle's lid, gulped down two tablets, and chased them with a generous helping of Old Crow.

As the pain in his leg dissipated slightly, he referred to an index card that bore an official address that he'd copied down from a book in the Dayton Public Library. He filled another glass with Old Crow and then slowly drafted a letter to the "Embassy of the Union of Soviet Socialist Republics" in Washington, DC. Smiling, he glanced at the fragments of *Argosy*'s rejection letter. If those damned magazines weren't interested in his UFO story, he mused, then the Russians surely would be, and probably would be willing to shell out considerably more compensation for his efforts.

Wright Arms Apartments, Dayton, Ohio
5:55 p.m., Tuesday, July 22, 1969

Grinning, Ourecky closed the door behind him, went straightaway to the fridge, grabbed a cold bottle of Schlitz, popped off the cap with an opener, turned on the window-mounted air conditioner, switched on the television, kicked off his shoes, and then sprawled out on the couch. He had arrived home just in time for the evening news. The Apollo 11 astronauts were on their way home; they were set to reenter the atmosphere and splash down on Thursday.

There was supposed to be a live telecast from Apollo 11 later this evening. The other big news item of the day concerned Senator Ted Kennedy—JFK's brother—of Massachusetts; he had left the scene of an auto accident this past Friday in which a former RFK campaign staffer, Mary Jo Kopechne, was found drowned in his submerged car near a bridge on Chappaquiddick Island. Ourecky found it ironic that the senator's tragic wreck had occurred in the very week that the greatest legacy of his brother—the late President John F. Kennedy—was coming to fruition.

He glanced at the clock on the wall; Carson would be here in thirty minutes, and he mused whether he had time to snatch a quick shower. After the Apollo 11 broadcast, they were heading out to dinner at a steakhouse to celebrate their latest assignment. On September 9, slightly more than a month away, they were going back upstairs.

In a whirlwind turn of events, he and Carson had gone from being summarily grounded to back-up crew status to finally being elevated to the prime crew for the September mission to intercept a Soviet radar surveillance satellite that threatened US and NATO fleets operating in the North Atlantic.

The ever-cautious Tew had wanted to punt the Navy's proposed mission to the next scheduled launch window—in December—but Admiral Tarbox was insistent that the Soviet satellite be disrupted at the earliest opportunity. Angry with Tew's reluctance, the scrappy admiral swooped straight from Ohio to Washington, where he spent the weekend lobbying high-level officials. As of this afternoon, since they were woefully unprepared to fly on such short notice, Jackson and Sigler were pulled off the mission, and he and Carson were ordered to immediately start training in earnest. As Ourecky had heard it, the unequivocal order to fly in September had come directly from the White House. While he was sure that some fear and trepidation would eventually set in, Ourecky was ecstatic at this point.

As he sipped his beer, the door started to open unexpectedly. "Drew?" he asked, sitting upright. Carson had developed an annoyingly bad habit of just popping in unannounced.

"No, dear. Sorry to disappoint you, but it's just me, the woman you're married to," said Bea, pushing the door open with her shoulder and dropping her small suitcase just inside the living room. She was dressed in jeans, a tie-died T-shirt, and leather huarache sandals. She wore no makeup, no bra, her hair was tousled, and her eyes slightly bloodshot.

"It's *Tuesday*. I thought you were scheduled to fly until Thursday, like normal."

"It's *so* nice to see you *too*, dear," she said.

"Hey, I didn't mean it like that," he replied, holding his beer as he stood up to hug her. "I'm just surprised that you're home today."

"Oh my God. *Beer?* Scott, *please* pour that out," she begged, nudging away his hand with the condensation-beaded Schlitz bottle. "I just can't bear the smell of beer right now."

Slightly bemused, he went to the kitchen, emptied the beer in the sink, and ran water after it. She was already sitting on the couch, removing her sandals, when he returned to the living room. He sat down beside her and asked, "Bea, are you all right? You don't look very good."

"I think you had better sit down, Scott. I have some news."

"Uh, dearest, I *am* sitting down." *She has news?* he thought. Even though he couldn't share it with her, he couldn't possibly imagine any news being more momentous than the news he had received today. But it only took a moment for him to realize that he was mistaken.

"Scott, I'm pregnant," she exclaimed, with a worried expression on her face.

Utterly stunned, he couldn't speak for a moment. Finally, he regained his composure, hugged her, and exclaimed, "Wow! You're pregnant! A baby! Bea, that's great news. We need to call my parents tonight. They'll be thrilled. And Drew's coming over in a little bit. We're going out to the Pine Club to . . . uh . . . have a steak. Hey, we'll all go together and celebrate. This is great!"

Her face suddenly turned pale; he wasn't sure if her change of countenance was brought on by his mention of red meat, Drew Carson or both, but it was fairly obvious that his plans for the evening—and for the next few months—were subject to rapid change.

She leaned away and put her hand on his forearm. "I'm happy, too, but it's not necessarily good news. I was so sick yesterday that I couldn't fly, so I laid over in Atlanta at Sally's place. I didn't get a wink of sleep last night. I flew in this morning and went straight to the doctor."

"Why didn't you call?" he asked.

"I didn't think it was serious and I didn't know if you were back in town. At first, I thought it was a stomach bug, like the Posts and the Sikes had last week. Boy, was I ever wrong."

"But, Bea, it's still *great* news."

"Scott, this baby is going to flip our lives completely upside down. We sure didn't plan on it. We'll need to move because this place is way too small for us and a little one. I'll have to quit flying, at least until the baby arrives and maybe a long time after that. I may never be able to go back to flying. The scheduling office is already making arrangements for me to work as a gate clerk, but I don't know how long I'll be able to do that."

Her lower lip quivered. She leaned against his shoulder and started crying softly. "Scott, I'm just not ready for all this. I wish my mum was here so I could talk to her."

He held her and said, "Look, Bea, we're okay for money, especially with this last promotion, so there's no need for you to keep working. And I'm sure that we can qualify for quarters on base, a two-bedroom or maybe even a three-bedroom for later when . . ."

Frowning, she looked up at him. "Please don't say that. We need to take this one step at a time. And Scott, I really don't want to move on base. I grew up in absolutely wretched military quarters all over the world, and that's not where I would choose to raise a child."

"Fine," he said abruptly. "We'll find a place in town. That's not a problem."

She pulled the last Kleenex from the box on the coffee table, wiped her eyes and then blew her nose. "Scott, that's not all. The doctor saw some things on my blood work that he wasn't happy with, so they want to run more tests tomorrow. He said I have an iron deficiency and that there may be other possible complications. He said I might not even be able to carry this baby to term. In fact, I may never be able to have a baby." She started sobbing again.

There was a knock at the door. It opened, and Carson stepped in, bearing a bottle of scotch. "Hey, I saw Bea's red VW roller skate outside," he said. "I thought she wasn't . . ."

Ourecky anxiously shook his head.

"Uh, I take it that this is not a good time?" asked Carson.

"I'll call you later," replied Ourecky. "Sorry, Drew. Hey, I'll see you in the morning, okay?"

Carson nodded. "Do whatever you have to do, Scott. I'll cover it with Virgil if you need to come in a little late tomorrow. We have a tight schedule, but I know that we can catch up."

Aboard Air Force One, arriving at Johnston Island
4:58 p.m., Wednesday, July 23, 1969

As the big jet rolled to a stop, Tew looked up from his note cards and glanced outside. A phalanx of khaki-clad, white-helmeted Army military police immediately assumed perimeter positions on the tarmac. Apollo 11 was scheduled to splash down tomorrow, and President Nixon was on his way to the aircraft carrier USS *Hornet* for the historic moment.

By sheer coincidence, Johnston Island was the closest US-controlled landmass to the recovery site, so it was being used as an intermediate stopover. On the president's return trip from the Hornet, Tew was slated to give him a quick tour of the Pacific Departure Facility launch site, if his hectic schedule permitted. The president had apparently expressed a tremendous interest in Blue Gemini after June's successful mission.

Given only a few days' notice, Tew had hustled to join the flight in San Francisco this morning. Wolcott had remained at Wright-Patt so that they could get a jump on preparing Carson and Ourecky for September's mission.

While Tew bore misgivings about sending up the wayward pair again, especially so soon, he knew that they were the only ones capable of successfully pulling off the rapidly impending mission. He had also learned that the president, a former Navy man, was apparently swift to defer to the Navy's position in any inter-service arguments; Tew's suggestion of delaying the launch was quickly squelched, apparently soon after the Chief of Naval Operations visited the Oval Office for coffee and a quick conversation on Monday morning.

Despite the opportunity to personally escort the new Commander-in-Chief through the Project's launch facilities, Tew had an ulterior motive for the impromptu venture. Given the opening, he felt confident that he could thwart Tarbox's plan. He was sure that he could convince the president that it was foolhardy to execute the Navy's mission on such short notice, and that they should either revert the launch date back to December or abandon the flight altogether. After all, Blue Gemini's charter was to intercept and interdict suspected Soviet nuclear weapons—Orbital Bombardment Systems—in space, and while the Gemini-I had proven effective at destroying other satellites, that wasn't the intent.

Tew had been ensconced in the rearmost compartment of the plane for the entire journey. He had seen the president one time during the flight, an encounter that lasted perhaps fifteen seconds, during which time Nixon casually mentioned that this would actually be his second visit to Johnston Island, since he had been on the atoll very briefly during his wartime service.

As a set of boarding stairs was moved into position, the head of the Secret Service detail announced over the PA system: "Everyone, please remain in your seats until the official party has departed the plane and you are granted clearance to stand up."

Looking out his window, Tew watched as President Nixon, Secretary of State William Rogers, Henry Kissinger, Apollo 8 astronaut Frank Borman, and a small party descended the metal stairs, where they were greeted by Admiral John McCain, the base commander, and some officers that Tew didn't recognize.

Despite the blistering heat, Nixon wore a dark gray suit complete with tie. He waved to a group of spectators—a mix of military personnel and some civilian workers—and then went over to shake hands and speak to them. Less than ten minutes later, he boarded the Marine One helicopter and promptly departed the island.

As the helicopter flew away, an aide tapped Tew's shoulder. Looking like he was barely out of college, he was one of several personal aides on the flight. "General Tew?" he asked.

"Yes?" replied Tew. His long-sleeved shirt was damp with perspiration. With the air conditioning switched off, the temperature inside the plane had already started to climb significantly, and he anticipated that it would be nearly unbearable in just a few minutes.

"General, I need to pass on the president's regrets. He had hoped to speak with you today, at least briefly. He'll return for a few hours after the splashdown, and if his schedule permits, he'll see your launch facility then. Until then, we've made arrangements for you to stay here on the island. There are VIP quarters, but there may not be any space available since most of the president's party is remaining here tonight. They are putting up some overflow tents, though."

"Of course," sniffed Tew, shaking his head as he loosened his tie. "But I'll be able to escort the president to the launch facility tomorrow, right?"

"Really, sir, I wouldn't absolutely count on it. I don't know if you've been watching the news this week, but Japan is very upset with us because they just discovered that we've been storing chemical weapons on Okinawa without their permission. They were moved there during the Kennedy Administration, and there was an accidental release last week."

"I'm *very* aware of that, young man."

The aide shrugged. "Well, sir, the Japanese are adamant that those chemical munitions be moved out of Okinawa, and there's a good chance that they will be coming here to Johnston Island. All of this just came up, and the president is supposed to receive a briefing on the options tomorrow. I'm guessing that will probably trump your sightseeing tour. Sorry, General."

As he watched the young aide stroll away to speak to someone else, Tew fumed. While the opportunity to fly on Air Force One was exhilarating, it looked unlikely that he would be able to brief Nixon tomorrow, and would lose over forty-eight hours of work as a result of this folly. And those were forty-eight hours that he could scarcely afford to squander at this critical time.

So here he was—an Air Force major general, due to receive his third star in short order—and he was effectively treated as a faceless nobody. He would probably spend the night on a borrowed cot in a sweltering tent and then depart tomorrow on a cargo plane crowded with support personnel. It was absolutely amazing; he personally oversaw the third largest manned space program in the world, and yet he could not be any more anonymous.

Filyovsky Park, Western Administrative Okrug, Moscow, USSR
9:35 a.m., Sunday, July 27, 1969

The morning air was cool, almost unseasonably so, and the sky sparkling clear. Climbing out of his sedan, Gregor Mikhailovich Yohzin mused that it was a truly excellent day for a jaunt in the park. A major general in the RSVN—Soviet Strategic Rocket Forces—Yohzin oversaw the initial testing of medium-range ballistic missile prototypes at the Kapustin Yar cosmodrome.

Presently, he was in Moscow on a temporary assignment with the GRU, at their "Aquarium" headquarters near Khodinka Airport, writing summaries of NASA technical reports concerning the American Apollo lunar program. He was still compelled to juggle his normal RSVN chores and the brief stints working for the GRU. Between the two organizations, he spent roughly one week out of every month in the capital city and three weeks at Kapustin Yar.

Reflecting on the NASA reports, Yohzin sighed as he contemplated the gross disparities between the Americans and himself. He envied the

Americans, particularly for the vast resources that were made available to them, as well as the tremendous intellectual freedom they were afforded. In contrast, he felt like Sisyphus, the Greeks' mythic prototype of futility, compelled to roll an enormous boulder uphill, only to watch it roll back down again, over and over and over.

If there was even the slightest benefit to his Sisyphean existence, it was that his workdays had a set pattern and structure. As a small child during the last gasps of the Russian Empire, Yohzin had experienced the turmoil and chaos of the Bolshevik Revolution; as a middle-aged man, he relished a calm, almost monotonous routine. Except when he was engaged in a unusually demanding test regimen, or when he was compelled to travel to Moscow to work with the GRU or participate in administrative busywork, his days at Kapustin Yar were just about as predictable as a worker's at a textile factory or farming collective. Virtually every day at six o'clock, he returned home to his humble apartment. Once there, he would join his family for supper, then retire to his study for an hour of tranquil solitude, and then accompany his wife for their daily stroll through the central commons. Content in seeming dullness, his placid life could almost be measured out by a metronome.

They lived in a town affectionately called *Znamensk* by its denizens but officially known as *Kapustin Yar-1*. The dreary settlement of apartment buildings lay in the center of the massive cosmodrome and had been purposely constructed to accommodate the personnel who worked there. An enormous number of military personnel, engineers, scientists, and their families called it home. They strived to make their lives there as normal as possible, despite the ever-present dangers in their midst. Like the top secret military facility that surrounded them, their little town appeared on no maps and was all but sealed to outsiders. It was a relatively comfortable existence, but Yohzin wished that his two sons could attend better schools. In his opinion, his boys weren't being challenged academically, at least not to the extent that they should be, and they would suffer for this shortfall once they matriculated to their post-secondary education.

As he strolled along the oxbow bend of the Moscow River that encircled Filyovsky Park, Yohzin recalled the unusual incident at the Paris Air Show last month, when the Germans made contact with him. He had attended the prestigious event at the behest of the GRU, which wanted him to scrutinize new American rocket and spacecraft technology on display. True to their nature, probably concerned that he might capitalize on the foreign travel to defect to the West, the GRU assigned a pair of operatives to keep watch over him. After distracting his GRU babysitters, an unseen German had briefly spoken to him before slipping him a note with a Moscow phone number. Although he hadn't seen the man's face, the voice was very familiar, and he was almost certain that he was one of the former V-2 engineers Yohzin had worked with just after the War.

Yohzin suspected that the brief encounter in Paris was some form of recruitment effort, perhaps a prelude to an invitation to participate in espionage, but he still held out hope that his old German friends were sincerely reaching out to him. Although he should have immediately reported the approach, he held onto the note as he mulled over the offer for several days, anxious to know if it was worth the risk.

Finally, last night, he was done vacillating. His curiosity overcame his reluctance and he called the phone number. After agreeing to a day and time, a calm voice hastily dispensed instructions for a clandestine meeting this morning in the park.

With tail wagging briskly, Yohzin's faithful Alsatian—Magnus—kept pace at his left heel. On a quiet stretch of the river, a light sheen—possibly an oil film—glistened on the water's surface. Yohzin walked for approximately fifteen minutes before locating a distinctive landmark that had been designated as a starting point. He carefully counted paces before spotting a dark-shaded stone that jutted out slightly from a stone wall, perhaps just a fraction of a centimeter.

"*Sitz, hund,*" ordered Yohzin, gesturing toward a spot next to the stone wall as he tugged lightly on the dog's lead. He crouched down and tightened his shoelaces. Making sure that no one was within the immediate vicinity, he slightly dislodged the loose stone and retrieved a tightly folded scrap of paper that had been wedged beside it.

He took a seat on a nearby park bench and unfolded yesterday's edition of *Pravda*. Pretending to read the newspaper, he memorized the terse instructions on the tiny note. The directions were simple enough: He was to continue walking on the path by the river until he saw a man wearing a gray jacket carrying a dark green umbrella under his left arm. The man would ask for directions to the nearest subway station, and Yohzin was to direct him to the Filyovsky stop on Minskaya Street. If it was not safe to meet, the man would carry the umbrella under his right arm. Likewise, Yohzin would carry his rolled newspaper in his right hand if it was safe to meet, or in his left if it was not.

After he digested his instructions, he digested the paper . . . literally. In accordance with his instructions, he slipped it into his mouth, chewed it up and swallowed the evidence. Surprisingly, the taste wasn't entirely unpleasant, even slightly sweet, and the scrap all but melted in his mouth. Yohzin suspected that the note was written on edible rice paper.

Following the specified security precautions, he walked on. Conscious that he might be strolling to his doom, Yohzin's heart pounded in his chest. His palms were sweaty, his hands trembled, his stomach was queasy, and a rivulet of sweat burned his eyes. He was well aware that the clandestine meeting could be an insidious trap orchestrated by the GRU or KGB, and in mere moments, he could be in a dire predicament, even more perilous than working around experimental rockets. Now regretting his decision, he almost pivoted around to walk back to his car. But he elected to press on, so strong was the luring prospect of seeing his German associates again.

The path narrowed as it traversed a low area that was a maze of dense thickets, lush undergrowth, and tangled vines. Slowing his pace, peering through a gap in the vegetation, Yohzin glimpsed a man who matched the physical description he had been given. The man had obviously selected a blind spot in the path for their meeting place. Still very apprehensive, Yohzin looked for potential escape routes. Also conscious that the stranger might be accompanied by confederates, lurking in the shadows, he vigilantly scanned the surroundings.

As he closed the gap, Yohzin shifted his newspaper to his right hand. He paused briefly, and looked towards Magnus; the dog's hackles were raised, as if he knew that he was in the midst of great danger.

"Pardon me, but I'm not from Moscow and I'm a little befuddled. Is there a subway station nearby?" asked the man in Russian, holding out a tourist map that depicted the Metro underground transit network. Slightly taller than Yohzin, the stranger was handsome, well groomed, and was graced with an athletic frame. He wore black-framed glasses that appeared to be of Soviet manufacture. His face bore distinct features that hinted of a Scandinavian lineage. He appeared to be in his mid to late thirties.

"Uh, the closest stop is the Line Four concourse on Minskaya Street," answered Yohzin in a faltering voice, using the pre-arranged safety phrase he had memorized earlier. Drawing in a deep breath, he gathered his composure. He gestured towards the southeast and added, "It's about half a kilometer from here, in that direction."

"*Spasiba*," replied the man. He then quietly asked a few questions to verify that Yohzin was legitimate.

Although this stretch of the path was largely deserted, they weren't entirely alone; even as they exchanged their verbal bona fides, a sailor swaggered by, wearing the insignia of the Black Sea Fleet. Likely home on furlough, the seaman was accompanied by a fetching young woman with brunette hair. Recognizing Yohzin's rank as he drew near, the sailor saluted stiffly, and Yohzin returned the gesture.

The "tourist" stooped down to admire Magnus. "What a handsome dog," he commented. "Obviously a purebred. What kind is he?"

Yohzin swiftly deduced that the stranger was not who he expected, since any German worth his strudel would have immediately recognized Magnus's breed. He wasn't Russian, either; his spoken Russian was *too* good. As if it had been entirely acquired in a sterile classroom, the stranger's speech lacked any peculiar nuances or inflections that might have associated him with a particular region or city. Gnashing his teeth, Yohzin could not believe that he had been so naive, and stringently hoped that he had not been lured into a trap. "You're not German?" he asked bluntly.

"No," answered the man, standing to his feet. He extended his hand. "American. I'm Smith."

Yohzin grudgingly shook the man's hand. "*Smith?* Don't Americans use first names as well?"

"No, just call me Smith, at least for the moment. Sorry for the deception, but we weren't sure you would cooperate if you knew it was us instead of the Germans." Just a few feet above them, a gray dove swished by, on its way to light in a nearby tree.

Glowering, Yohzin shook his head. He gestured for Magnus to lie down; the dog did so, and focused its gaze on a gaggle of geese frolicking at the river's edge. Sensing the canine's attention, the geese scattered.

"The message in Paris stated that I might have an opportunity to see my German friends again," said Yohzin. "Was that a lie?"

"Not entirely. In time, we can arrange that meeting with your old acquaintances," answered Smith, clearly striving to put him at ease. "But in the meantime, we have a favor to ask of you."

"You want information."

"Exactly," answered Smith. "So, General Yohzin, are you willing to consider an offer? Since you came here, you had to suspect that we were interested in purchasing information. We know of your work with the RSVN. We're willing to pay handsomely, depending on the quality and timeliness of what you might provide. It could be a very lucrative situation for you."

"*Nyet*," replied Yohzin emphatically. He was amazed at the Americans' cavalier willingness to engage in such an audacious gamble. "I am *not* interested in striking a bargain with you, Smith, regardless of your nationality. And money holds absolutely no interest for me, since I couldn't possibly spend it here without drawing undue attention to myself."

"Then perhaps you could spend it *elsewhere*," noted Smith. He seemed to be slightly off balance, perhaps because he was likely accustomed to dealing with greedy opportunists who were primarily motivated by money or other material reward. "Maybe you could spend it in America. We could eventually spirit you out of the Soviet Union, and you could retire in the United States, with a king's ransom to boot."

Yohzin smiled and asked, "How do you know that I'm not going to just call the GRU and have you arrested? How do you know that this is not a trap?"

The American chuckled and replied, "And how do *you* know that this is not a trap? I could just be a thug, biding my time until I bash in your skull, or I could be a GRU officer myself. How would you know? In any event, if the GRU or KGB was to swoop in and nab us right now, you would certainly have some explaining to do. Anyway, we're just chatting here, exploring possibilities. It's merely a conversation."

Yohzin shook his head. "Maybe, but I am still not inclined to sell my nation's secrets, and since I seriously doubt that you might entice me to do so, perhaps it's best that we just go our separate ways and forget that we ever met."

"If you insist," said Smith. "It was nice to make your acquaintance, even if we could not come to an agreement. That phone number will always be answered, in case you change your mind. If the notion of money offends you so much, perhaps you might think of a more palatable arrangement."

"I don't think so," replied Yohzin. "And I think it's futile to pursue this discussion any further."

"Circumstances often change, Comrade General."

"And just as often, they *don't*," replied Yohzin, turning to leave. He softly clicked his tongue, and Magnus took his place at his left heel. "*Do svidaniya*, Smith."

3

THE GHOST FROM WEST VIRGINIA

Forward Operating Base—Command and Control North
MACV-SOG
Da Nang, Republic of Vietnam
1:25 p.m., Monday, August 4, 1969

Sergeant First Class Nestor Glades patiently sat through the final debriefing with his team, recounting the intimate details of a six-day mission to locate an NVA communications relay site. The mission had been a resounding success and he had brought back his team—unscathed—as well as two high-value prisoners. Now it was time to go home, back to the States, yet again.

Glades was a "One-Zero," a recon team leader, assigned to the Military Assistance Command Vietnam Studies and Observations Group or MACV-SOG. MACV-SOG was a secretive special operations organization that conducted highly classified reconnaissance and strike missions throughout Southeast Asia. With his unprecedented string of successful missions and long-standing record for bringing his teams home intact, Glades was regarded as living legend by his peers in MACV-SOG.

In the tradition of his departure routine, a ritual that he had repeated six times in the past four years, Glades would meticulously clean his small arsenal of personal weapons—his CAR-15, an AK-47, an M79 grenade launcher, a Remington 870 twelve-gauge shotgun, an M1911A1 Colt .45 caliber pistol, a silenced High Standard .22 caliber pistol, and a Ka-Bar combat knife—before carefully wrapping each in oiled cloth and packing them into two plywood footlockers.

He would clean his personal equipment and web gear with the same degree of care, and stow them with the weapons. Then he would lock the footlockers and surrender them to the supply room, where they would be waiting for him when he returned in six months.

After the footlockers were securely stowed and his duffle bag was packed, he would compose a brief letter to Deirdre, his wife, to let her know that he was homeward bound. Then he and his team would retreat to the compound's small club, where they would knock back more than a few beers as they reminisced over fallen comrades. Then, as part of his close-out ritual, Glades would open a C-ration can containing sliced peaches in syrup, a can that he had carried through his entire tour. Like partaking in some peculiar form of communion, the men would pass the little green can around the circle, and each man would eat a slice and sip some of the sweet juice.

After sharing the peaches with his teammates, he would solemnly hand the team over to his "One-One," the assistant team leader. The One-One would take on the role of the team's One-Zero; in the days to come, he would mold them into his own vision of what a recon team should be. With luck, half a year later, if the new One-Zero did a good job, many of the men would still be alive and Glades would resume leadership of the team.

Like most MACV-SOG teams, his was a mixed element, composed of US Special Forces and indigenous—"indig"—soldiers, usually fierce

Nung or Montagnard mercenaries. The indig troops were intensely super-stitious, and despite his diligent efforts to dissuade them, they fervently believed he was endowed with supernatural powers that could be harnessed to protect them as well. On the camp, the indig troops fought over any vacancy that came available on his team. Some went so far as to bribe the camp barber for samples of his hair, swept from the plywood floor in the tent that served as a makeshift barbershop. The shorn hair was woven into amulets, which fetched a hefty price by local standards.

For the Americans, it was a standing joke that as recon men drew close to the end of their tours, they often found religion or Glades or both. If they were willing to submit to his relentless work ethic and incessant training regimen, then the odds were favorable that they would walk upright off the Freedom Bird, instead of being carried off horizontally in an aluminum box.

But this prospect of survival carried an exacting price. While they were in camp, the days were long and the routine was hard. Actual missions were a respite. Glades and his team intently rehearsed every aspect of every mission, from tasks so seemingly simple as stopping for a map check to how they would handle a wounded prisoner after an ambush.

When he was finally satisfied with a rehearsal, then he would toss in different variables and unknowns. How would they react if a key man was wounded? What if a vital piece of equipment failed to func-tion? Rather than rely on fads, fancy gimmicks or untested gizmos, he distilled missions down to simple plans that could be executed in even the worst of circumstances.

They practiced IADs—Immediate Action Drills—until they were rote. They practiced individual skills until they became second nature. One day during monsoon season, the rain came down so hard that even Glades would not take them out to the ranges; instead, they stayed in their team hut and practiced changing magazines—over and over and over, thousands of times—until their fingers bled. He was not content to merely go through the motions and had no tolerance for anyone who didn't share his point of view.

Glades excelled in two significant areas. First, he was unmatched in his ability to find objects and people in the most dire of conditions. Second, he was unrivalled in his aptitude to lead men to seek and kill enemies of the United States of America. So he felt very secure in his job, since his skills and natural talents were constantly in great demand. Invariably, there was always something missing that needed to be located. And in lieu of that, there was always someone who needed to be killed.

Because of his unique abilities, he had an unusual relationship with the US Army and MACV-SOG. He came and went just about as he pleased. Like a ghost of sorts, he was virtually invisible to the Army's normal personnel management system. Instead of being assigned at the impersonal whims of the Army, he rotated between MACV-SOG and the Florida Ranger Camp at Eglin Air Force Base.

At Eglin, he trained junior officers and NCOs undergoing the final phases of Army Ranger School, shaping them into fighters and leaders who would keep their men alive and effective. He also returned to his home in Milton, just outside Eglin, to get reacquainted with Deirdre and their three children. Life would be normal for a while, or relatively so, until he returned to Vietnam six months later. And the cycle would go on indefinitely, so long as the war was being fought, until he was dead or was too damaged to return to combat.

As he packed his weapons, he recalled his childhood in West Virginia, where he grew up as the son of a coal miner. If asked, Glades usually attributed his uncanny marksmanship and field savvy to his father, but in truth, his extraordinary abilities were borne from a peculiar mix of childhood hunger and his mother's love. His mother had given him many other gifts as well.

When he turned seven, his father handed him an ancient Remington .22 caliber rifle. For Glades, it was a memorable occasion, a red letter day, because he rarely saw his father above ground in daylight except on Sunday, when the miners and their families packed into the Methodist church on the outskirts of the soot-choked company town.

His father escorted him out behind their home, a ramshackle clapboard cottage exactly like the hundred other houses provided by the company. He gave Glades a ten-minute lesson on how to line up the rifle's sights and carefully pull the trigger to send the bullet true. Then, his father solemnly counted out five .22 caliber rimfire bullets from a faded yellow box. He told his son that the five rounds were his five chances to kill a rabbit or squirrel for the dinner pot, and if the boy failed to bring home meat for the family stew, there would be a hard lesson to learn.

His older brother had previously undergone this same ritual, and had failed at his five chances, so Glades already knew the harsh consequences: There would be a severe beating with his father's heavy leather belt, and then the cherished rifle would be taken away forever. The welts and bruises would heal, but the shame would linger indefinitely.

With liquor on his breath, his father sternly recited the other salient conditions of the arrangement. If he brought home game with his five bullets, he would receive one .22 caliber bullet a day from then on, and one extra bullet per month, with the understanding that he would bring home meat or suffer the repercussions. A miss would be punished with a beating, but the rifle would remain his.

As he wiped excess oil from the silenced High Standard automatic, Glades remembered that day as if it were yesterday. As his father explained the simple workings of the bolt action Remington, his mother tended to her hardscrabble garden in their tiny back yard. She picked fat wriggling slugs from vegetables and carefully pruned suckers from tomato plants. Even though autumn was already arriving, she walked barefoot on the damp soil.

She wasn't that old; she had married at sixteen, and had borne Glades and his four siblings before she was twenty-one. He was sure that she had once been beautiful, but it was hard to see that beauty through a countenance that was an unvarying mix of bitterness, resentment, and sheer exhaustion.

Her face was pale white skin stretched taut over high Irish cheekbones and dappled with a constellation of pinpoint black freckles. Her head was crowned with a thick mantle of jet-black hair, as dark as the anthracite coal that came up from the earth below. She was thin, painfully so. Today, such skinniness would become synonymous with beauty, glamorized by a waif-like model named Twiggy. But his mother's form wasn't by choice or vanity; her perpetual gauntness was the result of constant hunger allied with the wasting effects of a chronic infestation of hookworms. In the years to come, she would grow thinner still as tuberculosis settled deep in her lungs.

The following days were filled with disappointment as Glades missed one shot, then another, and yet another. But then he cornered a hissing possum, and dispatched it with his fourth bullet. Elated, he brought the pitiful creature to the door and called for his mother.

As God's creatures go, possums aren't particularly endowed with physical beauty, but his mother wept when she saw the gut-shot marsupial, dripping dark blood from an array of sharp teeth. For the first time in his life, he glimpsed an expression of joy in his mother's face, probably when she realized that there was a chance that she and her brood might eat more regularly. While he was bursting with pride from his successful hunt, that pleasure quickly faded as he vowed to never again waste a bullet, to always deliver meat to the table so that he could see that same expression on his mother's face again and again.

As his mother dressed the possum and sliced it up for their evening meal, Glades cut a small patch of skin, about an inch square, from the animal's gray hide. He scraped the underlying flesh and fat from the back of his souvenir, and gently rubbed salt into it to preserve it. From that day on, he carried the patch of gray fur with him, like some boys might tote a lucky rabbit's foot.

For the next ten years, every morning, rain or shine, snow or sun, Glades carried the Remington with him to school, swaddling it in sackcloth and concealing it in a thicket before he entered the single classroom. On the way to school, he cut through the woods, carefully scanning for tracks and spoor; on the way home, he found and killed his prey. And every day, his grateful mother met him at the door, patiently waiting for the fresh meat that would help nourish her family.

As the years passed, she doted on her young hunter and provider, treating him much differently than her other children. His siblings hauled water, split firewood, washed dishes, swept floors, shoveled snow, scrubbed laundry, pulled weeds, and performed a multitude of other chores. In stark contrast, Nestor had but *one* chore, to kill efficiently, and he grew more proficient at his singular task with every day that passed.

Sometimes, his mother sang to him as she chopped potatoes and vegetables for the evening stew. As he grew still older, she patiently struggled to teach him the Gaelic poems and Irish dances she had learned from her mother and grandmother. In his efforts to please her, he clumsily tried to follow along, stumbling through jigs and butchering the language of the ancients. But she persisted, insisting that it was important that he knew the ways of his ancestors, that he could never know when such things would come in handy.

From the day when his father first placed the gun in his hands, Glades endured only four "whuppins" for misusing his daily bullet. Ironically, he suffered one beating—a lighthearted one at best—for killing two fat raccoons with a single shot. It was wasteful, his father insisted, since it was far too much meat for their pot, and they couldn't afford ice for their wooden icebox.

Two weeks before his seventeenth birthday, his mother succumbed to the ravages of tuberculosis. At the funeral, as his stoic father silently wept and his siblings bawled aloud, Glades gazed into the plain wooden coffin and saw that his mother had become beautiful again; her pale face was a masque of serenity and relief. He was even sure that he saw the subtle smile that she shared only with him. And in a tiny church overflowing with grief and the grieving, he felt no need to cry in his mother's passing.

The next day, he hitched a ride to Morgantown. He sought out an Army recruiter, who allowed him to stay in a back room until he was seventeen. That day, as the grinning recruiter momentarily looked away, Nestor signed his mother's name on the parental consent form. Ironically, she had never learned to read and write; in her short life, she had left only one signature: an "X" hurriedly scrawled on her marriage license.

He found his place in the Army, a home where he received all the bullets he could ever hope for. He fought as a paratrooper in Korea, where he earned a Silver Star. He transferred to Special Forces in 1959, served a tour in Laos with Project White Star, and was then picked for an exchange tour with the British Special Air Service. After completing their brutal Selection Course, he accompanied an SAS squadron on an operational mission in Oman, despite explicit orders prohibiting him from participating in combat operations with the Brits.

Returning from Oman, he met Deirdre at Hereford. She was the nineteen-year-old daughter of the SAS Regimental Sergeant Major. The RSM was an unusual soldier by SAS standards; he was Irish and had served in an Irish Guards tank battalion in the War before coming to the Regiment in the early fifties.

Glades had first spied her—a petite spitfire with bright red hair and a personality to match—at the Regimental Christmas dinner. He couldn't take his eyes away from her, and chatted her up by the Christmas tree as she helped pass out presents to the children in attendance. He spent the next few weeks wooing her with the Gaelic poems and songs that his mother had taught him. In her, he found a woman as beautiful and kind as any man could ever hope for, but yet as mentally and physically tough as any man could ever hope to be.

Barely two months after meeting her, Glades asked the RSM for Deirdre's hand. The RSM grudgingly conceded, knowing that although the Yank sergeant would take her far away, perhaps never to return, there could be no better man for his daughter. So they were married in the chapel at Hereford, and she later accompanied him to the States. And despite the wars and long separations, they had been together ever since.

Glades was an unusual fit for Special Forces, even though it was an unconventional organization that traditionally drew the sort of men who found difficulties fitting in elsewhere. Although the nucleus of Special Forces was the tightly knit "A team," composed of twelve highly qualified soldiers, Glades typically worked most effectively by himself or with just two or three other handpicked men.

Additionally, although the Special Forces were renowned for their ability to blend into other cultures, Glades quickly found that he had virtually no capacity to learn foreign languages. He could spend days and weeks memorizing words and phrases, much like the Gaelic poems he had learned from his mother, but for some reason they just never seemed to sink in. Context and conjugation were just mysteries to him.

His inability to learn other languages was a bane upon him. Other Special Forces men joked that if Glades weren't so damned good at finding stuff and killing people, he would have been sent away long ago. Oddly though, he was extremely successful at working with foreign soldiers, communicating mostly through hand signals and pantomime. In the case of his MACV-SOG indig soldiers, once they learned that the surest path to survival was to stay close to Glades and emulate his actions, they actually spent a considerable amount of their own time learning as much English as they could, so that they could effectively communicate with *him* rather than the other way around.

For his part, he always contended that he could operate virtually anywhere in the world if he had just four simple expressions in his working vocabulary—*Yes, No, Please, and Thank You*—along with just a few other simple phrases, like the appropriate numbers from zero to ten. These, reinforced with some mission-specific phrases, like *"Be still or I will kill you,"* typically sufficed for most of his interactions.

Others could laugh at his inability to communicate, but they could not question his sometimes uncanny accomplishments. A few years ago, a U-2 spy plane had lost power over the Soviet Union, eventually crashing in southern Uzbekistan. Months after the CIA and other agencies had abandoned the task, Glades was called upon to recover the pilot's body and destroy sensitive equipment at the remote site. Although the Soviets knew that the highly classified aircraft was somewhere out there, they had yet to find it themselves, even though they were actively looking for it with over a thousand men drawn from a Motorized Rifle Regiment based in Termez.

Naked, pushing a "poncho raft" containing native clothes and basic supplies, Glades had swum the ice-cold Amu Darya river from neighboring Afghanistan to meet a CIA-contracted Uzbek guide in Soviet territory on the far shore. After making initial contact with the guide, Glades simply drew a cartoon-like picture of the spy plane, pointed to it, and shrugged his shoulders. Then he pointed north and asked, "*Where?*"

Accompanied by a small herd of goats, which provided shallow cover for their presence and a reliable food source for the treacherous quest, he and the guide trekked for over three weeks, eventually locating the crash site high in the rugged mountains east of Karshi. In the course of the twenty-three-day search, the two men had not found it necessary to exchange a solitary word.

Glades smiled as he reflected on the interlude in Uzbekistan. Even though he hadn't learned even the slightest smattering of the Uzbek language, had almost died, and had lost all desire to ever again taste spit-roasted goat, he had actually enjoyed his sojourn in the remote wilderness. For those three weeks, he was so far out there that his presence couldn't even be reflected as a colored pin on someone's map. There were no briefings or debriefings to attend, supply requisition forms to fill out, no equipment to account for, no radio frequencies or map coordinates to memorize. For that brief time, life was simple and good.

He sighed as he closed the hasps on the footlockers. He reconciled himself with the notion that it was time to go home, and while he longed to see Deirdre and their children, he would miss this place. And beyond that, he would miss the men who would die in the coming weeks and months, mostly because he would not be here to keep them focused and safe.

As he checked his watch, he heard his One-One's Bronx accent from outside the teamhouse, "Nestor, come *on*. The guys are waiting. It's time to eat peaches."

Yeah, mused Glades. *It was time to eat peaches.*

4

PIGEONS

Woodland Cemetery, Dayton, Ohio
3:30 p.m., Saturday, August 9, 1969

Even as he desperately labored to gather information about his primary target, the UFO research agency named Project Blue Book, Soviet GRU Major Anatoly Nikolayevich Morozov was also tasked to perform an initial evaluation of an American airman stationed at Wright-Patterson Air Force Base. Apparently anxious to do business with his nation's enemies, the airman—Staff Sergeant Eric Yost—had contacted the Soviet Embassy last month, desiring to sell information about UFOs. As with any potential source, using their embedded contacts within the Air Force personnel system, the GRU had already executed a fairly extensive check of his records.

Interestingly, the would-be turncoat possessed a Top Secret clearance, so he potentially had access to juicy information, which was why Morozov's bosses had authorized the evaluation and—if so warranted—recruitment. Otherwise, Yost's drivel-filled letter and four pictures, which showed several blue coffin-shaped boxes being unloaded from an official station wagon, would be smoldering ashes in the Embassy's basement furnace.

Besides Yost's security clearance, other issues raised some eyebrows within the GRU, particularly given the airman's proximity to Blue Book. Yost insisted that he could provide tangible evidence of alien space-craft captured by the US Air Force and stored at Wright-Patterson. Also noteworthy was the location where the photographs were taken, a hangar belonging to the "Aerospace Support Project." Although the GRU was aware of the organization, no one had a clue about what went on behind its doors.

If Yost's information could draw a line between the Aerospace Support Project and Project Blue Book, his recruitment could be fortu-itous, perhaps adequate to justify an extension of Morozov's Ohio mission. If Morozov remained clear of the Embassy long enough, then his intelligence career might be rescued from the monotonous doldrums of file work, making tea, and conducting surveillance on the bored mistresses of obscure US military officials.

On the other hand, it was equally likely—probably more so, in fact—that Yost would turn out to be just another self-inflated, greedy malcontent. His official records indicated a history of disciplinary troubles resulting from a proclivity for drinking to excess. Maybe Yost wasn't worth the effort, but then again, as Morozov's aged mother was apt to point out, mushrooms don't grow in pots of gold and silver. Consequently, Morozov was doled some operational cash and a substantial amount of leeway to determine whether Yost might be of potential value.

After going through the preliminary steps of the agent recruitment mat-ing dance and mailing a set of meeting instructions to Yost's post office box, Morozov was seated on an uncomfortable concrete bench, waiting for the American sergeant to arrive. He watched squirrels sprint anx-iously between the tombstones and listened to leaves rustling on tree branches.

As he waited, Morozov opened a local newspaper and placed it in his lap. That was the safety signal; per the instructions, if Yost saw the paper in Morozov's lap, he would know that it was secure to meet. If the safety signal was not correctly displayed, Yost should say nothing, walk by, leave the cemetery, and wait for further instructions to be mailed.

He heard a faint crunching noise and observed a slovenly man slowly approaching on the graveled walkway. The man had a very awkward gait, but Morozov couldn't determine if it was a pained limp, a drunken stagger, or perhaps a mixture of both. As he drew closer, Morozov saw that the man was unshaven and had short dark hair, was of medium height and very thin. Wearing khaki trousers and a red plaid shirt, he matched the general description of Yost and was wearing the identification colors specified in the instructions for the meeting.

"Is there a mausoleum near here?" asked the man, clearing his throat and speaking in a quiet monotone. "I'm looking for one of my ancestors."

"Have a seat," ordered Morozov.

"Eric Yost," muttered the man, slurring his words as he stuck out a hand.

Although it was not part of the protocol established for the meeting, the GRU officer shook Yost's hand. His grip was weak, like a woman's. It was like he imagined all American handshakes to be. Yost's foul breath reeked of liquor. "Mr. Yost, have you been drinking?"

"Is Pope Paul Catholic? Hell *yes*, I've been drinking! It's Saturday, isn't it? You want a nip?" Yost held out a silver flask. The flask bore an inscription engraved in cursive script: "To Eric, *mit liebe*, From Gretchen."

Straining to conceal his disgust, Morozov shook his head. *What kind of moron would imbibe before a potentially perilous clandestine meeting?* "Your identification, please," he said.

Yost took out his wallet and clumsily extracted his Air Force identification card. He tendered the plastic-laminated card to the Soviet officer.

Cupping it in his hand, Morozov scrutinized the credential and then looked at Yost. It was definitely a match, except that the frowning man in the photograph wore a moustache.

"It was a phase I was going through last year," explained Yost, obviously recognizing Morozov's concern with the discrepancy. "My wife didn't like it, so I shaved it off. I guess it really doesn't much matter anymore what that ungrateful bitch thinks."

The details also matched what Morozov had read in Yost's personnel dossier. He gave the card back to Yost. A flock of doves landed in front of the concrete bench, plaintively cooing as they jostled each other on the walkway.

"I hate pigeons," observed Yost, shooing one away with his foot before taking a sip from his flask. "Nasty things. They're like flying rats."

"They're not so bad." Morozov pulled out a brown paper bag from a jacket pocket and then cast bread crumbs to the ravenous birds. "I like to feed them. In Russia, the old people consider them sacred, a living symbol of the Holy Spirit."

"I didn't think Commies . . . I mean, Communists . . . believed in God. So you're religious?"

"Not really. Not anymore, anyway. I grew up during the War. We had to eat pigeons and a lot of other things not nearly as pleasant, or we would have starved. We thought of it as taking communion, which made some things a little easier to swallow. Listen—time is short, so let's get down to business. Do you have something for me, Mr. Yost?"

Yost nodded. "I have two sets of pictures. Here's the first set." He handed Morozov a small stack of color prints and then jammed his hands into his trouser pockets.

Setting his bag of crumbs on the bench, Morozov examined the prints. The sequence showed six identical pale blue boxes, perhaps a meter long by a half-meter wide, being unloaded from an Air Force station wagon into a hangar or warehouse. "You seem to have a fixation with boxes, Mr. Yost," he noted. "These aren't much different from the four pictures that you sent us by post. These containers are smaller, perhaps, but they're still just boxes."

"They're *not* just boxes," insisted Yost, still slurring his words. "Those are *obviously* caskets. So were the others I sent you."

"Hah!" sniffed Morozov, picking up his paper bag and sowing more crumbs. "If these are caskets, they're for children, maybe. They're much too small for—"

"*Look*," urged Yost. "There are armed men guarding them. You want to tell me that those are just simple boxes? And do you see those civilians standing here? I'm guessing that they're doctors, getting ready to do some sort of autopsy or dissection."

"Well, I'll concede that it appears a little unusual," commented Morozov, examining the details again. "But when we talked on the telephone, did you not assure me that you had photographs of an alien spacecraft?"

Pressing another set of photographs into Morozov's hands, Yost chuckled and said, "Feast your eyes on *these* snapshots, comrade."

Morozov sorted through the prints. They showed a series of crates, of various shapes and sizes, being unloaded from trailers. He was mystified why Yost seemed so convinced that they had anything to do with alien spacecraft. "These are just *crates*, Mr. Yost. Granted, they're unusual, but they are still crates." Morozov sniffed, handing the prints back. "Nothing more. Nothing of note."

Jabbing a stubby finger at a print that depicted an enormous flat circular crate being hoisted by a massive forklift, Yost declared, "Look, this crate is the perfect size and shape for a flying saucer! Do you think it's just a coincidence that I saw some sort of alien spaceship being moved out of here one night, and then a few weeks later, these circular crates are being moved in? And the coffins? You don't think that this hangar is a holding area for UFO stuff?"

"Keep your voice *down!*" snapped Morozov, glancing from side to side, almost as if he were concerned about inadvertently waking the dead. "Let's take this one step at a time, Mr. Yost, and not be so hasty to jump to conclusions. You claim that you saw an alien spaceship. Describe it to me."

Yost described the details of the bizarre vehicle he had witnessed in July of last year. "And that's it," he stated. "Obviously, it had to be powered by some sort of anti-gravity machine."

Morozov clicked his tongue quietly and shook his head. "I'm confused, Mr. Yost. You go to the trouble of documenting all of these mysterious boxes and crates, but when you saw *this* object, this space vehicle that you're so adamant about, you didn't see fit to photograph it as well? How do you explain that discrepancy?"

Yost groaned and blurted, "Because I didn't have a damned *camera* yet. I started carrying one after that. Plus it was nighttime when I saw it, so the pictures probably wouldn't have developed. But I'll bet you this: sooner or later, I will see that damned thing again, and when I do, I'll be ready."

Morozov sat quietly, pretending to scrutinize the photos as he contemplated the situation. Although Yost's claims were farfetched, Morozov didn't doubt him. The GRU officer considered himself an accomplished judge of character, and while the American was clearly a drunken buffoon of marginal moral fiber, he did appear to be entirely forthright about the strange object. He had obviously seen *something* that he sincerely believed to be an alien spacecraft.

Whether Yost's claims merited further pursuit was another matter entirely. Morozov was authorized a fairly sizeable purse for the initial recruitment if he genuinely thought that Yost could furnish something of value, but a more pronounced effort would require approval from his bosses back in Washington. At this point, the temptation to walk away was great, but the temptation to exploit Yost was greater. If he elected not to recruit Yost as a source, Morozov could probably accomplish his espionage chores and write all of his reports in short order, probably a couple of weeks at most. But if he convinced his GRU superiors that Yost was worthy of prolonged exploitation, then Morozov might be here indefinitely. Moreover, if the American sergeant's claims did pan out, it might be the coup that finally lifted Morozov's fortunes so he could matriculate to a more prestigious assignment, like Vietnam.

Even if his outlandish claims of aliens and UFOs were taken out of the equation, there was yet another reason to exploit Yost. Unlike most sources that the GRU ran, where they routinely solicited Americans under the guise of a false flag, Yost had made a conscious effort to be recruited by Soviet intelligence. If Yost ever eventually proved to be obstinate, Morozov could exert a tremendous amount of leverage; because a genuine traitor had something to truly be afraid of, they were subject to great manipulation. Since he had a Top Secret clearance, he had to possess access to *some* useful information, even if it didn't concern UFOs.

He had no doubt that the brash American was of value, but it was critical that he dominate the haggling that would ensue. He contemplated his negotiating options as he flicked more bread crumbs to the pigeons. A cardinal rule of developing unsolicited sources was to convince them that their information was only of marginal value, regardless of what they knew or suspected.

"And how about that pilot who went over the cuckoo's nest last year and then just vanished?" asked Yost, interrupting Morozov's thoughts. "The pilot who was working in this hangar? Do you suppose that was just some kind of coincidence also?"

"How can you be so sure that he was a pilot?" asked Morozov.

"My friend is an orderly at the base hospital. He saw this guy's records before he disappeared. There's a special code on the folder for pilots. That's how he knew."

"Who is your friend, the hospital orderly?" queried Morozov. "Tell me his name."

"I can't do that. If you contacted him, then . . ."

"I have no intent to contact him, Mr. Yost. I just want to verify that such a person exists and is actually assigned to the hospital. Certainly, you can understand that until we authenticate this information, it's all hearsay. I'll tell you what: give me his name, and if your information checks out, I will give you a two-hundred-dollar bonus the next time we meet."

Yost penciled a name on the back of one of the photo prints. "His name's Carr," he noted. "Bob Carr. He's been assigned to Wright-Patt for the past three years. He transferred here from Rhein-Main Air Force Base in Germany, near Frankfurt. That's where we met."

Morozov turned his head away and examined the photos again. Before them on the sidewalk, one of the pigeons walked around in a small circle and then keeled over. The others, seemingly indifferent to their fellow's plight, continued to peck at Morozov's crumbs.

"One of your holy pigeons is sick," observed Yost, chuckling. He tipped up his flask, emptied it, and replaced the cap. "Maybe your bread is moldy."

"No. Sometimes they become greedy and eat too much," said Morozov, placing the photos on the folded newspaper between the two men. "And then they become easy fodder for cats and foxes. Such is the nature of greed, Mr. Yost."

"So how much are you going to pay me for the pictures?"

"I'll tell you what, Mr. Yost, since I'm in a generous spirit today, I'll pay you one thousand dollars for these photos, under one condition."

"Just a *thousand* bucks?" groused Yost. "Whatever your condition is, it had better be good."

"I want *more* information about what's in that hangar," said Morozov curtly. "So here is what I propose: If you can provide me with a name and contact information for someone who works in that facility, I will pay you *another* thousand dollars, plus a substantial incentive if we're able to recruit them. They have to be reliable, though, and must have routine access to the hangar."

Grinning, Yost bent over to scratch his ankle. "I work close by," he answered. "I am pretty sure that I can scratch up some names."

As Yost pulled down his sock, Morozov noticed that the skin around his foot was pale and puffy, like that of a corpse that had been submerged under water. "Let me make something clear to you, Yost—you will be paid only if their information checks out. Do you understand?"

Yost nodded. "The names should be easy, but the phone numbers might be difficult. What could I tell them if they get suspicious?"

Morozov pondered the American's question, and then replied, "Let me think on that. I'll send some instructions to your post office box. One more thing: How about this agency called Project Blue Book? And the

Aerospace Support Project? Can you acquire information about people who work there as well?"

"I'll see what I can dredge up."

"Good," said Morozov, handing Yost two rubber-banded stacks of ten dollar bills. "As a show of good faith, I will pay you *two* thousand dollars for your photos today. Later, I might also issue you some special equipment, like a miniature camera, and special training. We will also compensate you for your time during training. Does that sound fair enough to you?"

"Now we're starting to talk turkey," declared Yost, quickly counting the cash.

Morozov handed Yost an envelope. "Here are instructions on how to contact me. Do *not* contact me unless you can provide me with reputable information about someone who has routine access to the interior of the hangar. Do you have any questions, Mr. Yost?"

Satisfied with his tally, Yost stuck the money in his pocket, grinned, and shook his head.

"You know your way out of here, I assume," said Morozov, tucking the pictures in the interior pocket of his jacket. "We'll be in touch."

Morozov waited patiently as he watched Yost slowly hobble away and then shifted his attention to the throng of pigeons. A few more of the gluttonous birds had fallen victim to his vodka-laced bread crumbs; they lay on their sides cooing in docile confusion. Morozov collected four of the fattest ones, snapped their necks, and then rolled them up in the newspaper.

Never knowing where his next assignment might be, the pragmatic officer made a point of maintaining the survival skills he had honed as a child of Stalingrad. Like Yost, most Americans perceived pigeons as nuisances, but Morozov knew that the birds' breasts were delicious when breaded and fried. He would enjoy these tonight, cooked on a hot plate in the dreary boarding room he rented by the week.

As he strolled out of the cemetery, he mused that few Americans would ever stoop to consuming pigeons, but probably fewer still would ever know the taste of human flesh. Recalling the horrific days when his family huddled in the shelled-out remains of their Stalingrad apartment, Morozov only wished that he could forget.

Simulator Facility
Aerospace Support Project, Wright-Patterson Air Force Base, Ohio
9:54 p.m., Sunday, August 10, 1969

Drenched in sweat, almost too numb to move, Ourecky sprawled on the lower steps of the stairs leading to the Box. His hands ached and elusive numbers danced in his thoughts, refusing to be confined in orderly equations. He had been tired before, but rarely this drained, even during their marathon runs earlier in the year. They had been at it all weekend, even after spending the entire week immersed in a seemingly endless round of intelligence and technical briefings in Colorado Springs.

Bea left for Atlanta this afternoon; he hadn't seen her all week and hadn't spoken to her since last Sunday. On a positive note, she was due to come off flight status after this round of flights; grounded, thankfully, she would work gates at the airport until the baby arrived in March. As yet, because of Bea's concerns that she might not even be able to carry the baby to term, Ourecky had not shared the news with anyone, not even Carson.

"Let's go again, Gunter," declared Carson, swaggering in from the suit-up room with a Coke bottle in his hand. "T-Minus ten minutes to orbit, and then let us cycle through IVAR and the post-insertion routine. Just another hour, and then we can call it a night after that."

"It's nearly ten o'clock, Drew," lamented Heydrich, looking at the clock on the wall. "And it's *Sunday*."

"But I still want to make sure we've polished off the rough edges." Carson sipped the last of his Coke before tossing the empty bottle into a nearby trashcan.

"There are *no* rough edges left to polish." Wearing a rumpled white shirt and coffee-stained tie, Heydrich looked even more disheveled than usual. "My guys are *exhausted*, Drew. They're on their last legs. I really don't see what we're going to accomplish with another run. You two are more than proficient on *everything*. We'll be back first thing in the morning. Please let my men get some rest."

Carson shook his head as he tore open a pack of chewing gum. "If my memory serves, Gunter, Virgil directed you to keep the Box operating until I felt confident that we were ready. I don't feel entirely confident, so I want to make another launch run. I don't want to pull rank on you, but if I have to call Virgil, I will. Please don't make me do that."

Heydrich nodded glumly, sighed, and slowly pivoted to face the far end of the hangar. A gaggle of controllers were gathered at the coffee urn, apparently grumbling over their lost weekend or perhaps drawing up plans for a mutiny against Carson's tyrannical reign. With a wave of his hand, Heydrich beckoned them back to their stations.

Zipping up his flight suit, Carson met Ourecky at the foot at the stairs. "Saddle up, buddy. We're going again."

"Gunter says we're ready," mumbled Ourecky, slowly clambering to his feet. "Don't you trust his judgment?"

"*Gunter's* fat ass isn't going to be strapped on top of that rocket on Johnston Island," answered Carson, speaking over his shoulder as he swiftly scrambled up the stairs. "*Yours* is. And I think we've covered this ground already. We're ready when *I* say we're ready, and not before. So quit griping and start moving."

"But . . ."

"Scott, I don't know if you're getting too cocky or whether you're just getting lackadaisical, but either one will get you killed," said Carson fervently. "One more round. Let's go. *Mush!*"

Following him, Ourecky slowly scaled the four metal steps like a condemned man headed to an appointment with the gallows. He observed Carson as he climbed into the simulator and squeezed into his seat. Ourecky *hated* the Box, but the spry pilot seemed to relish in it. If he didn't know any better, he might suspect that Carson was just a glutton for punishment, but there seemed to be something far deeper at play. Carson became an almost different person in the simulator, intensely focused and resolute. Now, as the pilot lay on his back cinching his harness while simultaneously verifying switch settings, his outward demeanor changed as he slipped into that determined persona.

Recalling their pre-dawn sessions at the base boxing gym, Ourecky realized that he had witnessed the same pained but resolute expression on Carson's face, when the pilot pummeled the heavy bag or endured a brutal sparring session against a much larger opponent. He suspected that the Box was like a penance or cleansing ritual that Carson felt compelled to inflict upon himself, a torture he must suffer until his personal demons were exorcized or at least temporarily diminished. Ourecky knelt down before his open hatch and allowed a technician to assist him as he wriggled into place in the snug cockpit. As he slipped on his headset and adjusted the microphone, he looked to his left to watch Carson review a pre-launch checklist. If the Box truly was a rite of purification, he only hoped that Carson was sufficiently purified before it also destroyed him and his marriage in the process.

5

THE FOURTH WORLD

Cap-Haïtien Airport, Haiti
2:35 p.m., Tuesday, August 12, 1969

Just arriving from Miami, emerging from the DC-3, Matthew Henson was immediately awestruck by the sweltering heat. It was almost overwhelming, like someone had thrown open the door to a massive blast furnace. Gabon had been hot, but not anything like this.

After the passengers collected their luggage at planeside, a pair of Haitian soldiers cajoled them into a file and walked them inside. The one-story terminal was constructed of unpainted concrete cinderblock. It was dirty and loud, jam-packed with waiting families and curious on-lookers. Shouting vendors hawked trinkets, wood carvings, fruits, vegetables, and unlabeled bottles of local rum.

The handful of disembarking Haitians melted into the bustling crowd; the soldiers paid them no mind, instead focusing their attention on Henson and the dozen Americans and Europeans arriving with him. As they queued up for their passport stamps, Henson breathed a sigh of relief as he realized the new arrivals were only subjected to a cursory inspection of their luggage and a quick screening of their travel documents.

"You American?" asked a man waiting in line behind Henson. Appearing to be in his mid-thirties, he was deeply tanned, chubby, of medium height, and had graying brown hair. He wore a wide-brimmed Panama straw hat, khaki chinos, old work boots, and a faded brown work shirt. His meager luggage consisted of a well-travelled gym bag and a banged-up cardboard box labeled "Re-Built Carburetor & Spark Plugs—Hold for CT."

Henson nodded. "I am. Matt Henson. I work for Apex Exploration Services in Ohio. We do mineral exploration and supporting work for mining companies."

"Sure you do. I'm Craig Taylor. Missionary Air."

"Missionary Air?" asked Henson. "You're a missionary?"

Taylor chuckled. "Not hardly," he answered, lifting the brim of his hat and wiping sweat from his brow. "I work a Maule M-4 bush bird out of here. I fly for missionaries scattered all over the country. All denominations. I shuttle people around, deliver supplies, mail, fresh food, medicine, you name it."

Nudging his gym bag forward with the toe of his boot, Taylor added, "Their stuff comes in here, I receive it, fork over the requisite bribes and handling fees, and then fly it around the island. It's a righteous gig, especially considering that I lost my flying ticket back in the States."

"How did you manage that?" asked Henson, whacking an iridescent green fly perched on his forearm. As he brushed away the flattened pest, it was immediately replaced by two more.

"I was suspected of moving some illicit cargo, if you know what I mean. They could never make a smuggling rap stick, so they just nailed me on a string of safety violations."

Hearing his name shouted from the crowd, Taylor smiled and waved at a pair of elderly Americans who were obviously missionaries. Turning back towards Henson, he said, "I wasn't being nosy when I asked if you were from the States. It's just that I've been down here long enough to see most people come and go at least a few times, and I just didn't recognize you."

"No problem. Look, once I settle in, I'll be shopping around for a small building to work out of, maybe a truck, odds and ends like that. Any ideas?"

Taylor nodded. "Well, personally, I like to stay close to my plane. Spare parts are hard to come by down here, so I don't like things to come up missing. I have a shed out by my airplane, on the south end of the airport."

Henson nodded. "That's how I like to work also. Once I find a suitable place, I'll probably just string up a hammock and stay there. In the meantime, do you have any suggestions for lodging? Any decent places to eat?"

"There are a couple of really good hotels in town. The Hotel Roi Christophe is top of the line and relatively affordable. The local grub is really good, but they don't consistently enforce sanitary standards, if you catch my drift. Avoid the mom-and-pop bistros that cater to tourists."

"Inflated prices?" asked Henson. "That's common with tourist traps."

"Yeah, but that's only part of the problem. They'll serve you up a heaping plate of gristle and rice dressed up as haute cuisine, and then you'll spend the next few days camped on porcelain. My recommendation is to hire yourself a cook if you're going to be here a while. The fresh fruits and vegetables are really good, and they're safe if you give them a thorough scrub, but I would be cautious about eating any meat."

A throng of barely clothed ebony children scurried through a gap in the flimsy barriers. With their hands outstretched, the waifs flocked around the Americans and cried out, *"Blan! Blan! Blan!"* A Haitian soldier shooed them away, swatting a couple with the butt end of his M1 rifle.

"Blan?" asked Henson. "That sounds almost like *"blanc."* That's French for white."

Taylor nodded. "Correct. *Blan* is Haitian Creole for white, but it's also what Haitians call *any* foreigner, regardless of their skin color. You had better get used to that notion quick, because if you think your black skin will yield you any special treatment here, you're mistaken."

"Interesting," noted Henson, shoving his suitcase forward with his foot. "So how do you get on with the missionaries? You don't strike me as the religious type."

"It's not unlike any other business," replied Taylor, lighting a cigarette. "You just have to know who you're dealing with. Most missionaries are very practical people, and they recognize that they would be between a rock and a hard place without my services, so they don't waste a lot of time beating me over the head with a Bible."

Taylor continued. "We occasionally have some hardcore Southern Baptists come in here, and they can be a pain. They swing in here preaching fire and brimstone, quoting the book of Leviticus at every turn, and they have it in their heads that they're going to change this place for the better. The first thing they intend to do is to expunge voodoo from the cultural landscape."

Henson laughed. "Voodoo? Now, we're talking. Man, I grew up in New Orleans. There's *nothing* you can tell me about voodoo that I don't already know."

"Trust me, it's not the same thing," asserted Taylor. "Not by a long shot. Down here, voodoo is not just a few fakers in fancy shops selling trinkets and potions to tourists. It's deeply engrained in the Haitian culture. It comes from their African traditions. Voodoo is a powerful force in this country. If you don't understand that or can't at least appreciate it, you might as well climb back on that plane and head back to Miami."

"It's *that* serious?" asked Henson, shooing away an aggressive mango vendor.

"Yeah. If you're going to work here, you have to understand the linkage between the people and voodoo. This might strike you as a bit odd, but Haitians are a deeply religious people. The population is roughly ninety percent Catholic, but they're also one hundred percent voodoo. A lot of missionaries just can't comprehend how the two things can exist simultaneously. They can't comprehend how the average Haitian can tank up on rum on a Saturday night, participate in a voodoo ceremony at the local peristyle, and then appear for Mass on Sunday morning. I guess the simplest explanation is that they're hedging their bets by covering all the spiritual options available to them."

"Makes sense," noted Henson. "I'll try to keep an open mind about it."

Taylor stepped closer, lowered his voice and tersely asserted, "Let me warn you: While you're here, don't run afoul of the voodoo people. A lot of what they do is homegrown hypnotism mixed in with hocus pocus fakery, but there's a few of them that conjure up some powerful magic as well. If you hang around long enough, they'll make a believer out of you."

"So the voodoo doctors run the show?" asked Henson.

"*Houngans.* A voodoo priest is a *houngan.* A voodoo priestess is a *mambo.* A *houngan* or a *mambo* oversees the rituals at a peristyle, which I suppose you would call a voodoo temple. A *bokor* is a houngan who practices black or evil voodoo. The most powerful *houngans* are the *houngan asogwe*, who are kind of like high priests. Some *houngans* are also *loup-garou.*"

"*Loup-garou?* I know that word. It's French for werewolf."

"Sort of," said Taylor. "But here, *loup-garou* means shape-shifter. A *loup-garou* is someone who is able to change into another form, like a snake or a bird. Anyway, to answer your question, the *houngans* control the people, but even they fear the *Tonton Macoutes.*"

"*Tonton Macoutes?*"

Taylor subtly motioned towards two hulking Haitian men lurking in a corner. Dressed in dark suits with open collars, the scowling men wore mirrored sunglasses and carried identical bags slung over their shoulders. "The *Tonton Macoutes* are secret police who answer only to Papa Doc Duvalier," he explained. "He's granted them blanket amnesty for any crimes that they commit, so they're none too shy about abducting or torturing anyone who might be an enemy of Papa Doc. Everyone fears the *Tonton Macoutes,* even the *Fad'H.*"

"*Fad'H?* What's that?" asked Henson.

"*Fad'H. Forces Armées d'Haïti,*" explained Taylor. "The Haitian military. Anyway, *Tonton Macoute* literally means *Uncle Satchel*. It's from a folk tale where a boogeyman prowls the streets after dark. If he catches a kid out late, he stuffs them in his satchel and the kid is never seen again. And that's pretty much how they conduct business."

"I'll watch my step."

"Do that," advised Taylor. "See how heavy those satchels hang? They all tote an Israeli Uzi submachine gun and a short machete in their bag. If they ever snatch you up and start hacking into you with the machetes, then you'll be begging for the Uzis in short order. And there's more."

"More?" asked Henson.

"Yeah. Most Haitians believe that the *Tonton Macoutes* have been granted special voodoo powers that make them even more potent than a *houngan asogwe*. They believe that the *Tonton Macoutes* can take possession of a person's soul and leave them to walk the earth as something less than human."

"A zombie?" smirked Henson, rolling his eyes.

Taylor closed his eyes and grimaced. "Yeah, Henson, you think it's funny *now*. Wait until you've been here a while. You'll probably be a little less skeptical then."

Henson looked furtively at the two *Tonton Macoutes*. One glared back, grinned menacingly, and then nudged his mirrored sunglasses down onto the bridge of his broad nose. Despite the stifling heat, a chill froze Henson's spine when he glimpsed the man's intimidating eyes; they were jet black, both inert and vibrant at the same time, like a shark's.

Shuddering, Henson quickly averted his gaze, wondering if it was really possible that the man could wrest his soul from his body. Looking ahead, he saw a young woman behind a desk wave him forward.

"You're on deck, Henson," said Taylor, offering a business card. "Come find me if I can ever be of assistance. Don't bother with the telephones; they don't ever work. Just flash my card to anyone at the airport, and they'll take you to find me. Good luck."

Henson nodded and stuck the card in his shirt pocket.

Sighing, Taylor shook his head. "Matthew, let me give you your first lesson about doing business down here in Haiti. These people have *nothing*, so anything and everything is of value to them. If someone hands you a business card, it's a hugely significant gesture. It implies that they accept and trust you and want to establish a relationship. If you jam it in a pocket without looking at it, like you just did, you're making a subtle statement on how little you respect them."

"Thanks." Henson pulled out the card and placed it carefully in his wallet. He stepped forward to the desk and tendered his documents to the woman. A listless *Fad'H* soldier, barely conscious in the infernal heat, lolled on a stool by the door. Two yellow chevrons, sewn sloppily on his uniform sleeve, identified him as a corporal.

The woman compared Henson's passport to a typewritten list of names. Turning her head, she nodded at the *Fad'H* corporal. Suddenly animated, he stood up and snapped his fingers. Two other soldiers whisked through the door and seized Henson by his upper arms, while another grabbed his luggage. Swiftly escorting him outside, they jammed him and his confiscated belongings into the backseat of a waiting Willys jeep.

The jeep jolted into gear, and they left the airport grounds, heading roughly north on Highway Three. Henson had endured similar drills in other countries and had learned to remain calm and observant. In the week since Grau had directed him to come down here, he had studied Haiti, so he was relatively familiar with the geography.

He had also read an old Marine Corps guidebook, published when the immortal leatherneck hero Chesty Puller fought here as a corporal in the early twenties. It contained a basic Haitian Creole translation guide that Henson had gleaned for rudimentary vocabulary. Now, he eavesdropped on the conversation between the corporal and the driver, diligently trying to pick out words and context.

Nothing he had read could adequately convey the harsh reality and hopelessness of this place. The outskirts of Cap-Haïtien were dominated by sprawling expanses of slums. The people lived in squalor like he had never witnessed in Africa or elsewhere.

Most of the dwellings were shacks fashioned of wood scraps and cardboard. Now and then he saw a corrugated metal roof, but more often the roofs were thatch. Naked children frolicked in puddles of fetid water; women knelt in the same stagnant pools to wash clothes, and Henson also witnessed others squatting there to relieve themselves. A noxious stench, a mixture of odors from charcoal cooking fires and heaps of rotting garbage, permeated the air.

Shortly after leaving the airport, they came upon an accident that had brought traffic to a standstill in both directions. A porter had been pushing a massive cart, heavily laden with household goods—quite likely the entire furnishings of a modest middle-class home—when an axle broke, flipping the conveyance and scattering its contents. The porter's arm had been fractured in the mishap, and a sliver of white bone poked out from a profusely bleeding gash in his right forearm. A cloud of black flies had already descended on the wound; the swarming pests jostled for their share of fresh blood and newly torn flesh.

Residents of the surrounding neighborhoods converged on the scene. While none helped the injured man, most rummaged through the spilled goods. Cradling his shattered arm, the barefoot porter—who essentially hired himself out as a human truck, a common occupation in Haiti—pleaded for them to stop.

Most of the looters were content to grab an item or two, but scuffles broke out over the more prized plunder. In the trickle-down manner of wealth distribution in an anarchic society, the strong took the choicest items from the weak, then the less-weak took from the more-weak, and so on and so on, until an uneasy equilibrium was reached.

The traffic jam grew progressively worse as drivers and passengers left their vehicles to join the spree. The lanky corporal eventually grew impatient, cursed, and rushed into the swirling crowd. Prodding with his M1 carbine, he ordered the passersby to cease their pillaging and to push the broken cart off the roadway. Begrudgingly, they shoved it to the shoulder, and then the frenzy of looting and squabbling resumed.

Carrying a box of silverware, the *Fad'H* corporal swaggered back to the jeep and climbed in just as traffic started to flow again. As they passed the derelict cart, Henson saw the hapless porter squatting by the side of the road, holding his broken forearm between his knees as he simultaneously wept and swatted at the merciless flies.

As they surged past the aftermath of the accident and picked up speed, Henson's attention was drawn to an incandescent flickering glow in the distance. As they drew closer, it grew so intensely bright that he had to squint and shield his eyes. Finally, he realized the source of the mysterious light. Two Haitian men had strung a pair of wires to a power line running beside the roadway and were using the bare ends to weld together pieces of a broken bicycle frame.

As they rolled into Cap-Haïtien's central district, the gray teeming squalor was supplanted by picturesque rows of quaint stucco buildings in peaceful pastel shades. It was as if a giant hand had neatly sliced blocks of ice cream—strawberry, pistachio, vanilla, peach—and fashioned an entire town from them. The streets were reminiscent of the whimsical images of Cap-Haïtien that Henson had seen in the tourist brochures.

The jeep pulled into a fence-enclosed compound dominated by a stately white building. The *Fad'H* headquarters was guarded by at least two squads of soldiers; the stern-faced black troopers wore starched fatigues and whitewashed webbing, and were armed with vintage M1 rifles. The rifles' wooden stocks glistened with hand-rubbed linseed oil.

Henson's two attendants gruffly pulled him from the jeep and brought him inside. "Colonel Roberto will see you, *blan*," announced the corporal, nudging him down a hall and into an office. A wooden sign—hand-lettered in inlaid gold leaf—identified the office's occupant as Colonel Roberto Hector Gonzalez, Commander of the *Fad'H* Northern District.

It was a strange quirk of Haitian culture that while the nation was overwhelmingly populated by the descendants of African blacks previously enslaved by French plantation owners, its aristocracy was predominately of Hispanic or mulatto heritage. Just like so many glaring contradictions of this country, Henson found it puzzling that the slaves would cast off the yoke of one white oppressor only to be effectively dominated by another. After all, a prevalent joke—although not entirely humorous—was that the Haitian flag's background was the same as the French tri-color, with the white part cut out, just as the revolting slaves had literally chopped the French out of Haiti in the early nineteenth century. But despite the traumatic excision of the French colonists, high offices and military ranks were rarely held by Haitians of African descent.

The corporal shoved Henson into a wooden chair in front of Colonel Roberto's large mahogany desk as the driver deposited his baggage on a table in the corner. Wearing an elegantly tailored dress uniform that showed not the slightest dampness of perspiration, the fastidiously groomed colonel was almost absurdly handsome, as if he were the thoroughbred product of countless generations of careful breeding.

The *Fad'H* officer's thick mane of black hair was meticulously combed, with not an errant strand to be seen, and the pencil-thin moustache that adorned his upper lip was absolutely symmetrical. Looking almost too perfect, like a malevolent version of Ricky Ricardo cast as a Latin dictator, he probably could have made himself at home in any steamy, intrigue-festering banana republic in the western hemisphere.

The corporal proffered Henson's passport to the colonel, snapped to attention, saluted, and briskly exited the office. Colonel Roberto intently scrutinized the document and then removed a small ledger from a desk drawer. Still silent, he scanned a list of names in the ledger, obviously attempting to find Henson's. Frowning, he slowly shook his head and laid the passport on his desk. "Just what is the nature of your business here?" he asked, getting straight to the point.

Henson answered calmly, "I'm with Apex Exploration Services, based in Dayton, Ohio. We do exploration work for mining companies."

Henson then recited the cover story he had told so many times in the past few months, adding the embellishments specific to the local situation. In this instance, he described how Apex was on the hunt for bauxite, which was used to manufacture aluminum, and elaborated on how he intended to hire locals to assist him in the search.

Roberto listened patiently to Henson's explanation, then politely clicked his tongue and commented, "Ludicrous. There is nothing of value here. Nothing on top of the ground, nothing beneath the ground, nothing in the air or water. If there was, outsiders would have come and plundered it long ago. So, Henson, since you're obviously trying to pass yourself off as someone innocuous, perhaps you should refine your flimsy story to fit the circumstances."

"It's not a story, sir," retorted Henson firmly but quietly. "I *do* work for Apex Exploration. If you wish, you can contact them in Ohio to verify my status. I am here to look for bauxite."

Roberto chuckled. "Let's cut to the chase, as you Americans are fond of saying. When that plane comes in from Miami, every American who climbs off is either a tourist, a missionary or a CIA operative. I know that you're not a tourist or a missionary, because you're not whimpering or praying or demanding a lawyer right now. You're much too sedate. So why don't you just admit to working for the CIA, so we can establish a more cooperative relationship? I think that once you contact your superiors, you'll determine that we share mutual objectives. Wouldn't it be preferable to work in concert, instead of stumbling around each other?"

"I don't work for the CIA," reiterated Henson. "I'm employed by Apex Exploration Services."

"So, Henson, you disavow any connection with the CIA?" asked Roberto, smiling as he drummed his fingers on his desktop blotter. His fingernails were sparkling clean and perfectly manicured.

"I do. Sincerely, sir, I don't work for the CIA."

"If you insist, I'll play along with your ruse. Open your suitcases, Henson," ordered Roberto. "Let's see what you have."

Henson stood up and strolled over to the table that held his luggage. He unlocked the bags and flipped them open for Roberto's perusal. Roberto totally disregarded the wooden case that contained his sampling tools, camera, notebooks and other sundry tools of the exploration trade, and focused instead on Henson's suitcase.

Within the suitcase, on top of his clothes and personal articles, Henson had carefully arrayed the universal currency of enticement: three cartons of cigarettes, a fifth of Jim Beam, a small envelope of cash, and the latest edition of *Playboy*.

Roberto set those items aside and then carefully rifled through the rest of the suitcase. Smiling contentedly, he extracted a hardbound edition of William Manchester's *The Arms of Krupp*. Henson had purchased the recently published volume during his layover in Miami just hours ago. "This should suffice for a gratuity," he declared, very matter-of-factly. "But if I take this from you, will you have anything else to read?"

"No," replied Henson. "That's all I brought. I figured it would last me through this trip."

"Oh," muttered Roberto quietly, frowning as he leafed through his new acquisition. "Then look in my bookcase there and take anything that interests you. I'm done with all of those. I hope you're not offended, but it's just so hard to acquire anything decent to read down here."

Henson looked at the bookcase to the right of the table; its three broad shelves strained under the weight of purloined literature. Above the case hung two hand-carved wooden picture frames; one contained a Yale diploma and the other displayed a black-and-white photograph of a much younger Roberto, dressed in a graduation robe and grinning broadly, receiving the diploma. Henson smiled to himself, realizing that Colonel Roberto was obviously one of the more educated, refined, and eloquent thugs in the Third World. Scanning the titles in the bookcase, he selected a French book, *Papillon*, by Henri Charrière, that detailed the author's odyssey after being sentenced to life at the French penal colony on Devil's Island.

"Ah," observed Roberto. "*Bon*. Excellent choice." Gesturing for Henson to take his seat, he picked up his phone and spoke in Haitian Creole. Setting down the receiver, he said, "My driver will take you to the Hotel Roi Christophe for the evening. Please enjoy the accommodations tonight, with my compliments. The manager there is an old friend of mine; I'm sure that he will see to your every need. I assume that your employer will facilitate the remainder of your stay?"

Repacking his bribe offerings before latching the suitcase, Henson replied, "I appreciate your kindness, but I really doubt that I'll have much need for a hotel after tonight. I work in the field, mostly, and make my bed wherever I happen to be at the end of the day. I'll be here for the next couple of weeks. Depending on what I find, my company might deploy some additional equipment. If that's necessary, then I'll try to stay close to it."

"Hopefully, you'll find what you seek, although I stringently doubt that there is anything of value to be found. In any event, Henson, I bid you the best of luck, but I also ask a favor of you."

"Gladly."

"You are free to travel as you wish. I will instruct my soldiers not to detain you or otherwise obstruct your passage. But wherever you go, don't make the mistake of giving these people false hope. There is *no* hope here. Americans and Europeans flock to this place with grandiose notions that they can change things. It's like they're passing through a revolving door; they're here, and then they're gone. They arrive with lofty aspirations and then go home with tragic stories to tell at dinner parties and fund-raisers, but *nothing* is changed. The truth is that there is nothing that *can* be changed. The sooner that you accept that, the better."

Henson nodded solemnly.

"But Henson, be forewarned: if you're here for some form of mischief that has nothing to do with scrabbling around for bauxite, it's better that you tell me sooner than later. Think about that tonight, and if you feel like paying me another visit tomorrow, then I'll be happy to receive you."

"Thank you, Colonel," replied Henson, closing his suitcase and clicking the clasps shut. "But I don't think there will be a need for that."

Roberto reached into his jacket. "My card, Mr. Henson. Please don't hesitate to call on me if I can be of assistance, particularly if you find yourself in difficult circumstances."

Henson accepted the card like it was a delicate and valuable banknote, examined it, and then carefully placed it in his wallet. "Then I'm free to go?" he asked.

"But of course," replied Roberto, gesturing at the door with a flourish. "My men will take you back to the airport, if you wish, or to the Roi Christophe. Be sure to visit the Citadelle while you're here. It's the big castle in the mountains overlooking the city. Fascinating place: a lot of history there, but not much of it pleasant."

"I'll be sure to see it. *Adieu*."

"*Adieu*, Mr. Henson."

6

FALSE FLAG

Gulf of Mexico, Sixteen miles south of Panama City, Florida
12:35 p.m., Thursday, August 14, 1969

It couldn't possibly be a more beautiful day over the Gulf of Mexico. The air was sparkling clear, the water was crystalline blue, and the sun shone brilliantly. It was an afternoon tailor-made for a glistening image in a glossy tourism brochure.

As perfect as the conditions were, Carson and Ourecky weren't here to frolic on the pristine white beaches, but were closing out an entire week of practice water landings. Presently, they were seated in a special water drop training vehicle being hoisted aloft by a massive Army CH-54 "Flying Crane" helicopter. Consisting of little more than two enormous Pratt and Whitney turboshaft engines bolted to a skeleton-like airframe, the Sikorsky cargo hauler looked like an enormous insect.

The water drop trainer was a new addition to their pre-flight training regimen. Before their flight in June, he and Ourecky had undergone rudimentary ditching drills in the infamous "Dilbert Dunker" contraption at Pensacola Naval Air Station. Compared to the live water drops, the Dunker experience was like riding a cheap carnival attraction. A mock-up of a helicopter cockpit was mounted on a 25-foot metal rail. Once they were securely strapped in, a klaxon blared and the mock-up quickly slid down the greased rail and into the deep end of an indoor swimming pool, where it abruptly plunged underwater and rolled over. In the lopsided and confusing environment, the principal takeaway lessons were to not panic and to use rising air bubbles as a ready reference to orient themselves so they could find their way to the surface. *Just relax and follow the bubbles to safety* declared their Navy instructors.

Like the paraglider training vehicle they flew to practice for normal landings, the water drop trainer was a stripped-down mock-up of the Gemini-I spacecraft. For simplicity's sake, in order to prepare the crews for the most hazardous of the potential water landing scenarios, the three skids were permanently locked down and welded in place. The control panels were simply painted replicas of the real panels; very few of the instruments actually functioned.

Fabricated from sheet metal and spare parts, the water drop trainer was expressly built to be repeatedly immersed in brine. Consequently, it was furnished with only the most basic of appointments; the rest were simulated with carefully placed ballast to ensure the appropriate weight and balance characteristics. There was no environmental system, so their suits became hot and uncomfortable in very short order. But despite the ersatz quality of most of the equipment, the two hatches functioned exactly as they did on the real spacecraft; after all, one of the primary objectives of the training was to evaluate their ability to evacuate the vehicle if it was in danger of sinking. Rigid booms replaced the inflatable boom components of the paraglider, so the sail's deployment was considerably faster and smoother than their live drops from a C-130 cargo aircraft.

Glancing to his right, Carson watched as Ourecky removed his helmet and tugged a rubber "neck dam" over his head. Shaped like a big gasket, the purpose of the tightly fitting diaphragm was to prevent water from seeping into the suit once they were dunked in the drink. Since the neck dams instantly rendered their suits into oppressively hot saunas, they typically didn't don them until just a few minutes before the drop. As aggravating as the rubber collars were, they were essential: a suit with a little leakage was a problem, but a waterlogged suit could be a death sentence. After Ourecky replaced his helmet and locked it into place, Carson fitted his own neck dam and re-donned his helmet.

He heard the helicopter pilot's voice over the intercom: "We're steady at altitude. Angels twelve. Five minutes to drop. Ready, gentlemen?"

"We're good," replied Carson. "Angels twelve. All secure and ready for drop."

"Good luck," said the pilot. "Have a safe ride down. Make sure you steer clear of any oil rigs out there."

"Will do. Have a safe flight back to Rucker," replied Carson, consulting the pre-drop checklist and turning his head towards Ourecky. "Arm pyro and seats."

Ourecky threw a series of toggle switches. "Paraglider jettison pyro armed. Hatch pyro armed. Ejection seats are armed."

Carson ensured that the check lights were showing green. "I verify that paraglider jettison and hatch pyro are armed. Seats are armed. Beacon transponder is activated. Check hatch pawls."

"Hatch pawl is neutral."

"*My* hatch pawl is also neutral," answered Carson, running his hand along the hatch locking mechanism. "Both hatch pawls in neutral position. Tighten and lock restraint harness."

"Restraint harness is locked."

"Roger. My restraint harness is locked as well. Ready for this, buddy?" asked Carson. He spit out his chewing gum and stuck the gray lump to the "circuit breaker panel" to his left. "Last drop before we head for Wright-Patt."

"I'm ready as I'll ever be," answered Ourecky. "But I'll be even happier when I'm back home."

"Me too." Carson had been concerned with Ourecky's behavior this week. Although the engineer consistently executed his procedures correctly, it didn't seem like he had his head fully in the game. He didn't fear that Ourecky was losing his confidence; if anything, the right-seater had recently become almost annoyingly cocky at times. Still, something was off; Carson felt sure that he was distracted, but couldn't divine the source of the distraction, and Ourecky didn't seem too willing to share.

Now, it was time to focus on matters at hand, because even in a training environment, they could swiftly find themselves in tremendous danger. Although they had practiced landing the paraglider-equipped Gemini-I on virtually every imaginable surface—snow, ice, desert, sand, gravel, packed earth, rain-slick asphalt—in any conceivable environment, this intensive training session prepared them for the most dangerous post-reentry scenario of all, an emergency splashdown at sea.

Why could water landings be so dangerous? The first problem was the physical nature of water. As high divers, water skiers, and Golden Gate suicide jumpers could readily attest, although water might appear to be a kind and forgiving medium, it was brutally incompressible. The fluid may yield ever so slightly to the knife edge of a diver's hands, but it didn't readily surrender the right of way, particularly to large blunt objects.

The second problem with water landings lay in the highly utilitarian design of the Gemini spacecraft. From the very outset of the program, the Gemini was ingeniously designed to touch down on dry land. NASA's water-landing configuration was actually an afterthought. Had the designers known that every NASA Gemini mission would eventually terminate in a splashdown at sea, they could have readily omitted the landing gear skid wells from their blueprints even before they left the drawing board, and the astronauts could have ridden into orbit in a cockpit considerably more spacious and accommodating. Why did NASA shelve the concept? The crucial paraglider could not be refined in time. After expending four years and over 165 million dollars in development, NASA abandoned the concept of returning to terra firma when the paraglider proved troublesome during initial testing. Luckily for the Project, the primary contractor—North American Aviation—was well on its way to perfecting the paraglider even as NASA's ten Gemini flights were drawing to a close.

But an object exquisitely configured to touch down on dry land did not necessarily do so well on water. In an absolutely perfect world, as a paraglider-equipped Gemini-I made contact with the water, its skids would skim along the surface of the sea like a giant water bug in a Disney nature documentary, eventually losing momentum and slowly settling into the waves to bob gently like a massive fishing cork. But reality was much harsher. Numerous experiments had shown that upon contact with the surface of the water, instead of dissipating energy by hydroplaning, the three skids effectively acted as a giant speed brake.

If the Gemini were designed like a more conventional aircraft, where the landing gear could be lowered and retracted by a hydraulic mechanism, this might not be an issue. Theoretically, if the gear were raised and lowered by hydraulics, then the crew could simply retract the skids if they saw that they were coming down over the water. But alas, the Gemini was contrived with the most spare of functionality in mind. The tripod landing gear's design reflected this simple philosophy. Once the gear were down, there simply was no means to raise them again. The skid struts were actuated by an extremely reliable compressed gas system, and once the gas forced the struts to their full extent, a simple set of spring-loaded detents latched them irrevocably in their extended position.

So, almost without exception, even in the most tranquil of sea states, their water landings were typically much more traumatic than sliding in on gravel or even packed earth; as soon as the skids bit water, the two men were usually subjected to a bone-jarring wallop that slammed them forward against the windows and instrument panels.

Worse yet, combined with the braking effect of the skids, the spacecraft's forward momentum had a profound tendency to swiftly propel the nose underwater. Once its blunt nose planed underwater, the Gemini-I acted with all the hydrodynamic characteristics of a concrete cinder block. Consequently, flaring and jettisoning the paraglider became that much more crucial. Even then, it was extremely difficult to gauge the spacecraft's height above water in daylight, even in the best of conditions, and virtually impossible at night.

If the paraglider was not stalled and was still driving forward with any appreciable speed at all, the effect was not unlike an unconscious skier being dragged behind a very powerful speedboat. More often than not, they would go underwater *fast*, a consequence which presented them with an entirely new set of potential hazards. Within a matter of seconds, as they sank, water pressure would prevent them from manually opening the hatches. If they were quick on the draw and not incapacitated from impact, they could disable the low-velocity ejection seats and fire the hatch pyro. Then it was a simple matter of scrambling out as quickly as possible as seawater gushed in. Of course, the spacecraft sank even faster as it progressively took on water.

The hatch pyro was the fastest route to safety, but it was ineffective once the spacecraft had submerged a few feet underwater. Actually, the pyro charges weren't just ineffective, they were downright dangerous. Given the incompressible nature of water, the explosive force of the pyro charges had to be dissipated *somewhere*, and the only *somewhere* that was available was the spacecraft's cabin. So, the crew had to detonate the charges at exactly the right time to make their escape, because if they fired them too late, they would likely die in the process. And if by chance they inadvertently skipped the all-important procedural step of disabling their ejection seats, it really didn't matter when they fired the pyro, because the ejecting seats—even though they were low velocity models and not the standard Weber seats fitted in NASA Geminis— would smash their heads against the unforgiving hatches, which would simultaneously shatter their necks.

If there was even the most remote possibility that they might overshoot or undershoot a dry landing site and come down in water, they learned to delay lowering their skids until the last minute, even though there was a risk that the skids would not fully deploy. After all, a dry ground belly landing with skids up was far preferable to a water landing with skids down. If the skids weren't deployed, there was a fairly good chance that the spacecraft would remain afloat almost indefinitely, provided they landed in a decent sea state. That was the ideal outcome, since the spacecraft could be recovered with all the intelligence data they had collected on the mission. On the other hand, the spacecraft could not be economically recycled after immersion in salt water; some parts could be recovered, but for all intents and purposes, the vehicle would be counted as a lost asset.

If the skids were already deployed and they saw that they were definitely going down in the drink, the most prudent recourse was to jettison the paraglider and eject. If they did elect to ride the spacecraft down with skids deployed, the underlying theme was to hope for the best but prepare for the worst. The best case scenario ended with them floating peacefully on blissfully serene seas, patiently awaiting the arrival of stalwart pararescuemen from the 116th Wing; the worst case scenario saw them helplessly descending to the ocean floor as their precious air gradually dwindled.

Carson's visions of nautical oblivion were interrupted by the helicopter pilot's voice: "Ready for release?"

"We are *ready* for release," answered Carson, bracing himself. Like most of their live training, the drop would be tremendously exhilarating, if it weren't also so damned frightening.

"Five, four, three, two, one, Mark. Release," stated the pilot. There was a loud clicking noise followed by a slight jolt, like they were starting off on a roller coaster. They lurched forward, almost nose down, and fell a short distance as the paraglider unfurled. After plummeting earthward for a few seconds, the paraglider gathered lift and then they were soaring. As part of their regular routine, Carson executed his controllability and stall checks to verify that the paraglider had deployed correctly. After that step, he set a course for the designated compass heading and kept a wary eye on the altimeter.

Unlike landing at a contingency recovery site, manned by an LSO team, an emergency water landing would be flown blind; there would be no TACAN beacon to guide on and no updated information concerning local weather conditions. To effectively simulate the contingency, they were literally on their own after being released from the helicopter.

Actually, as Carson well knew, they weren't entirely alone; a HH-3A "Jolly Green Giant" helicopter, loaded with several pararescue PJs and specialized recovery equipment, trailed approximately a quarter-mile behind the training vehicle. The PJs were tasked with prying open the hatches and extricating them from the vehicle if they were knocked unconscious or otherwise incapacitated on landing. As an additional safety net, a pocket fleet of six Boston Whaler safety boats were holding station in the vicinity of the predicted touchdown area.

Their first order of business, if they determined that a water landing was inevitable and ejection was not a viable option, was to retrieve their survival kits and position them so they were immediately at hand for an abandon ship drill. In addition to being crammed with essential survival goodies—emergency rations, drinking water, fishing kits, signaling devices and their all-important one-man life rafts—the bulky stowage bags also served handily as cushions to prevent them from slamming face-first into the instrument panels.

"Initiating contingency water landing checklist. Retrieve your kit," said Carson, watching the magnetic compass as he kept them on heading.

Ourecky hurriedly popped the quick-release catches on the survival rucksack, disconnected the lanyard that connected it to his parachute harness, yanked it out from its storage compartment beside his seat, and wrestled it into his lap. "I'm up, Drew," he announced, cradling the bundle like a small child.

Carson checked the compass, nudged the hand controller slightly, and then said, "Take the controls."

"I have the controls," answered Ourecky, slipping his hand under Carson's to grip the center-mounted hand controller.

"You have the controls," stated Carson, relinquishing the piloting duties to the right-seater.

He quickly retrieved his own rucksack and wedged the pack into place for landing. He verified their altitude—just a hair over four thousand feet—before resuming control of their descent.

Minutes later, he announced, "Passing through two thousand feet. Check your harness and brace for landing."

Ourecky double-checked his harness and replied, "Two thousand feet. Restraint harness is tight and locked. I'm braced."

"Harness is tight and locked," stated Carson a few seconds later. "Passing through one thousand feet. Watch the contact light and stand by to jettison paraglider." Peering through his window to gauge their height, he wasn't too fond of the final landing phase. Their altimeter wasn't particularly reliable. Neither was the contact light; the sensor, which closed a relay when it was immersed in salt water, worked roughly fifty percent of the time. Watching the rippling surface of the Gulf, he estimated that they were about two hundred feet up. He gently tugged the hand controller to begin flaring the paraglider in order to reduce their forward drive. Timing was crucial; ideally, if he managed the landing correctly, he should bring the paraglider to a full stall—with no forward movement—at almost exactly the time they made contact. If he stalled it too early, they would plummet straight down to a very hard landing. Too late, and the nose would plow under the waves.

"Stand by . . . stand by . . . stand by," said Carson, manipulating the stick with a light touch. "Brace . . . brace . . . brace!" The two men were jolted forward as the skids collided with the waves. Grunting, Carson winced as his harness restraints bit sharply into his shoulders. The contact light blinked green. "Contact! Jettison paraglider!"

Ourecky threw a switch and responded, "Paraglider jettisoned!"

Carson knew that the trailing HH-3A helicopter was hovering nearby, about a hundred yards behind them, and four pararescuemen were already spilling out to act as lifeguards. Afterwards, it would pull away to orbit the landing site at a comfortable distance, ready to drop a flotation collar and other personnel if reinforcements were required.

Looking out his window, he saw that he had timed it almost perfectly. The nose was not awash and sinking; the mock-up spacecraft was afloat and stable, at least for the moment, so there shouldn't be a pressing need to abandon ship.

Although the splashdown was certainly realistic, he reminded himself that if they had to flee over the side, an actual spacecraft would have an abundance of potentially dire hazards. Besides the ejection seats and various pyrotechnics, the RCS—Reentry Control System—thrusters could spew very toxic residual fuel. The Gemini-I's exterior skin would be painfully hot to the touch, and if they went into the water, they had to be especially mindful to remain well clear of the heat shield, which would still be plenty hot enough to boil water and scorch flesh. The paraglider itself could become a treacherous hazard if it came to rest over the spacecraft; besides the potential danger of becoming tangled in various lines and cables, it would be as if they were suddenly draped with a wet, impermeable blanket of nylon fabric.

"Disable hatch pyro and seats. We'll open the hatches manually," declared Carson, unlocking his neck ring. "Helmets off."

Ourecky threw switches on his panel before tugging a lanyard at the base of his ejection seat. "Pyro is disabled. Seats are disabled. Removing helmet."

Carson paused as he watched Ourecky doff his helmet and stuff it down into his footwell, and then ordered, "Open hatches." Ideally, to keep everything in trim, they would work in unison to swing open the hatches simultaneously. He smoothly unlatched his hatch and flung it open. Buoyed by a faint breeze, the cool salt-scented air quickly began to dispel the stale hot air of the cabin. Glancing to his right, he saw that Ourecky was having problems with this hatch's ratchet locking mechanism.

"My hatch pawl is jammed!" blurted Ourecky, frantically straining at the hatch lever. "It's not budging!"

"Keep working on it," answered Carson calmly as he released his shoulder restraints. He leaned hard to the right to compensate for the slight shift in their center of gravity. "We're stable. Relax. I'll sit tight until you crack it open."

Suddenly, everything went out of kilter as the capsule unexpectedly teetered to the left. Briny water sloshed in, tentatively at first, and then gushed in like a tidal wave.

The vehicle was foundering, much faster than Carson had witnessed on previous runs. An orderly exit was no longer an alternative. "We're taking on water, Scott," he declared. "We have to egress *now*."

Coughing and sputtering, Ourecky defiantly persisted at the hatch lever even as the water surged over his face. He sucked in a deep breath at the last second, but was off slightly on his timing; he obviously drew in a considerable amount of salt water as he inhaled.

As the mock-up rolled completely underwater and started sinking, Carson glimpsed a SCUBA-equipped PJ swimming into view. Clad in khaki UDT shorts and a dark blue T-shirt, the wiry PJ braced himself in the hatch opening, reached in, snatched him by the shoulder harness, and started tugging him free of the vehicle.

The PJ obviously expected Carson to panic, but the pilot was not so easily rattled. With the mock-up capsule sinking, this was definitely an emergency situation, but since they were here to train for emergencies, Carson decided that there was still training yet to be done. After all, what would they do if this happened after an actual flight and there were no PJs to lend a helping hand? He flashed the "okay" sign to the PJ, which was intended as a signal for the rescue swimmer to back off, but the tenacious PJ continued his aggressive extrication efforts.

Carson vigorously shook his head at the SCUBA-equipped PJ, reached up, and broke his grasp on his shoulder harness. Relinquishing his grip, the PJ apparently got the message and retreated slightly to observe. Leaning to the right, Carson tugged Ourecky's survival ruck-sack out of his lap, heaved it overboard, and then wrestled Ourecky out of the seat and through the open hatch. The kit was positively buoyant, so they would find it on the surface when they got there.

Ourecky was limp and virtually unresponsive; Carson was sure that he had likely swallowed a lot of seawater when the mock-up keeled over unexpectedly. Kicking with all of his might, he made sure the right-seater was clear of the mock-up before groping for the short lanyard to inflate his life raft. Since he was supporting Ourecky, he could only search for the lanyard with his one free hand. Unable to immediately find it, he decided to just swim for the surface.

After a few seconds of concerted stroking, he felt cold water flowing in around his neck and into the suit; his rubber neck dam had probably been displaced either by increasing water pressure or his physical exertions or both. Hugging Ourecky tightly around the waist, he kicked—*hard*—but realized that he just wasn't making any headway towards the surface. He recalled the oft-repeated mantra from their Dilbert Dunker training ordeal: *Just relax and follow the bubbles to safety.* Unfortunately, as much as he wanted to just chill out and casually follow the bubbles, they were rising and he was sinking. *This really wasn't good.*

At least two minutes had elapsed since he had sucked in his last breath; the burning sensation in his lungs and pounding sensation in his temples were really disconcerting. Most people would become frantic at this point, but as Carson knew from extended training sessions in the Pensacola pool, as anxious as he felt right now, he could stay under for at least another two minutes before he blacked out. He had learned that little tidbit through painful experience; to reinforce their teaching point, as they timed his endurance with a stopwatch, the Navy survival instructors had physically held him at the bottom of the deep end until he passed out.

Now, as he slowly descended past the cold thermocline, still kicking vigorously, he tilted his head back and looked at the rippling surface; about five feet above him, the PJ hovered, obviously ready to respond if he saw any indication of distress. With a start, Carson suddenly realized that they were making upwards progress towards the surface. The saved had become the savior: Ourecky was now supporting him with one arm around his waist, flailing wildly with the other, and kicking like Johnny Weissmuller as Tarzan, on his way to rescue the toothsome Jane from a toothy crocodile. Carson smiled to himself. He shook his head at the patiently hovering PJ, flashed an okay sign, and allowed Ourecky to drag him to safety.

Sputtering, they broke the surface and gasped for air and then finished the drill by inflating their little one-man dinghies and climbing aboard. As a team of frogmen brought the mock-up to the surface and rigged it for recovery, the two men gratefully clambered into a net-sided "Billy Pugh" basket dangling from a hovering HH-3A helicopter. The helicopter flew for just a short distance before depositing them on the deck of a Boston Whaler safety boat.

2:10 p.m.

Later, stripped out of their waterlogged spacesuits, the two men reclined in the spare shade of the Boston Whaler's center console and lunched on Army C-rations as they reviewed the exercise. Soaking wet but obviously dehydrated, Ourecky guzzled water from a green plastic Army canteen.

"Thanks for hauling me up out there," noted Carson. A flock of petrels flew by, low over the waves; apparently spotting a school of fish close to the surface, two of the birds broke off from the formation and dove into the water.

"You're welcome, but I'm sure you would have done the same for me. Besides, I owe you for yanking me out of that damned can."

"A team effort, so I suppose we're even," replied Carson, furtively sampling a C-ration officiously labeled "Chopped Ham and Eggs." Deciding that it was edible, if not half-bad, he dug into the remainder with a white plastic spoon. "Hey, you remember how we wanted to approach Tew and Wolcott to convince them to let us fly stripped down, in regular flight suits instead of the pressure suits?"

"Sure," replied Ourecky, twisting the lid closed on the empty canteen. "But I seriously doubt that they will ever buy off on it."

"Well, I think we have another strong argument to support our case."

"If you say so, but I doubt that they will approve it. So, any chance we can grab a little time off? It's been a mighty hectic week."

Stripping off his soaked T-shirt, Carson replied, "Time off? Nope. Let's zoom back to Wright-Patt, grab a night's sleep, and spend the weekend tuning up our procedures in the Box."

"But we're solid on procedures, Drew," grumbled Ourecky, slowly shaking his head.

"Getting a little cocky, are you? We still have plenty of work to do and not much time to do it. Anyway, what's gotten into you lately? I've never had to kick you in the butt to train before. This is no time to be loafing, Scott. What's up? What's with all the sniveling?"

"It's Bea," replied Ourecky, staring at the decking.

"Hey, I *get* it, Scott," said Carson, wringing out his white undershirt. "I'm not the smartest guy in the world, particularly when it comes to women, but you want to log some time with Bea. That's natural, but we still need to be ready. We've got just a little more than three weeks. You kids will have plenty of time to get reacquainted after we return from orbit."

"Look, Drew, there's more to it than that," answered Ourecky. "Bea's in a pretty fragile emotional state right now. She needs me to be there for her, but she begged me not to say anything to anyone until she was further along . . ."

"Further along? Further along with *what?*" demanded Carson.

"Bea's pregnant," confessed Ourecky quietly.

"Well, congratulations!" Carson slapped Ourecky on the shoulder. "You two kids didn't waste any time, did you?"

"We sure didn't plan it this way. We really wanted to wait a couple of years. Personally, I wanted to hold off at least until we were done flying missions."

"With things as they are, that might be quite a while," replied Carson. "And I guess that you're aware that Virgil is pushing hard to have the program extended for more missions."

Ourecky nodded glumly. "Anyway, I've known for a few weeks now and I really wanted to tell you, but Bea was having a lot of complications, and the doctors were really concerned that she was going to have a miscarriage. As it is, she's not entirely out of the woods."

"Oh. Sorry."

"So, can you cut me a little slack, Drew? Do you understand why I need to spend this time with Bea?"

"Sure," answered Carson. "I *fully* understand."

"So since we're so solid on procedures, can we take the weekend off?"

"*Nope*. I can sympathize with your plight, Scott, but since I'm still the boss on this mission, and I think we need some additional polish, then we're spending the weekend in the Box. No questions, no ifs, no buts, and no whining. I promise you that I will cut you loose as soon as I can, but not until we've dotted all the i's and crossed the t's. Okay?"

Ourecky nodded.

"Besides, now that Bea has a baby on the way, you have that much more reason to knuckle down and train. It may be a pain in the ass, Scott, but training gets you home."

"I know you're right, Drew," admitted Ourecky, dipping a cracker into a flat can of C-ration peanut butter. "But that sure doesn't make it any easier. Look, can you do me one favor? Can you keep a lid on this thing until I tell you otherwise?"

"Done, but in the meantime, let's keep our focus on flying. Fair enough?

"Fair enough."

The Falcon Club, Dayton, Ohio
9:32 p.m., Saturday, August 16, 1969

As he strolled through the poorly lit entrance of the Falcon Club, Jimmy Hara had a lot on his mind. For one thing, his health had been off lately. For the past several weeks, he had been suffering frequent headaches and sharp pains in his joints. His wife finally convinced him to see a doctor at the base hospital, so he went on Thursday and was subjected to a regimen of tests that showed nothing conclusive. The docs had drawn a significant amount of blood, which was being tested for various conditions; the results were supposed to be back this coming week.

If his health issues didn't concern him enough, there was also a significant development in the Project's security operations. Only twenty minutes ago, Hara had been comfortably parked on his living room couch, eating popcorn and watching *Adam-12* with his wife and kids. Then he received a call from his office notifying him that a Blue Gemini worker had filed an urgent report from a payphone in the Falcon Club. During their monthly counterintelligence briefings, the Project's personnel had been primed to aggressively report any suspicious activities, particularly anything that might involve a suspected magazine or newspaper reporter.

The concerned worker was Stan Hubbert, an aerospace engineer who worked for Gunter Heydrich in the Simulator Facility. In his terse call, Hubbert conveyed that he was being approached by someone who expressed more than a passing interest in his workplace. Hara and his counter-intelligence operatives had already prepared for this contingency and were now putting their plans into play. So they could better understand the potential threat, Hara's intent was to gather more information on the curious interloper, ideally before handing the case over the OSI and FBI. After all, the man might merely be collecting scuttlebutt for a local newspaper or—worse case—gathering hard news for one of the national weekly magazines.

He cleared his thoughts to focus on the immediate task at hand. Orienting himself to his surroundings, he quickly spotted Hubbert at a booth in the secluded lounge adjacent to the bar. Tonight was Luau Night at the club, so the bespectacled engineer wore an orange Hawaiian shirt, allowing him to blend in effectively with the tiki torches and other glintzy Polynesian décor. As Hara watched, Hubbert chatted with a balding man who appeared to be in his late thirties, wearing dark chino trousers and a white short-sleeved shirt. While still on the phone with the Hara's office, Hubbert had been briefed on the response plan, and he was obviously doing his utmost to keep the man engaged while not acting furtive or suspicious.

Hara leaned against the bar, munched a half-handful of stale peanuts, lit a cigarette, and then gestured towards the barmaid, a heavyset, middle-aged woman dressed in a brightly colored muumuu. The woman's bottle-blonde tresses were set in a massive bouffant, lacquered board-stiff with hairspray, spacious enough to conceal a covey of quail.

As he waited on the barmaid, Hara looked at the paneled wall behind the bar and studied the portraits of former Wright-Patterson pilots who had been killed while flying. He hadn't been in the club in months and noticed that the most recent additions to the wall included several pilots shot down over Vietnam, as well as Tom Howard and Pete Riddle, who had perished in a "T-38 training accident," according to a typewritten label under their photographs.

The barmaid placed a coaster and napkin in front of Hara and said, "Jimmy . . . long time, no see. How have you been, baby?"

"Busy." Hara slipped a ten across the bar. "Hey, Sally, I need you to do me a big favor. I think that guy over there is trying to run some sort of scam on my friend, so I need to keep a sharp edge. Can you fix me a gin and tonic, extra heavy on the tonic, just enough gin to pass the whiff test? I'll probably be ordering them for a while, so keep the same thing coming. I'll run a tab, and there's a tip in for you, also."

"Sure thing, Jimmy," she replied, grinning. "But I had better not see you two walking out hand in hand at closing time."

"I only have eyes for you, Sally," he said, winking as he reached out to take her hand. "You know that."

"Oh, you're such a sweet, smooth-talking man, Jimmy," she replied, fluttering her eyelashes as she poured his faux drink. "I sure hope that your wife appreciates the treasure she's found."

Hara smiled and then walked over to the booth where Hubbert was sitting. "Stan the *Man!*" he declared. "Forgive me for acting surprised, buddy, but I don't see you out very often. Did Gunter unshackle you from your oar?"

"For the moment," answered Hubbert, standing up and grasping Hara's hand. "Hey, Jimmy, care to join us?" He motioned towards the stranger in the booth. "This is Eric Yost. He works on base, in that warehouse right down the road from us."

"Pleased to meetcha," said Yost. On the other side of the club, the house band had started into a new set of dance tunes, so Yost had to speak loudly to be heard over the music.

"Eric, this is Jimmy Hara," said Hubbert. "He's an A & P guy."

"*A & P?*" asked Yost. "You work at a grocery store?"

"Airframe and Power plant," explained Hubbert, nudging his black-framed glasses up on his nose. "A & P. Jimmy's an engineering technician."

"Engineering technician," sniffed Hara, sliding into the booth opposite Yost and taking a sip from his drink. "That's just a glorified description for a wrench-turner. You could train any monkey to do my job, provided that you could dumb him down first." He looked at Yost—if that was his real name, which was highly unlikely—and suspected that he might be seriously ill. He was painfully skinny, his face was drawn and his skin was pale and pasty. His hands trembled slightly, like he might have some sort of nervous disorder, and his left eyelid twitched erratically.

The three men made polite conversation for a few minutes, before Hubbert announced, "Gents, if you'll excuse me, I gotta go hit the hay. Early day tomorrow."

Yost and Hara watched Hubbert walk away. Several minutes of uncomfortable silence elapsed before Yost warily observed, "Pardon me, Jimmy, but I don't recognize you."

"So why would you?" asked Hara, holding up his empty highball glass and motioning towards Sally for a refill.

"Oh, I work a couple of doors down from Hangar Three, so I see most of the guys coming and going, and I get accustomed to the faces."

"Then it makes sense that you haven't seen me. I'm TDY a lot. Nevada, Arizona, you know the drill," explained Hara. Now he understood how Yost had picked Hubbert out of the crowd at the club. The unanswered question was whether he had become casually familiar with the Blue Gemini workers, or whether the familiarity was the product of deliberate surveillance.

"Yeah, I understand. So if you don't mind me asking, what is it that you guys do in that hangar of yours?"

"I *do* mind you asking, but to be honest, it's not anything too exciting. I spend most of my days dismantling stuff," confided Hara. "The engineers look at it, and then I piece it back together. Like I said before, any monkey could do it."

"I suppose," replied Yost, obviously sensing that it was futile to pursue any further line of questioning. "Anyway, I'm sorry for being so nosy. Just too curious, I guess."

"Curiosity killed the cat," quipped Hara, accepting a glass from the barmaid. "Sally, dearest, could you be so kind as to bring a fresh one for my friend here? Put it on my tab."

"Sure, Jimmy," said Sally, wiping the tabletop with a damp rag. "Another rum and coke, sir?"

"Please."

Hara was sure that if he spent enough time with Yost, he could slowly and methodically peel back the layers of the onion to determine what he was up to and which publication he was working for. He was fairly certain that Yost was pretty much as he presented himself. He clearly wasn't a trained intelligence operative or a cagey reporter with any measure of street smarts. He had the air of someone who had landed himself in a jam and was presently clawing his way out. That didn't make him any less despicable, but perhaps a little more manageable.

With decades spent in counter-intelligence work, Hara was an avid adherent to the ancient Latin adage "*In Vino Veritas*." Although Yost might be reluctant to recruit Hara because he hadn't witnessed him going in and out of Hangar Three, a substantial amount of alcohol, liberally applied, might overcome his aversions. Hara felt confident that he could drink Yost under the table on any given day, and it shouldn't be much of a momentous feat tonight since he was just sucking down diluted drinks at two bucks a pop.

Finally, after almost an hour of throwing back hard liquor—at least on Yost's part—the two men had become as intimately close as old war buddies. With his inhibitions loosened, Yost dropped what little was left of his guard. He leaned towards Hara and quietly conveyed, "Look, Jimmy, you're a great guy and one of my best friends, so I need to come clean with you. I have a friend who's very interested in your hangar. He's with an Israeli company, and they want to compete for Air Force contracts."

"Israelis?" mumbled Hara, staring at the ice cubes slowly melting at the bottom of his glass. While he presented a calm façade, he was growing physically sick. His stomach was literally churning. *Israelis?* He seriously doubted that Yost's acquaintance really was an Israeli, and now it was extremely apparent that this caper had little to do with gathering information for some slick article in some glossy tell-all magazine. How could he have been so stupid?

"Yeah," said Yost. "Israelis."

"Israelis?" Hara smirked, raised his eyebrows and laughed. "Then I *really* don't think that your friend would be too interested in what we do."

"Don't be so quick to think that. To be honest, it could be worth your while to talk to him. I'm sure that he would be willing to come off his wallet if it helps his company to land some American contracts. Also, I don't know how close you are to retiring from the Air Force, but you might be able to parlay this situation into a very lucrative gig later on."

"Hey, look, I don't know," answered Hara. "You seem like a pretty decent guy, but I just met you and I don't want to put my security clearance in jeopardy."

Yost persisted for several minutes, until he finally persuaded Hara to provide his contact information. "C'mon, Jimmy. The Israelis are our allies. What could it hurt?"

Feigning reluctance, Hara eventually scrawled his name and phone number on a napkin and slid it across the table. "Here," he said. "I won't promise that I'll talk to your guy, but if he calls, I'll listen to what he has to say."

After downing another drink, Yost announced that it was time to leave. "Maybe I'll run into you again down here, or see you on the base," he said.

"Maybe," replied Hara. He watched as Yost limped towards the door, and then observed one of his operatives leave the bar and nonchalantly follow him outside. Another of his sleuths was waiting in the parking lot, waiting to follow Yost home. Over the next few weeks, until he was arrested or the matter otherwise resolved, the sergeant would be under constant surveillance. A "flaps and seals" specialist would surreptitiously read his mail and his phone would be tapped. Since they were now aware that Yost was in a position to routinely observe Hangar Three, they would make subtle adjustments to limit what he could see. More so than anything else, they didn't want to tip him off that they were conscious of his activities. As loathsome as he was, Yost was just a small fish; the prize catch was his handler, whoever that was.

"You okay, Jimmy?" asked Sally, approaching the booth. "You look like hell."

"I'm okay," he replied. "Do me a favor, please. Clear out my tab and bring me another one."

"Jimmy, you know I'm charging you full price for these watered-down drinks, right?" asked Sally, picking up his empty glass and rattling the ice. "That's what you asked for."

"I know. Bring me full octane this time."

Hara seethed with anger. He *despised* traitors. It took all the restraint he could muster not to follow Yost into the parking lot and immediately kill him. At this point, his anger for Yost was only exceeded by his ire for himself and his own actions. It was virtually impossible to guess how much damage had already be done if a foreign intelligence service, even one theoretically benign as the Israelis, was manipulating Yost. He was responsible for safeguarding the Project's secrets, and he felt like he had failed miserably.

Aerospace Support Project
8:32 a.m., Monday, August 18, 1969

"Virgil, got a minute?" asked Jimmy Hara.

"Barely," replied Wolcott, looking up from a mass of paperwork. "What's on your mind?"

"Plenty. Is Mark around?"

Wolcott glanced towards the door to Tew's private office and said, "He's indisposed. I don't want to wake him unless this is awfully danged important. He has way too much on his plate right now; he doesn't need any distractions. Savvy?"

Hara nodded. "Do you remember when we briefed everyone on a potential security threat last month?"

"About someone pokin' their nose around and gettin' too curious about what we do over in the hangar?" asked Wolcott. "Didn't you imply that you thought it was some sort of reporter?"

"I did," replied Hara. "But I was wrong. We threw out a hook for this guy and we got a bite. A big one. Virgil, I'm pretty sure that it's more serious than we thought. He might be working with a foreign intelligence service, maybe even the Soviets."

Hara looked at the floor. "I've bungled this one, Virg, and I'll accept the consequences." Swallowing, he bent forward and closed his eyes, like a shamed samurai offering his head to his master.

"Straighten up, Jimmy. It can't be so bad that you won't be able to handle it."

"It is. I talked to this guy at the Falcon Club on Saturday night. His name is Eric Yost. Not only was Yost watching Hangar Three, but we also initially misidentified him as a guy named Dan Kroll because he was driving Kroll's car and living in Kroll's apartment. What's so stupid is that Yost gave me his real name in the Falcon Club, and I was too quick to jump to the conclusion that he was trying to pass himself off as someone else."

"So what's the connection to this Kroll?" asked Wolcott.

"They work together here on base. Kroll is TDY in Thailand, so it looks like Yost is just housesitting for him. Virgil, I know you want me to handle things internally, but if there's a foreign intelligence service involved, especially if they're trying to actively recruit our people, then I'm obligated to notify the OSI and FBI. I should have already . . ."

Wolcott gritted his teeth, shook his head, and declared, "*No.* The last danged thing we need is for the OSI or FBI to inject themselves into our operations right now. We're at a crucial juncture and there's just way too much at stake. We danged sure don't need outsiders pokin' their noses into our business."

"But . . ."

"Jimmy, we've got another launch in less than a danged month. I'm up to my ears in the minutia, and it's all that I can do to keep Mark from having a damned coronary frettin' about it. So don't trouble me with all the burdensome details from your end. If we've got a problem, just *fix* it."

"Fix it?"

"That's right, hoss," replied Wolcott. "Just *fix* it."

Waffle 'n' Egg Diner, Dayton, Ohio
8:19 a.m., Monday, August 25, 1969

Morozov abhorred the notion of conducting meetings in uncontrolled locations, but Yost had all but insisted on this place, ensuring him that they could be assured of a booth in a quiet corner. At least he had been right on that account; aside from the cooks and wait staff, there were only seven people in the diner.

Four elderly men occupied one booth, playing dominoes and swapping war stories. At the other end of the diner, a middle-aged businessman in a poorly fitted blue suit ate a late breakfast and perused a trade magazine.

The remaining two souls were a couple—both apparently in their early thirties and obviously married but not to each other—who clearly didn't know what constituted appropriate behavior in a public setting. Judging by the tables that had yet to be bussed and the stacks of unwashed dishes waiting by the sink, the regular breakfast crowd had come and gone.

Morozov was desperate. Besides Yost and his friend—Bob Carr, the medical orderly who worked in the base hospital—he had not developed any reliable sources at Wright-Patterson. The Washington GRU office was clamoring for more information about Project Blue Book and UFOs, particularly in light of Yost's claims that the Air Force physically possessed and studied alien technology at Wright-Patterson, but Morozov had yet to provide anything substantial. In a note he had deposited at a dead drop, Yost implied that he was finally successful in obtaining contact information for a technician who actually worked inside Hangar Three, who was also apparently involved in flight testing activities.

Although he had been specifically dispatched to seek information about Blue Book, Morozov was becoming increasingly convinced that the answers lay within the Aerospace Support Project. Since Yost had not yet been forthcoming with useable contact information about anyone working in that organization, and it was yet to be seen if his new "contact" would be of any value, Morozov submitted a formal research request to the GRU's Department of Archives and Operational Research—commonly known as the "Encyclopedia"—to identify personnel currently assigned to the Aerospace Support Project.

The GRU had been recruiting and cultivating sources within the Air Force and Navy personnel systems for several years, mainly to harvest background information on POWs being held by the North Vietnamese, particularly to identify any prisoners who might have access to classified projects or compartmented information. Unfortunately, Morozov's request had not yielded results; the Aerospace Support Project's personnel records were kept separate from most records and were not yet accessible to their sources.

Despite this administrative snag, Morozov did have at least some slight insight into some personnel assigned to the Project. Although Carr demanded a king's ransom for his labors, the orderly had proven considerably more reliable and industrious than Yost. Through his connections at the base hospital, Carr was able to determine that six test pilots had been assigned to the Aerospace Support Project. The clue to the puzzle was the hospital's internal protocol for labeling medical records. Through some patient investigation, Carr was able to identify six sets of medical records with a unique code similar to Agnew's. The special code and extensive lab work signified that the six had been subjected to an unusually thorough physical examination, as had Agnew.

All six were unmarried, which was apparently a prerequisite, and they were all relatively small in stature. He was also aware that two—Howard and Riddle—had been killed in an alleged training accident earlier in the year. Another—Agnew, the pilot that Yost had first identified—had been removed from the Project for psychiatric reasons, and had subsequently vanished. That left at least three—Carson, Sigler and Jackson—still assigned. Carr had also identified a seventh man, an engineer assigned to Eglin Air Force Base, who had undergone the same exhaustive physical, but Morozov didn't feel he was as relevant as the pilots.

Morozov suspected that the pilots were probably involved in flight tests of reverse-engineered alien spacecraft. The tip-off was their physical size; certainly, a captured UFO could not accommodate anyone any larger than the six pilots. He strongly felt that he was on the verge of a breakthrough and hoped that Yost's latest revelation would be the key that finally unlocked the door.

He heard a loud noise from the parking lot and turned to watch Yost arrive in his flashy red Mustang. He couldn't believe that the American could see fit to drive such a conspicuous vehicle.

Whistling, Yost shuffled in, wearing blue chino trousers, a white T-shirt, and new brown work boots. "How are you?" he asked, sliding into the booth opposite Morozov. "Long time, no see."

"Hush! Pay attention," urged Morozov quietly. "If anyone asks, we've just met. You answered a want ad for a part-time watchman job at Acme Tool and Die in Riverside. I am interviewing you for it. My brother—Sergei—owns the business. Do you understand?"

"Got it," replied Yost quietly, grinning. "Cover story, right?"

Morozov nodded. He was beginning to doubt that this idiot was worth the effort. A waitress walked up to take their order. She was in her mid-thirties, slightly overweight, with curly auburn hair and sharp features. Morozov guessed that she was of Irish or Scottish descent.

"Alice," said Yost cheerfully. "It's been a while, hasn't it?"

Frowning, the waitress studied his face, as if trying to place him, and then grinned, revealing teeth long in need of a dentist's care. "Oh, Eric! You've gotten so skinny that I barely recognized you. Where have you been lately?"

"I'm working night shifts now. I usually sleep during the day. Alice, you look like you've lost a lot of weight yourself. You look *great*."

"Oh, you're always so sweet. I could just bottle you up and pour you on pancakes," she said, laughing as she retrieved a pencil from behind her ear. "So, what can I bring you, Eric?"

"You buying?" asked Yost, looking across the Formica table top at Morozov.

Morozov nodded reluctantly.

Yost perused a catsup-spattered menu. "Alice, I'll have the Hungry Lumberjack breakfast. Eggs scrambled with cheese. Can I finagle some biscuits also? Along with the waffles?"

She nodded in assent. "Biscuits? Sure. Bacon, sausage or hash?"

"Yeah," he replied.

She repeated the question. "Bacon, sausage or hash?"

"All three, darling. After a full night at the shop, I'm famished," he answered, slowly standing up from the table. "And if you two don't mind, I need to go wash up."

Watching Yost limp away, Morozov asked, "So you know Mr. Yost?"

Still jotting down the order, the waitress answered, "Eric? Oh yeah. He's in the Air Force. He works in some warehouse on the base."

"Do you think he's trustworthy?" asked Morozov. "I'm interviewing him for a part-time security guard job at my brother's machine shop. Would you vouch for him?"

"Oh, I'm sure you can trust him. He's as good as gold. He's a real sweet talker, though. He always used to keep after me to go stepping out with him sometime."

"And would you? Can't you see he's wearing a wedding band?"

"His wife left him. Poor thing, he just can't bear to come completely to grips that it's over. So the answer to your question is yes, I would go out with him, if he ever asks again. Who knows? It might be fun."

"So you obviously trust him," he observed, glimpsing Yost on the way back from the restroom. "This job is only part-time, and mostly he'll just be sitting at a desk. It's an unnecessary nuisance for us, but we're working on a government contract and they insist we hire a watchman. But it's still very important that we know he's trustworthy."

"If you say so," she replied. "And what will you be having, sir?"

"Just coffee, please," answered Morozov. "Black."

"So where are you from? You have some sort of accent. Poland, maybe?"

Poland? *Poland?* Morozov wanted to guffaw. "I'm from Michigan," he said. "My parents came here from Slovakia right after the First World War."

"I suppose it's all about the same thing, isn't it?" she said, laughing. "We're all kind of mutts in this country, aren't we?"

Mutts? Truer words were never spoken, thought Morozov. As the waitress walked away, Yost slowly walked up, whistling the theme from *Hawaii Five-O*, and slid back into the booth.

"Do you have something for me, Mr. Yost?" asked Morozov, casually sliding his newspaper across the table as if to point out an article.

"Indeed, I do," replied Yost, slipping a note under the newspaper as he feigned interest in a headline. "That's contact information for a guy named Jimmy Hara. He's an engineering technician who works in Hangar Three. He's an airframe and power plant mechanic."

"Airframes and power plants?"

"Yep. To be honest, he didn't seem too enthusiastic about talking to you, but who knows? You should give him a ring sometime. I fed him the Israeli line, just like you told me." Yost looked up from the tabletop and added, "You do remember our deal, don't you? That's a thousand dollars up front, and a bonus if he's cooperative, right? You still pay me a thousand bucks regardless of whether he agrees to talk, right? Isn't that the bargain that we struck?"

Sliding the newspaper back, Morozov nodded as he sighed. "Yes. That's what I agreed to. I will leave your money folded in this paper, and I will leave first. Understand?"

Yost nodded and then casually touched a finger to his lips, as if he was scratching under his nose. Bearing a heavily laden tray, Alice walked up to the table, placed two large plates before him, and said, "Here you go, gentlemen. Hungry Lumberjack and two coffees. Catsup?"

Yost looked down, nodded, and exclaimed, "Man, that looks good and I am *starving*. Would you mind if we finished this interview after I've finished this grub? I don't want it to get cold."

"We will, Mr. Yost," said Morozov, slipping out of the booth. "Enjoy your meal. If you'll excuse me, I'm going to call my brother, and I'll be right back." As he walked towards the pay phone located in the front of the diner, he glanced back over his shoulder. It was revolting, like watching a voracious swine feeding from a swill-filled trough. As he watched Yost slurp his coffee, Morozov longed for the day when he would no longer have to rely on the American sergeant.

7

BABY BLUES

Aerospace Support Project
9:30 a.m., Wednesday, September 3, 1969

Chewing on a well-frayed toothpick, Wolcott leaned back in his chair, folded his arms across his narrow chest, and studied the men seated around the table. Waiting for Mark Tew's final decision on the impending mission, Carson and Ourecky were obviously anxious to finish their final preparations before heading west on Saturday.

Heydrich looked bored, as if this was just another routine day back at Peenemunde. The three men looked to Tew. To Wolcott, it was painfully obvious that his friend was diligently searching for even the slightest excuse to scrub the flight. Tew believed the whole effort was a rush job to placate the Navy, and he didn't like rush jobs.

Finally, closing a binder, Tew made his declaration. "Try as I might, I can see no reason that we can't execute as planned. Gunter, are you confident that these two are ready?"

"I am," said Heydrich.

"Carson? Ourecky?" asked Tew. "Are you up to this?"

"We are," answered Carson. Ourecky nodded in silent agreement.

"Then we go," stated Tew. "*No* monkey business this time. No arguments or discussion. Don't expect me to be lenient if you feel compelled to violate orders again. Understood?"

"Yes, sir," replied Carson and Ourecky in unison.

"Are we still good for launch next Tuesday, Gunter?" asked Wolcott.

"We are, Virg. Crew Three is at the PDF for the pre-launch checks. Major Jackson called me this evening and reported that everything is in line."

"And the Navy?" asked Tew. "Are they happy with everything?"

"They are, Mark. They're extremely anxious to see this new Soviet platform knocked out."

"That's fine, Gunter," noted Tew. "But you need to make it abundantly clear to them that we will do nothing to place our crew at risk to accomplish their objectives. We're doing this in a hurry, but we won't be in such a rush that we abandon safety and sound operating principles."

"They understand," said Heydrich, gathering his paperwork and maps.

"Okay, gents," interjected Wolcott. "As the boss said, we go on Thursday. Anything else on the agenda?"

"Well, Virgil, Scott and I have a recommendation," answered Carson. "We'll write it up formally, if need be, but for future missions, we would like to fly in standard flight gear. No pressure suit, but just a regular Nomex flight suit and flight helmet. With an oxygen mask, of course."

"*No!*" blurted Tew, vigorously shaking his head. "*Out* of the question."

"I concur," said Wolcott, spitting out his toothpick. "Shucks, Carson, I know you boys hanker to be as comfortable as possible, but we're not ridin' down that trail. The suits stay on."

"It's not a matter of comfort, Virgil," claimed Ourecky. "The suits are restrictive and awkward. We have to move around a lot more than the NASA crews did, especially when we unstow and re-stow equipment in the storage bays behind the seats. It's hard to swivel around to reach stuff without inadvertently bumping into switches and instruments. That could be disastrous if we're operating in close proximity to a target."

"Well, pardner, I concede that's a valid point," noted Wolcott.

"Not so fast," snapped Tew. "How about an ejection scenario?"

"General, our low-velocity seats are only viable during the post-reentry phase," asserted Carson. "They're of absolutely no use on the pad or on the way up. Mode Two—salvo-firing the retros and coming down on the paraglider—is the only abort mode available during boost phase. Since we're remaining in the vehicle, one way or another, standard flight gear should suffice."

"Okay," replied Tew, wringing his hands. "Then let's focus on post-reentry. You're implying that you wouldn't need the suits if the paraglider failed and you had to punch out?"

"That's correct, sir," interjected Ourecky. "With a normal paraglider deployment, we're already sufficiently low that we wouldn't need a pressure suit."

"There's still a possibility we would need oxygen if we went out too high," said Carson. "But all that would require is a modification to the existing oxygen flow manifold block on the seat, so that it feeds the oxygen to a mask rather than the suit. And there's more. The suits could be extremely dangerous if we go down over water. If they take on water, regardless of whether we eject or evacuate the spacecraft after landing, they're dragging us under."

"Good point. But what about a micrometeorite strike?" asked Wolcott. "If you're in orbit and your pressure vessel was perforated, hoss, you ain't survivin' without a suit."

Ourecky answered, "Virgil, the Apollo astronauts fly to the moon and back without wearing their pressure suits. I doubt that anyone is going to convince me that that they would fare any better or any worse than us if they caught a micrometeorite."

"Okay. I'll cede that, hoss," replied Wolcott. "Sounds like you've pondered it thoroughly, but I'm still straddlin' the fence. Gunter, what's your take on this?"

"I agree with them," replied Heydrich. "If they're not leaving the spacecraft for extravehicular activity, the suits are more of a hindrance than anything else. There's still a risk of the spacecraft being damaged while they're maneuvering during close proximity operations, but if the vehicle's pressure vessel is punctured, it's highly unlikely that they'll survive reentry."

"Personally," said Carson. "I would much rather have my lungs sucked out through my nose and die instantly instead of being slowly broiled alive."

"Duly noted, Major Carson," said Tew. "We'll consider your idea. I'll talk with the Life Support people to see if it's a viable option. Even if they concur, the earliest we could fly in shirtsleeves would be Mission Four. That's still a long way down the road, gentlemen."

"Sir, speaking of Mission Four, aren't Jackson and Sigler still slated to fly it?" asked Ourecky. "It's none of my business, but will they be ready to launch in March?"

"Since you're so concerned, yes, we're fairly confident that they'll be able to fly the next sortie, pardner," said Wolcott. "We're anxious for them to jump in the saddle, especially if it'll take some of the pressure off you two, particularly since Four is stacking up to be a relatively simple flight. Now, Ourecky, would you care to tell me why it's any of your concern?"

"Uh, it's a personal issue, Virg, but I would rather not fly in March or April, if possible."

"He's going to be too busy handing out cigars." Carson smirked.

Tew looked up from his paperwork, and his jaw dropped abruptly. "*What?*" he blurted.

"Major Ourecky's going to be a father," announced Carson. "Bea's pregnant."

"*When?*" demanded Tew, reaching for a nearby desk calendar.

"Her doctor says that she's due towards the end of March," answered Ourecky.

As Tew traced his finger on the calendar, his eyes opened wide and he swallowed deeply.

"Is there a problem, sir?" asked Carson.

Pushing aside the calendar, Tew slowly shook his head and declared, "Virgil, I want you and Gunter to absolutely ensure that Crew Three is prepared to go up in March."

"Will do, boss," replied Wolcott, stubbing out his cigarette in a glass ashtray.

"Thank you, sir," noted Ourecky gratefully. "I really appreciate it."

"Not a problem," said Tew. "And how is your wife, Ourecky? Bea, right?"

"Oh, she's fine, sir, but she had some questionable results on some initial blood work and other tests, so the doctors are monitoring her closely. They were initially worried that she might miscarry, but it looks like we might be beyond that."

"So the pregnancy is proceeding normally?"

"I hope so. We have an appointment tomorrow, so we should know more then, but it seems like everything is going normally. She's sick a lot in the mornings, but apparently that's very common, and she doesn't like being stuck on the ground, either. Delta has her working a gate at the airport until she's able to fly again."

"She's seeing a doctor at the base hospital?" asked Tew.

"No, sir. Her doctor's at Grandview Hospital, in Dayton."

"Grandview?" asked Tew. "That's in midtown. Don't you live closer to Wright-Patt? Wouldn't it make sense for her to see a doctor on base?"

"Agreed, sir," replied Ourecky. "But she's been seeing this same doctor since she's been working for Delta. On top of that, sir, I don't think she's too keen on using the base hospital."

Sighing, Tew placed his hands flat on the table. "Ourecky, we'll do our *utmost* to make sure you're here when she goes into labor, but you know we can't make any promises. If you're out of town when that day comes, then I can send someone to fetch her to the base hospital. Talk to her, and see if you can't persuade her to see a doctor here instead of at Grandview."

"Thank you, sir," said Ourecky. "I'll do that."

Wolcott waited until the others filtered out before he confronted Tew. "Okay, Mark, do you care to tell me what that was all about? Since we've been hangin' around each other for the past couple of decades, I know when you're antsy about something. So what are you frettin' about?"

Tew spun the calendar around so that it was facing Wolcott. "Look," he said, jabbing the calendar with his finger. "If Bea's due at the end of March, then she probably got pregnant . . ."

Wolcott looked at the calendar. Immediately grasping the potentially dire ramifications, he groaned. ". . . at the end of June, right after Ourecky got back. *Tarnations!* I can't believe we ain't caught this sooner."

"Well, Virgil, how could we? For starters, we never anticipated that we would have any married men flying, so we never considered this potentiality. On top of that, if she had been going to an OB/GYN here on base, we would have known immediately."

Shaking his head, Virgil muttered, "I can't believe that this slipped by us."

"I'm hoping for the best, but who on earth can know what will happen?" asked Tew. "How could we have been so naive? This is uncharted territory. Ourecky spent time in weightlessness, was exposed to cosmic rays and God only knows what else, and there's not sufficient research to know what's going to happen to that baby. It's as if we've found ourselves in the middle of a giant science experiment, only it's been going on for a few months."

Grandview Hospital, Dayton, Ohio
10:45 a.m., Friday, September 5, 1969

"So you leave tomorrow morning?" asked Bea, trying to make herself comfortable in one of the awkward plastic seats in the crowded waiting room. Musak played quietly in the background, barely audible over the voices of expectant mothers and anxious teenagers.

"I do," replied Ourecky.

"How long this time?" she asked, fanning herself with her hand. "Or should I even ask?"

"Oh, I think we won't be gone much more than a week. We'll do prep work, then we do the flight testing later in the week. I an͏ʳ that I'll be home on Monday or Tuesday of the following week. Will you be okay?"

"Of course," she replied.

"Look, I know that you're happy with Doctor Blakely, but have you considered seeing a doctor at the base hospital? I've checked into it, and they have a couple of really qualified obstetricians there. We could book you an appointment, and . . ."

"*No*," she said adamantly. "You're right. I'm *very* happy with Doctor Blakely and I'm *not* going to see anyone else."

"Well, the subject came up earlier in the week, and General Tew mentioned that he would prefer that you saw a doctor at the hospital on base."

Rolling her eyes, Bea asked, "Since when is *our* baby General Tew's business?"

"It's not. He was just concerned about your welfare. The base hospital is a lot closer than this one. There's always a remote possibility that I won't be here when you go into labor, so he wanted to make sure that you got to the hospital safely."

"I thought we discussed this," she said. "You said you would be here when the time came."

"And I'll *try*, but I can't promise anything. You know that. General Tew and Virgil Wolcott are doing everything they can to make sure I'm here in March and April. They're really trying, Bea. Why are you being so stubborn? Can't you at least *consider* seeing a doctor on base?"

"No," she muttered, closing her eyes and grimacing. "Scott, did I ever tell you about my brother Charlie?"

"Charlie? I thought that was going to be *your* name if you were a boy. I thought you didn't have any brothers or sisters."

She reached out and took his hand. "That's not entirely true, Scott. I had a brother, Charlie. He was born in 1951, on base, about three weeks before my dad shipped out to Korea."

"But you never told me. What happened?"

"Charlie was turned the wrong way as he came out, and got tangled in his umbilical cord. He died during the delivery, and my mum almost did also. She told me years later that the hospital did an investigation afterwards. The doctor could have saved Charlie if he had been paying attention, but apparently he was distracted. So, Scott, don't ask me to have our child on base. I have enough bad memories about that place."

"I understand now," he said. "I wish that you had told me before. And I'll try to be here when the time comes."

"Don't *try*. Just be here, Scott. That's all I ask: just be here for this baby."

Dayton Airport, Ohio
8:30 a.m., Monday, September 8, 1969

Tew glanced at his watch; just as planned, he was fifteen minutes ahead of schedule. He took a seat opposite the Delta Airlines gate where Bea was checking in passengers for a flight to Logan Airport in Boston. Next to the gate, a mass of anxious businessmen lined up at a bank of ten payphones, waiting impatiently for their turn to make a hurried call or two before boarding.

He watched Bea as the line gradually dwindled. Although striving to be cheerful and attentive to every passenger, it was obvious that she was harried, even though her workday had barely begun. Her pregnancy was just starting to show; she wouldn't be able to wear her form-fitting stewardess uniform for more than another couple of weeks at best.

As he waited, he pondered tomorrow's planned launch of Mission Three. He desperately wished that that he could leap forward in time so that the agonizing wait could be over, the mission would be complete, and Carson and Ourecky would already be safely home.

Although the last mission had theoretically been a success, Tew wasn't confident that the massive battery failure had been a one-time fluke, just as he wasn't sure that there wasn't another gremlin lurking deep within the guts of their overwhelmingly complicated machines.

He closed his eyes, mentally grappling with the seemingly endless details necessary to prepare for each mission. *Were they ready?* With ten more missions yet to fly, and possibly more if they were granted an extension, he wasn't sure how many more times he could suffer through this stressful ordeal. Although he had earned his third star, which he was due to pin on in October, he strongly considered dropping his paperwork for retirement if the Project was extended. Unlike Wolcott, he intended to go home and stay there after he retired. No consulting or government work, just endless rounds of golf and doting on grandchildren yet to arrive.

He opened his eyes as Bea greeted the last few passengers, and walked over just as the last traveler passed through the boarding door. Smiling as she recognized Tew, she counted ticket stubs and compared her tally against the seat assignment sheet.

With the figures reconciled, she picked up a red phone and spoke into it. "Gate Eighteen, eighty-two boarded for Boston. Plus the flight crew, so seven more is a total of eighty-nine. Straggler? What's the name? Smith? Okay, I'll hold the gate open five more minutes."

"Tea?" asked Tew, offering a cardboard cup from the airport's lunch counter. "There's a teaspoon of honey in there. I think that's how you like it, right?"

"Oh, you're such a dear," she replied, gratefully taking the steaming cup. "What brings you this way, Mark?"

"Same as everyone else. I have a plane to catch. I'm early, so I thought that maybe we could chat if you have a moment or two to spare."

Checking the clock, she nodded. "The next flight is due in eighteen minutes. So are you headed to the same place where Scott and Drew went? They left this weekend."

"No. I'm not that fortunate. I'm bound for Washington. Budget meeting this afternoon, and then I fly right back this evening. So, how do you like working at the airport? Scott implied that you were grounded until the baby comes."

Leaning against the counter, she sighed, rolled her eyes and said, "I'm thrilled to be home every night, but I can't wait to fly again. This gate work is tedious; same thing, day in and day out. Everyone is in such a rush, and they're so rude and obnoxious."

"Sorry." He slipped his hand inside the inner pocket of his jacket to check his ticket folder. A shrill voice blared over the PA speaker, requesting that a passenger named Mr. Knowles pick up the nearest courtesy phone.

She sipped the tea, smiled, and said, "So is it still okay for me to call you Mark even though I'm officially an Air Force wife now? I certainly don't want to land Scott in any hot water."

"No need to worry about that, Bea."

"Mark, you might be here to catch a flight, but I suspect that this isn't just a casual social call. Do we need to talk about something? Is Scott okay?"

Tew nodded. "Scott's fine. He's just very busy, but you know that. He's a hard worker; my life would be a lot easier if I had three or four guys just like him."

"Well, if not Scott, is there something else on your mind?"

"There is," answered Tew. "Look, Bea, to tell the truth, I'm concerned about you and your pregnancy. Scott tells me that you're reluctant to go to the hospital on base. It's sure a lot closer to where you live than Grandview, and the doctors are . . ."

Frowning, she shook her head vigorously. "No, Mark, I'm not *reluctant*. I *refuse* to set foot in that hospital and that's final. Scott may be in the Air Force, but me and this baby are not," she said emphatically, gently patting the slight swell of her stomach.

He leaned towards her and said quietly, "Bea, I know what happened to Charlie."

She swallowed. "You do? How could you know?"

"I was between tours in Korea when it happened, back here on leave. Your parents were devastated. I can understand why you don't want to have your baby on base." Tew refrained from mentioning that he convinced Bea's father to go to Korea in the same timeframe.

"So then you know why it's futile for us to discuss it any further."

He grimaced and said, "Look, Bea, it's not that simple. From the first time we talked, I swore I would be honest with you and share everything I could, to the extent that I can. I owe you that."

"Mark, you owe me *nothing*. I don't know why you feel that you're indebted to me. Look, I'm really snowed here. Can we dispense with the pleasantries so you can just get to the point?"

He nodded solemnly. "Bea, the truth is that Scott is exposed to a lot of potential hazards in the course of his work, and while I'm not concerned about him being harmed as a result, there's not a lot of research to tell us how it might affect your baby."

"So that's what this I really about," she fumed. "You're not really concerned about me or the baby. You're more concerned with covering your tracks, aren't you?"

"Bea, it's not like that . . ."

"It's not? Well, Mark, put yourself in my shoes. I know men are fighting and dying in Vietnam. I see it every single night on television. It breaks my heart. I feel so sad for their wives and families, and I'm grateful that Scott's not over there, but sometimes I suspect that he would be safer there than . . . doing whatever it is that he does for you. When you told me he was going back to school, I was so confident that we would finally have a normal life. Then suddenly he's back here working on some classified project that he can't talk about. And a week rarely goes by when he doesn't come home with strange scars or marks that he can't explain."

She continued. "I hear rumors about secret things that go on at the base. There are people who swear that the Air Force keeps crashed UFOs in some hangar, and that they're trying to make them fly. Can you tell me that Scott's not caught up in that? Is that what this is all about? Is *that* why you're so worried about my baby?"

He shook his head. "We've talked about it before, Bea. I can't tell you what Scott does, but if it will dispel any of your fears, I can assure you that he's not involved in any form of UFO research. I *swear*. Does that help *any*?"

"If not UFOs, then *what*? Does he work on atomic bombs? Poison gas? Death rays?"

He tried to shrug off her question. "Bea, there's no simple answer, except to say that he's exposed to a lot of things that might be hazardous.

We have a fairly good understanding of what's harmful, but it's not an exact science. It's like radiation. Years ago, people didn't have even the foggiest notion how dangerous radiation was. Now we know that even X-rays can be dangerous, but back around the turn of the century, they were a novelty at carnivals and fairs."

"So he *is* involved in atomic stuff," she declared, slapping her hand on the counter.

"Don't be so quick to jump to conclusions. Scott is routinely around a lot of potential hazards: strong radio waves, chemicals, all sorts of things. By themselves, they're probably not harmful, but we don't know what the cumulative effects might be. I doubt that he'll suffer any ill health as a result, but for you and your baby, I would rather err on the side of caution."

"I appreciate your concern, Mark, but I am *not* going to the base hospital. Period."

Tew capitulated, loosening his tie. "Okay, Bea, we're obviously at an impasse, but I think we can resolve this situation, if you're willing to make some concessions. Here's my perspective: I want to do everything possible to ensure that you have a healthy baby. I can't make promises, but I will do my *utmost* that Scott is here when the time comes. Fair enough?"

Bea nodded and said, "Plenty fair, but I think you're wasting your breath if you expect me to see a doctor on base. I really don't think that there's anything else for us to discuss."

"Let's not be too hasty, Bea. As I see it, there's no need for you to go on base if you don't want to. It will always be an option that's available to you, especially in an emergency, but no one is going to force you to go there."

Lugging a duffle bag strapped over his shoulder, a young soldier sprinted up to the counter. Wheezing for breath, he said, "Smith. PFC Walt Smith. I'm on the Boston flight. Am I too late?" Painfully thin and darkly tanned, the soldier was dressed in Army khakis that seemed two sizes too large. He furtively glanced to the left and right, as if he wasn't comfortable in crowds.

"Mr. Smith?" asked Bea, smiling as she examined the passenger's ticket. "Right through that gate there, sir. They're expecting you."

Without replying, the soldier rushed to the door. Bea paused to amend her paperwork and call the revised numbers to the airline's operations desk. Setting down the phone, she said, "Sorry, Mark. You were saying?"

"I've talked with a couple of doctors at the base hospital. They can arrange with your physician for some extra blood work and additional tests. You might have to go in for a few extra visits, but we'll pay for everything."

"How lovely!" muttered Bea. Finishing the tea, she crumpled the white cup and dropped it in a waste can. "More poking, prodding, and needles. You men just don't know how lucky you are."

"No argument there," replied Tew. "We definitely won that coin toss. If men had to carry babies, the world's population would fit in a broom closet. Bea, I hate to rush you, but I have a plane to catch. Do we have an agreement?"

"Well, it sounds like a workable plan, except the airline is not going to let me off for any additional visits. I'm stretching my off-time as it is, and all the other girls are covering for me."

"I'll take care of Delta," replied Tew, recalling that one of his B-17 bomber squadron mates from England was now a senior vice president in the airline's corporate headquarters. "I'll make a few calls, and you'll have all the time that you need."

"And you'll let us know if there's anything wrong with the baby, right?"

Tew nodded. "Immediately."

"Promise?"

"Promise," replied Tew.

"Then I think we have an agreement, General Tew."

8

KILLING A DRAGONFLY

Mission Control Facility, Aerospace Support Project
6:05 a.m. Eastern, Wednesday, September 10, 1969 (GET: 31:05:03)

Although Mission Three was well underway, the atmosphere within Mission Control was calm, almost casual. Only five of the controllers were actively manning their consoles; the rest mingled to smoke and chat. Their real work was accomplished over three weeks ago, when they assembled and polished all aspects of the flight plan. Like skilled clockmakers of olden times, they wound the spring and set the intricate gears in motion, but once the action was started, there was scarcely little they could do except to maintain a vigil and hope for the best.

Unlike NASA's flights, where mission controllers analyzed an almost continuous stream of telemetry from spacecraft, the Gemini-I transmitted only snippets of data during the periodic contact windows. The intermittent communications were by design. The Soviets' capacity to track, assess, and catalog orbiting objects was not nearly as sophisticated as the American space surveillance network. Consequently, the Blue Gemini flights were planned to launch, execute their mission, and return to Earth long before the Soviets became aware of them. Additionally, radio transmissions—both voice and data telemetry—were kept to a bare minimum, and the short communications windows were restricted to remote areas where eavesdropping—even by the Russians' infamous "fishing trawlers"—was highly unlikely.

Even after the telemetry was received at the remote sites, it had to be relayed to Wright-Patterson for analysis, so the controllers typically looked at information that was at least thirty minutes old. If they spotted an anomaly, their options were limited to recommending that the mission continue or be curtailed. In most instances, the crew would have already diagnosed the problem and arrived at a similar conclusion. At this point, there were probably no two men as intimately familiar with their machine as Carson and Ourecky. Their actions on the last flight clearly demonstrated that they knew the Gemini-I spacecraft probably as well or even better than the engineers who designed and built it.

Gazing out over the room, Heydrich wished that he could relax and adopt the nonchalant attitude of his subordinates, but he had information that they were not yet privy to, and it weighed heavily on him. Leaning back in his chair, he was sure that a heart attack was imminent. The veins in his temple throbbed, his breathing was labored, and his chest pounded like a kettledrum in the pivotal crescendo of a Wagnerian opus. He removed his headset as he reviewed a report on projected fuel consumption but finally had to set his clipboard aside when he could no longer compel his eyes to focus on the numbers.

On one hand, he thought, massaging his aching temples, *things could not be any better*. Unlike the first two flights, the current mission had proceeded without the slightest flaw or delay. Carson and Ourecky were due to complete their close proximity operations in less than two hours. They were in the homestretch: all that remained was for them to deploy the Disruptor, descend to a lower orbit, loiter for few hours, reenter, and then land at Edwards Air Force Base. Their flight was stacking up to be the most perfect mission in the relatively short history of manned space flight, even though it would never be annotated in any official record books.

On the other hand, he mused, gazing at the nicotine-stained and well-chewed nails of his trembling fingers, *things could not possibly be any worse*. Since they were chasing a suspected ocean surveillance satellite, this was a Navy mission. Certainly, it was an Air Force crew in an Air Force spacecraft launched from an Air Force launching pad, supported by Air Force resources scattered all around the globe. Logically, then, the Air Force should call all the shots.

But in forcing the hasty mission to knock out the nemesis Soviet satellite, Admiral Tarbox had vigorously tugged on some significant political strings, some leading into the Oval Office itself. That probably should have been enough, but the die was definitely cast when he played the most sinister card in his hand: based on Tarbox's recommendations, the Navy committed to fund the entire mission—lock, stock, barrel and booster— so that not a solitary Air Force nickel would be expended in the effort.

Heydrich had toiled within bureaucracies—American and otherwise—long enough to appreciate the unique power of money, so he knew that there were few venues where money spoke louder than in the military budgeting process. The Navy's largesse carried a significant consequence: they were also granted considerable authority over the execution of the mission. To that end, eight Navy officers presently encroached in Heydrich's domain.

To their credit, the Navy overseers were largely satisfied with the Air Force's plan and had no desire to meddle, so the mission was proceeding as envisioned. But there was a *ninth* member of the Navy delegation— Ed Russo—and therein lay the problem. In the waning phases of an otherwise flawless operation, the headstrong interloper felt compelled to make changes.

Worse still, his proposed changes were by no means subtle, and he had succeeded in capturing the ear of Tarbox, the senior member of the Navy contingent.

If matters couldn't be any more infuriating, Russo and Tarbox were now talking to Wolcott. Judging by Wolcott's enthusiastic nods and broad grin, it was clear that Russo had won another convert. The mission's objectives were to intercept the suspect satellite, inspect it, and then deploy the Disruptor. According to the plan, the Disruptor's needle thruster would destabilize the Soviet satellite over the course of a three-week period, before it was gradually de-orbited.

Merely selling the Navy on the plan had been a monumental struggle, but they were eventually won over to the merits of incremental failure versus a catastrophic failure. As Tew had successfully argued, it was far preferable that the Soviets convince themselves that their design was faulty rather than arbitrarily blowing the satellite to smithereens.

Now, Russo had concocted a scheme in which the Gemini-I would back away a safe distance before firing the Disruptor's powerful main charge. The main charge was a ten-pound shaped explosive penetrator, similar to the warhead of an anti-tank bazooka rocket, which would blast through the exterior shell of even the most thick-skinned satellite. Carson and Ourecky would photograph the event, in order to provide a graphic record of the destruction.

In a sense, Ourecky's infamous photograph of the Soviet satellite's data plate, snapped during the June mission, had been their undoing. In classified briefings describing Blue Gemini, Tew and Wolcott frequently concluded their presentations with the startling image of Object 2368-B, much to the delight of their high-ranking audiences. Thus, it made perfect sense if the Navy was paying the freight to neutralize the satellite, they would desire a keepsake of similar dramatic impact to trumpet the success of their mission. Never mind that it was two Air Force officers who would be placed at risk to capture a glossy photograph of a Soviet satellite being obliterated, so long as a suitable trophy could be nailed over Tarbox's mantel.

But there were three significant flaws in Russo's shortsighted plan. First, the standing mission rules stringently dictated that the Disruptor's main charge was to be not employed unless there was substantial evidence that the target was an Orbital Bombardment System. And even under the most exigent circumstances, the main charge would not be triggered until the Blue Gemini crew had returned safely home.

The second drawback was that the mission change would add three hours to the proximity operations. Taken by itself, the delay was negligible, but that three hours would subsequently force a twenty-eight hour hold in returning the spacecraft to the planned landing site at Edwards Air Force Base under the cover of darkness.

Although he would never admit to it, Heydrich wasn't overly concerned about the potential dangers the crew might face as they fired the main charge. While he didn't relish the notion of contributing more scrap metal to the growing constellation of space junk girdling the Earth, he felt confident that Carson could back away to a safe stand-off distance that would enable them to fire the charge and get a good record of the shot. But tacking on twenty-eight additional hours was just begging for trouble.

Time was their constant enemy; the longer they lingered upstairs, the more likely that something would fail. Consequently, they planned the missions to finish the job and return home as swiftly as possible. The extended delay would entail at least a partial power-down to conserve batteries, and their power-up procedures—used to awaken the spacecraft from its slumber—were far from foolproof. The possibility always remained that if one switch or breaker was in the wrong position at some point during the complicated sequence, the finicky spacecraft might not return completely to life.

The third shortcoming was that while Tarbox could request the deviation, it still had to be blessed by Father Tew, and once a mission was underway, Tew wasn't inclined to tinker with an established flight plan. And he certainly would be leery to approve a revision of this magnitude.

Watching the three men stand up and walk towards the glassed-in office space at the rear of the center, Heydrich knew that Russo had convinced Wolcott that the concept needed to be presented to Tew. "Hey, Gunter!" said Wolcott, beckoning with a wave as he paused at the second row of consoles. "Wander up to the aquarium and hear this, pard."

Heydrich grimaced as he pushed himself up from his chair. "I've already heard it, Virgil," he said quietly, looking back over his shoulder to watch Tarbox and Russo entering the office.

The worn heels of Wolcott's cowboy boots clicked on the scuffed linoleum. He grinned and said, "Well, though I don't cotton to most crap that spews from Russo, even I have to admit that this is a mighty interestin' concept. It's a perfect opportunity to evaluate the main charge with the side benefit of getting' a good snapshot to verify that this stuff works as advertised."

"I don't like it, Virgil," uttered Heydrich, straightening his black tie and adjusting his belt. "I don't like throwing changes upstairs with short notice, and I don't like violating the mission rules after we've worked so hard to establish them. If this was an OBS, I might—"

"Rules are made to be broken, pard," observed Wolcott. "The fact is, Gunter, you worry too danged much. We only have nine more flights after this one, unless the program is extended. I don't know about you, but I'm bankin' on that extension. A dramatic shot of a Russkie satellite being pulverized in orbit would travel a long ways towards securing it. And let's face it, hombre, we ain't going to parlay into this sort of lucrative opportunity again. Mark would never buy off on it under normal circumstances. The only reason we're gettin' this chance is because the Navy is payin' the freight. So, ride with me on this, why don't you? Just this once, Gunter?"

Heydrich didn't respond but followed Wolcott into the office and closed the door.

With the decision-makers assembled, Tarbox spoke tersely, "Mark, Russo here has a proposal. Time is of the essence, so I want you to hear him out, and if there's no significant reason why we can't execute his concept, then I want it done."

"Yes, Admiral," replied Tew politely.

Russo took over, articulating his plan in roughly two minutes.

"Gunter, you've apparently already heard this plan and have had some time to digest it," said Tew. "What's your take?"

"I would strongly caution against it," said Heydrich abruptly, vigorously shaking his head. "It will add over a day to our profile. And besides violating our mission rule on indiscriminately firing the main charge, we have no substantial data on what constitutes a safe stand-off distance when it's fired. Granted, it's a shaped charge that would be oriented away from our vehicle, so it should be relatively safe, but there's still the potential of a secondary explosion."

Tarbox whirled to face Heydrich and fixed him with an icy glare that could readily render tropical seas into icebergs. "Since you feel so compelled to nitpick, *Gunter*, what do *you* recommend for a safe stand-off distance?" he asked. His voice was a high-pitched squeak that grated deeply in Heydrich's ears.

"At least a hundred and twenty-five miles, Admiral."

"Ludicrous!" sneered Tarbox. "They're up that high as it is."

"That's my point, sir. General Tew asked my opinion. I don't recommend firing the main charge until our crew is back on the ground. That's what is stated in our existing mission rules, and I think we need to abide by them."

"But we wrote those rules to be almost unduly safe," interjected Wolcott. "There's ample room for flexibility, pardner, and I think the potential payoff here is worth the risk. Don't you?"

"We're squandering precious time," snapped Tarbox, obviously losing his patience. "Rather than continue this discussion, I will make this very simple for you gentlemen. General Tew, unless you can immediately offer a substantial reason why we cannot execute this plan, then we need to send up the appropriate instructions to the crew on their next contact window, or we're going to lose this opportunity. Now, do you have *any* particular reason why we can't execute?"

"I don't," said Tew. "But I don't think it's safe, and I won't place my crew at risk."

Given the unusual circumstances, Tarbox seemed absolutely determined to flaunt his fleeting authority over Tew. "It's not *your* crew, General. They're flying *my* mission, a *Navy* mission, and the risk is tolerable to *me*, so we waive your exalted mission rules and *execute*. And while I don't want to formally order you to act, I will, and if you're not so disposed as to follow my orders, then I'll just pick up the phone and call someone who can and will order you to comply."

Anxiously watching Tew, Heydrich felt as if he were trapped between two hulking behemoths, squared off and staring at each other, anxious for battle. He had a sudden intense desire to be invisible; in lieu of anonymous transparency, he simply desired to be somewhere—*anywhere*—else right now.

Squelching any opportunity for Tew to reply, Tarbox abrasively added, "I'll remind you of something else, General. I'm very conscious that we had to come begging because you have the only platform flying, but that's subject to change. Don't be surprised if the tables are turned in a couple of years. In the meantime, make this happen and make it snappy."

"Yes, sir," replied Tew.

Heydrich was stunned; he had never seen Tew cave in. It was as if Superman's cape had been snatched away and a glistening lump of green kryptonite jammed down his throat.

"Good," sniffed Tarbox, ending his abrupt tantrum. "I'm glad you came to your senses, Mark. Since we don't need to discuss this any further, I'll leave you gentlemen to your work. I have phone calls to make."

As Tarbox left the office, Tew opened his desk drawer to locate a nearly depleted bottle of Pepto-Bismol, took a swallow, and then replaced the cap. Looking to Russo, he said calmly and quietly, "Let me offer you a little career advice, son. Shifting allegiances to suit your immediate interests is a dangerous game. It can have disastrous consequences. Suffice it is to say, when you're done currying favor with the Navy, you're eventually coming back to the Air Force. From this point on, I would strongly recommend that you adjust your behavior accordingly, so that the *remainder* of your career is not *entirely* miserable. Do you understand?"

"Yes, sir," replied Russo meekly.

"And one more thing. After this mission is over, when you leave here, *don't* come back. You're not to set foot on these premises again, under any circumstances. You are no longer welcome here. I'm officially declaring you *persona non grata*, and if I'm compelled to tattoo it on your forehead in red letters, I will. Am I making myself clear, Colonel?"

Russo nodded grimly.

Shifting gears, Tew looked to Heydrich and asked, "How long until their next contact?"

Glancing at his watch, Heydrich replied, "GET 32:03:00. Forty-six minutes from now. They'll talk to a land-based station in the Azores. That's out in the Atlantic, about nine hundred miles west of Portugal."

Tew sighed. "I know where the Azores are, Gunter, but thanks for the geography lesson. Since this is time-sensitive, let's focus only on the immediate changes. Their reentry will be delayed, but your controllers will have plenty of time to read up those plans later. Correct?"

"Correct, sir," answered Heydrich.

"Okay, Gunter, go confab with our opportunistic friend here and make the necessary changes to coincide with his brilliant new plan. Shoot them to the Azores as soon as possible."

Tew added, "And set up a phone call. Immediately. I want to talk *directly* to the Azores controller who will be communicating with the crew, to make it absolutely clear that this new set of instructions are coming from me. I want Mister Carson to emphatically understand that he doesn't have any leeway for deviation if he doesn't concur."

"Got it, boss," replied Heydrich.

"And does all of this meet *your* approval, Commodore Russo?" asked Tew sarcastically.

Russo stammered in reply, "Uh, it does, sir, but I want to apologize for—"

"No need for apologies, Russo. Just make sure that your ass is on the far side of that door before I ask Virgil to strangle you."

As Russo hastily left, Heydrich asked quietly, "Mark, am I missing something here? What exactly did Tarbox mean about the tables being turned in couple of years?"

From the way that Tew and Wolcott looked at each other, it was painfully obvious that they shared a secret. "None of your concern, Gunter," answered Wolcott, swallowing as he nervously twisted the silver tip of his bolo tie. "Never you mind."

"Make that call, Gunter. Get the Azores on the horn pronto," said Tew. "And let's make sure these boys come home intact, despite this idiotic damned scheme. *Macht schnell.*"

"*Zu befehl, mein Herr,*" replied Heydrich with a smile.

Tew and Wolcott observed Heydrich as he rallied his controllers in the front row of consoles. Wearing a satisfied half-smile, Tarbox stood close by the huddle, listening in.

"Sorry, buddy," muttered Wolcott. "I should have talked to you first. Now I feel like a real jackass for allowin' Tarbox to highjack our stagecoach."

"I'm sure you meant well," replied Tew. "Next time, Virgil, please talk to me first. I'm sure this will all go well, but I just can't stand the notion of putting those two boys at additional risk."

"Noted. So when are you going to spill the beans to Gunter about Tarbox's new show, Mark? Don't you think Gunter has a right to know?"

"Not yet. I don't want to cloud his focus with things that aren't important at the moment."

Behind them, the phone jangled on Tew's desk. Turning around, he answered it, and then said, "It's the Azores, Virgil. I'll handle this. Why don't you head down to the floor and keep an eye on Gunter and his crowd?"

"Will do, boss."

On Orbit
**6:43 a.m. Eastern, Wednesday, September 10, 1969
(REV 21 / GET 31:43:08)**

Although he certainly didn't fancy himself the artistic type, Ourecky liked shooting photographs in waning light, because the low sun angle seemed to add a dramatic aspect to what would otherwise be a boringly sterile image. Gazing through the camera's viewfinder, he thought that the high contrast and crisp shadows lent the Soviet maritime radar satellite a somewhat sinister appearance.

For a moment, he paused to admire their prey. The Navy had ample reason to be concerned about its recent appearance in orbit. The surveillance satellite was impressively huge, about the size of a house trailer, and exceptionally well designed. Dominated by two massive radar antenna arrays protruding at right angles from its cylindrical fuselage, it was also festooned with a complicated collection of smaller antennas, apparently connected to signal intelligence gear probably intended to sniff at different radio frequencies.

And to top it all, the thing was absolutely stable; it could not have tracked any truer than if it had been riding along polished steel rails in orbit. From this angle, it looked like an enormous dragonfly skimming effortlessly over the crystalline blue surface of a mountain lake. It was so startlingly beautiful, it seemed almost a shame to kill it.

This was their last photo pass before orbital twilight and would be the last before they deployed the Disruptor after the sun came back over the horizon. As Ourecky took pictures, Carson slowly maneuvered the Gemini-I around the target. It was a painstaking process, but the multiple photo passes ensured that they captured virtually every detail of the Soviet satellite.

Ourecky pointed the camera, adjusted the focus, and snapped the last three frames. "That's it for this roll," he commented, thumbing the film advance. "I hope our Navy friends are happy."

"And not a moment too soon," said Carson. He used the hand controller to make slight attitude adjustments to ensure that the Gemini-I had sufficient clearance from the Soviet satellite. "Twilight in two minutes. You've squirreled away some extra film, right? We still need to document the Disruptor placement."

"Oh, sure," replied Ourecky, removing the film spool from the camera. He let go of the camera, ignoring it as it floated a few inches from his face. He slipped the exposed roll into an aluminum film can, tucked the can into a pre-labeled slot in a film bandolier, and then neatly cribbed annotations into his photo log. "I'm holding at least five rolls in reserve. There's also still a hundred feet or so in the movie camera. We're just swimming in film up here. We could probably shoot a wedding or bar mitzvah on the way home to Ohio."

"Good idea. Maybe with the extra cash, they could buy us some decent in-flight food. Maybe something with some taste. Speaking of which, did you sample that new stuff they packed in the chow bins?

I think they're supposed to be processed fruit, but they're so hard and tasteless that I suspect they're really wood. Try any yet?"

"The pine wafers or the oak tidbits?" asked Ourecky. "Not too appealing to me. I'll just stick with my peanut butter, if it's all the same to you."

Carson chuckled and then switched on the floodlight for the forty-five minutes of station-keeping in orbital darkness. "Are you okay over there?" he asked, glancing to the right. "You look kind of glum."

Tucking his photo log into the side wall pocket, Ourecky nodded his head, smiled and said, "Well, as strange as this might sound, it's sort of a letdown that nothing's gone wrong. After months of fixing glitches in the Box and then the big battery failure on the last mission, it's kind of monotonous when there's no serious problems to fix."

Carson laughed. "Buddy, I'm sure that the time will come when we'll yearn for boredom. In the meantime, let's enjoy the moment." He sipped water from the dispenser and added, "Man, I would *kill* for a Coke right now."

"If only I had known," said Ourecky, adjusting the cabin lights before he stowed the camera. "I would have stashed one away for you."

"Yeah, and Tew would have a guillotine waiting on us at Edwards. Seriously, I don't think anyone's going to be drinking a Coke in orbit anytime too soon." He glanced up through the view port, making sure that the two vehicles were safely positioned. "Looks good. So, are we ready for the next contact window? Azores is next up, right?"

"Right," answered Ourecky, adjusting the radio controls. "We still have a few minutes. I'm going to switch everything on to make sure it warms up adequately. We're not exactly lacking for battery power on this cruise, so we might as well take advantage of the surplus."

"Sounds good in my book." Carson gazed out the window and caught the fleeting green flash of light as the last vestige of the sun disappeared into the horizon. "Hey, Scott, you know that with this stack flying so early and with Parch and Mike flying the next one in March, we'll be off the lineup for several months. Almost nine months, to be exact."

"And?"

"I'm thinking about going back to Wolcott and Tew to ask for a combat rotation in Vietnam. Just a quick tour, two or three months, and I'll be right back to fly the next hop with you. What do you think? Cool, huh? Surely they'll let me go after we've nailed these first two."

"Honestly?" asked Ourecky, adjusting the cryptographic gear. "I don't see that happening, Drew. Don't get me wrong, but I don't think they're willing to assume that kind of risk. Virgil might, but Tew would have to sign off, and I don't think you're going to convince him. You just *won't* let this one go, will you?"

"I like this," declared Carson, gazing out at brilliant stars in the darkness. "But this is never going on our records. I need *combat* time. There's no substitute for combat experience."

"If you say so. Hey, we should hear the Azores in another three minutes."

"Got it," answered Carson, adjusting his headset. "I'm ready."

Anxiously standing by with a sheath of pre-printed note cards, Ourecky prepared to copy the normal onslaught of contingency reentry data. Unexpectedly, he heard distorted chatter in his headset and looked towards Carson.

Simultaneously, both men looked at the mission clocks in their instrument panels; their GET—Ground Elapsed Time—was 32:00:24. "If that's them, they're over two minutes early on the contact," observed Ourecky. "They're *never* early." Although they were in adequate line-of-sight range to communicate with the island station, the established procedure was for the Azores to transmit at a reduced power setting, because of the potential of Soviet intelligence trawlers lurking offshore. What could be so important that they would risk breaking this protocol?

As a former fighter pilot, Carson was accustomed to extracting context and content from otherwise unintelligible fragments of garbled radio transmissions. "Scott, I'll cover this one. Prepare to copy whatever they read up. Crypto ready?"

"Crypto's loaded," declared Ourecky.

"I know we're early but go ahead and switch it on."

Ourecky toggled the switch on the secure voice controller, and the power light blinked green. Immediately, the transmission was much clearer but still barely audible: "Scepter Three, this is Azores Station, over."

Slightly baffled, Ourecky studied the mission clock again: GET 32:01:21. They still had over ninety seconds before the contact window. *Why was Azores transmitting early?*

"Okay, I'll bite," mumbled Carson. He keyed the mike and succinctly stated, "Azores Station, this is Scepter Three. Go ahead."

"Scepter Three, bump *immediately* to VHF Four and disable your voice recorder."

"Drew, what's that about?" asked Ourecky. "What do they mean by *bump?*"

Carson laughed. "It's fighter pilot lingo. Bump means to switch the radio to a bootleg frequency. Apparently, someone has some back-channel traffic for us." Dialing the radio to Channel Four, he added, "Let's see what they have to say. Turn off the tape recorder."

They heard a voice over the radio: "Scepter, Azores, are you on Four?"

"Azores Station, this is Scepter Three on VHF Four," said Carson.

"Roger, Scepter. I have orders *directly* from Tew. How copy?"

"I understand you have orders directly from Tew. Go ahead," replied Carson. Raising his eyebrows and shrugging his shoulders, he looked at Ourecky.

At a rapid-fire pace, Azores transmitted: "Orders follow. On this contact, indicate that you are not receiving voice communications from ground and are transmitting blind. When I read up new instructions, disregard them. State that you will proceed with Disruptor deployment and remainder of proximity operations and mission as planned. Resume normal communications on your next scheduled contact window. How copy, over?"

"Good copy, Azores," replied Carson. "I will indicate that I am unable to receive, will transmit blind, that I am continuing mission as planned, and will resume normal communications on next contact window. These are orders from *Tew?* Over."

"Roger. *Direct* orders. Switch back to VHF One and stand by for scheduled contact."

Just as the Azores controller had alluded, they received an entirely new set of instructions after the usual glut of contingency reentry guidance. Just as he had been directed, Carson feigned that their receiver wasn't functioning as Ourecky copied down the new information. For his part, the Azores mission controller did exactly what he should have done in such a situation; he continued to read up the instructions as if he could not hear Carson.

After the contact was over, Ourecky switched off the cryptographic equipment as Carson emitted a low whistle. "Well, isn't this just hugely ironic," noted Carson, unwrapping a stick of Juicy Fruit.

"Ironic? How so?"

"After the last go-around, General Tew was hell-bent to crucify the two of us for violating orders, and now he's personally directing us to do the same damned thing."

"Obviously, he had a good reason."

"That goes without saying," observed Carson. "I don't know who cooked up this halfwit nonsense, but I don't like it. I sure can't picture Gunter and his guys foisting something this asinine on us. I'm glad that the boss told us to ignore it."

"You and me, both."

"So are you ready to snare this critter?"

"Ready."

"Well, the sun will be up shortly, so let's deploy the Disruptor and head for the house."

Mission Control Facility, Aerospace Support Project
2:18 a.m., Thursday, September 11, 1969 (GET 51:18:06)

Hours later, as they waited for the news that the Gemini-I had successfully landed in California, Tarbox and Tew sat at the table inside the glassed-in back office. As he had been doing for over an hour, Tarbox crouched over a cassette tape recorder, listening intently to the radio communications between the Gemini-I and different communications stations as it passed overhead. Studying a hastily prepared transcript of the messages, he rewound the tape and played again the disjointed communications between the spacecraft and Azores Station.

"I don't know what you're struggling so hard to hear, but it seems fairly obvious that they had a communications failure," observed Tew. "A simple equipment malfunction. With all your years of flying, Leon, you've never experienced that?"

Rewinding the tape yet again, Tarbox cursed under his breath.

"It happens, Leon. Carson couldn't hear what was being transmitted up from the ground, so he did precisely what we had trained him to do. He continued the mission based on the last orders he had received. That simple. I'm woefully sorry that they weren't able to execute your plan, Admiral, but they did deploy the Disruptor, and now they're on their way home. Can you not just accept that? It was a successful mission. Let's just leave it that way."

"Sure," sniffed Tarbox, switching off the tape recorder. "Their receiver conks out at precisely the right moment to scuttle *my* plan, but it's miraculously resurrected in time for the next contact window? That's just a bit hard to swallow."

"You forget who's up there," answered Tew. "Those two are very adept at fixing things. It might have taken them a while, but they got the receiver repaired."

"Yeah, right," squeaked Tarbox. "Trust me, Mark, I'll dig to the bottom of this in due time."

There was a knock at the door. "They're down," announced Heydrich, opening the door as he wagged a cigar. He handed each officer a fat Montecristo to match his own. "Textbook landing at Edwards. Mission accomplished, Admiral. We'll fire the main charge at your order, on the next pass overhead, if you so desire."

"Plan on it," answered Tarbox, lighting his cigar with a wooden match. He puffed deeply, savoring the fragrant smoke. "But *I'll* be the one to push the button. And I want some cameras in here when we do it, Mark, so we have a record for posterity."

Tew set his cigar aside and looked out one of the windows to see Wolcott waving his cowboy hat, whooping it up with the jubilant mission controllers. "I'll make that happen."

"And just because your boys are safe on the beach, Tew, don't think the debacle is finished," said Tarbox, extinguishing the burning match head by squeezing it between his calloused fingertips. "Because our conversation is far from over."

9

IN DEEP

Dayton, Ohio
6:30 p.m., Thursday, October 2, 1969

S tanding outside an Esso gas station, Hara sipped coffee from a
paper cup and checked his watch as he waited for one of his operatives, Terry Smith. Smith and one of his men had been pulling
surveillance on Yost's disused house on Elm Street and wanted to talk to
Hara about recent developments.

After correctly identifying Yost as a security threat back in
August, Hara had taken a close look at his personnel file. His records
indicated a recent spate of disciplinary problems, most stemming
from a propensity towards excessive drinking. Oddly though, even
though knocking him down two pay grades, his commander had
neglected to submit the paperwork to downgrade Yost's security
clearance. Yost had once been hailed as an outstanding performer,
so maybe his boss clung to the thought of reforming him. But
since he held a Top Secret clearance, mostly by virtue of handling

cryptographic equipment, he could potentially cause some damage if he was not adequately monitored.

On a positive note, Yost's current assignment, effectively a punishment tour driving a forklift in a warehouse at night, lent him no routine access to any classified materials. Beyond his work, Yost lived a bizarre and rather miserable existence. He spent virtually every waking hour on base, either at work or camped out in his van in Parking Lot 20. As best as his surveillance team could determine, Yost stayed at Kroll's apartment long enough to sleep and clean up, and that typically was just a few hours at most. His behavior puzzled Hara. It seemed as if he was hiding from something, and Hara was curious to find out what that might be, since that might offer some insight as to why the sergeant was so willing to betray his country.

Smith arrived in his dark blue Impala. "What's up?" asked Hara, sliding into the passenger seat and shutting the door. He handed the coffee to Smith, who gratefully accepted it.

"We've been watching Yost's house for the past week. We have it entirely wired. To be honest, Jimmy, I think it's a dry hole. Yost hasn't set foot in the place in months."

"Interesting. I assume that you've also checked his van on base. Find anything?"

"Just a grubby damned mess. We did find exposed film that he hasn't processed yet, along with a handwritten log of when he watched the hangar. Wait'll you read his notes. He believes that old Dyna-Soar mock-up was some sort of captured UFO, and that the hangar is in someway associated with Project Blue Book. I know that you suspect he's some sort of spy, but he could just be a harmless nutcase."

"Okay, Terry," replied Hara. "So if there haven't been any substantial new developments, would you mind telling me why you called me out of the office? I have reports stacked to the ceiling, and they're not going anywhere right now."

"Because I think I know why he's so shy about leaving the base," explained Smith. "Two guys have been lurking around his house for the past couple of days, one in a car in front and one on foot in back. We've called the plate in to the locals." Smith handed Hara a slip of paper with a name and address. "We also checked with federal and state law enforcement, and I also called my private eye contacts. Nothing. These guys appear to be hooligans of some kind."

"Well, that would certainly explain why Yost isn't returning to this neck of the woods," observed Hara. "My guess is that Yost owes someone big, and they're looking to collect. Tell you what, Terry, why don't you drop me a block away? I'll check them out."

"Sure thing," replied Smith. "Hey, Jimmy, I don't want to make you feel self-conscious, but you haven't been looking so hot lately."

"Well, I don't feel that great, either. I've lost my appetite, I've been shedding weight, and I keep getting nosebleeds. I've been going to the doctor for the past two weeks, and they don't seem too damned inclined to tell me anything conclusive. They think it's some sort of blood disorder. They sent blood specimens down to Walter Reed, so right now they're waiting on the results. In the meantime, they have me choking down vitamins and iron pills."

"Sorry to hear that. I'm sure that you'll snap out of it soon enough," said Smith, sticking the key in the ignition. "Ready to roll?"

Hara nodded.

7:05 p.m.

Remaining cautiously out of sight, Hara observed the late model Dodge Charger for several minutes before making his approach. He stood by the driver's side door, but the man inside ignored him.

Hara loudly cleared his throat, and the man slowly swiveled his head to look at him. With greasy black hair tied back in a ponytail, he was clean-shaven and wore a black leather jacket, a black T-shirt, and black-framed sunglasses. The bucket seat next to him was littered with food wrappers, soft drink cans and empty coffee cups. "*What?*" he demanded.

"I'm looking for Elm Street," said Hara, stamping his feet. "Can you help me?"

"Sorry. I'm not from around here," replied the man, obviously agitated by the interruption.

"Not from around here?" Hara spotted a blood-spattered baseball bat in the car's back seat. "So what are you doing? Something going on? Are you a cop? Maybe I can help you if you're working a case."

"*Nothing's* going on," replied the man. "Just waiting on a friend to come home."

"So where does your friend live?" asked Hara, pointing at Yost's house. "That one?"

"Okay, buddy, I'm busy here, so I think you need to make yourself scarce," said the man, pulling back his open jacket to reveal the butt of a revolver protruding from a tan leather shoulder holster. "Am I making myself *clear?*"

In the blink of an eye, without the slightest hesitation, Hara leaned into the car window, reached into the man's leather jacket, unsnapped the holster, and snatched out the gun. With his left hand, he grabbed the man's ear and twisted it sharply.

"*Nice,*" commented Hara as he examined the blue steel revolver. "Smith and Wesson Chief's Special. Not my personal choice, but very reliable and simple. You definitely look like the sort of man who favors simplicity."

Hearing the cartilage in his ear pop and tear, the man grimaced.

Hara thumbed back the revolver's hammer and slowly pulled the trigger as he pointed it at the man's forehead. He caught the hammer just before it fell on a cartridge. "Wow, man, this action is smooth as glass. You must have filed it down. Nice work."

Mashing the gun's snub barrel against the man's sternum as he thumbed back the hammer again, he asked, "Care to explain what you're doing here, friend?" To punctuate the question, he applied a few more foot-pounds of torque to the man's ear.

"I'm here to collect some money," divulged the man, now considerably more attentive to Hara's inquiries. Fifty feet away, a rotten tree branch snapped loose from a dying oak and crashed to the pavement. Startled, the man jumped like he was coming out of his skin.

"Collect money? From *who?*"

"A guy named Yost. Eric Yost. He owns that ratty little dump over there."

"Oh," noted Hara. "Yost, huh? Well, gee, maybe I can be of assistance paying his debt. How much is Yost in the hole to you?"

"Twelve thousand bucks," muttered the man.

"*Huh?*" asked Hara incredulously, forcing the pistol harder against the man's chest. "What did you say?"

"Twelve thousand bucks."

"Oh. That's what I thought you said. Twelve thousand dollars? Sorry, man, I don't carry that kind of cash on me." With his left hand clamped like an unyielding vise on the man's reddening ear, Hara slipped the revolver into a pocket of his old field jacket and then yanked off the man's sunglasses.

"Ow! Man, don't break those shades! They're real Foster Grants."

"Shut up. I want you to take a look at something," said Hara, holding out the slip of paper bearing the driver's name and address. "Does this information look familiar?"

"Yeah," uttered the man, squinting to view the paper. "That's me. You a cop?"

"Nope. But I'm going to offer you some advice. *You* need to make yourself scarce. As of right now, Yost is off limits to you. He owes me big, much bigger than he owes you, and he pays *me* before he pays you. After that, I don't much care what you do with him, but if you touch him before I collect, twelve thousand dollars will seem like a very paltry sum when you're trying to catch your breath at the bottom of the Great Miami River. Understood?"

Grunting, the man nodded compliance.

"Here," said Hara, thrusting a small notebook and a pencil towards the man. "I'll make you a deal. Jot down a phone number or two where I can call you, and I'll let you know when Yost and I are squared up. Once he's settles up with me, it's open season. Got it?"

The thug nodded and furiously jotted down a phone number in the notebook.

"Thanks. I'm a man of my word, so I'll ring you up when the time comes. To make life even easier for you, I'll even tell you where I find him. In the meantime, don't ever let me catch you lurking around here again," hissed Hara, releasing the man's ear as he simultaneously drove the heel of his right hand into the man's chin. Unconscious, the thug lolled in his seat, with his head coming to rest against the doorframe.

Dayton, Ohio
11:04 p.m., Saturday, October 4, 1969

Jimmy Hara placed the call from a payphone two blocks away from Yost's apartment. There was no answer at the first number, but someone picked up the second number almost immediately. Judging from the sounds in the background—loud music and several people talking—the phone was in a nightclub or bar.

The conversation was almost entirely one-sided and was over in less than a minute. Hara concluded with, "Yeah, he should have at least a thousand bucks, maybe a lot more. But listen to me, you can smack him around all you want, and hurt him as you see fit, but don't kill him. If he ends up dead, then you and I have big problems, and I don't want that. Understood?"

Satisfied that his information was received, Hara hung up the phone, pulled off his gloves, and then walked two blocks north to occupy the vantage point he had selected. Taking a seat on an old milk crate concealed in the shadows by the Dumpster, he verified that the spot had a good view of the parking lot and the front of the apartment building. Now, it was just a matter of waiting until the show began.

Forest Park Apartments, Dayton, Ohio
11:48 p.m., Saturday, October 4, 1969

Yost was sound asleep when he was roughly yanked out of his bed and hurled to the floor. The lamp on the bedside table clicked on. Lying on the carpet, he looked up to see the same two thugs who had attacked him in December.

"How did you find me?" he asked, slowly crawling towards the wall.

"Doesn't really matter," said the larger of the two goons. "A little birdie whispered to me that you just came into a fairly sizeable windfall. Here's the deal, Yost—you're in for some pain and suffering regardless of what happens, but you're going to be hurting much less if you tell us where you stashed the cash. Do that, and we won't break your legs."

"We're going to find it," added the second thug, tapping his palm with a ball peen hammer. "No matter where it is, because we aren't leaving until we find it or you tell us where it is."

Cowering in a corner, Yost knew that trying to retain the money was a lost cause. "In the dresser over there," he confessed timidly. "In the top drawer, rolled up in a sock."

Brandishing a baseball bat, the second thug walked over to the dresser, opened the drawer, and quickly found the money. He took off his gloves, licked his fingers, and counted it. "Three grand and some change," he noted.

"That's all?" asked the first thug. "Every cent? You *swear* that you don't have anything else hidden in here?"

Yost shook his head.

"Good," said the first thug, jamming the sock into Yost's mouth. "That will do for an interim payment on your debt. Now, it's time for us to settle some other accounts."

12:25 a.m., Sunday, October 5, 1969

The thugs were fast but efficient. When the mauling was over, after they departed with their loot, Yost was delirious with pain. He was fairly sure that his right arm and several ribs were broken. Incapacitated, sprawled on the floor, he looked anxiously at the beige Princess telephone on the nightstand, but couldn't move toward it to call for help.

He heard the front door creak open, followed by faint footsteps.

"Yost?" asked a soft voice. "Yost, are you in here?"

"Jimmy?" mumbled Yost, glimpsing a familiar face. "Jimmy Hara? What are you doing here?"

"Just stopped by to pay you a visit," answered Hara, kneeling next to him. "Looks like I showed up right at the nick of time. Man, you look like shit! Rough night?"

"Can you help me?" stammered Yost. "I think I need to go to the hospital."

Nodding, Hara answered, "Sure, buddy. I'll take care of you. I'll get you right to where you need to be." Hara reached into the lower pocket of his faded field jacket, pulled out an object bound in a white handkerchief, carefully unwrapped it, and slid it under the bed.

"What was . . ."

"Listen to me, Yost," whispered Hara, leaning down so his lips were only inches from Yost's pulverized right ear. His warm moist breath smelled like rotten meat. "Do you know the penalty for treason?"

Yost paused, then cringed as he quietly croaked, "Death?"

"Correct answer, sunshine," whispered Hara.

Aerospace Support Project
5:45 p.m., Monday, October 6, 1969

Wolcott chuckled as he read an article concerning an incident near Cleveland yesterday. The Cuyahoga River had caught fire again. Actually, a slick of oil and debris had burned for about thirty minutes, but he thought it bizarre that a waterway could become so horribly polluted that it could actually burn. Scanning further down the front page, he read another headline and gasped. He punched the intercom button and ordered, "Winters! Run downstairs and drag Jimmy Hara up here right now. I don't care what the hell he's doin', I want him *now!*"

Just a few minutes later, Hara entered the office unannounced and took a seat at the conference table. "Did you call, Virgil?" he asked. "Something I can do for you?"

"Did you know about this, Jimmy?" demanded Wolcott. He held up the newspaper, pointed at a small headline in the Metro section, and read aloud, "*Wright-Patt Airman Believed Murdered.*" He folded the paper and slapped it on his desk blotter. "This Sergeant Yost they're talking about, is that not the *same* Yost you were investigating? Have you read this?"

Hara nodded, yawned, looked at the ceiling and said, "Oh yeah, Virg, I read the article. I also read the official reports at Dayton PD this afternoon. Want a quick summary? Yost's 1957 Chevy van was found abandoned on a bridge over the Great Miami River early yesterday morning. Dayton police found a bloodstained Oriental rug and two empty cement bags in the back of the van. There was a circular imprint on the carpet, about two feet in diameter, that looked like it was made by a metal washtub. They strongly suspect Yost had been killed in the van and then dumped in the river."

Like a bored teenager reciting a tedious English assignment, Hara droned on. "The Dayton cops are still dragging the river. Yesterday afternoon, their detectives searched an apartment where Yost was known to be staying, based on an anonymous tip called in by a neighbor. The caller said he heard strange noises from the apartment just before midnight and then saw Yost's van and another car—a dark-colored Dodge Charger—leave the parking lot immediately afterwards. He wrote down the license tag on the Charger.

"The detectives found a Smith & Wesson .38 Special revolver on the floor of the apartment, underneath a bed. It was assumed to have been lost in a struggle. They dusted Yost's van, the apartment, and the gun, but found no prints except Yost's. It looks like the bad guys were wearing gloves and were very careful."

Hara yawned, stretched and continued his recitation. "Dayton PD's big break came when their crime lab identified fingerprints on the cartridges loaded in the revolver. It appears that our culprit was astute enough to wipe down the outside of his gun and wear gloves during the crime, but he apparently didn't abide by the same precautions when he loaded it.

"The prints matched a known criminal who was believed to be a payment enforcer for a local loan shark. They picked him up at his apartment, and he just happened to be in the possession of over three thousand dollars in cash. Coincidentally, his fingerprints and Yost's fingerprints just happened to be on the bills. The suspect also owns a 1968 Dodge Charger. The description and the license plate match what was reported by the anonymous caller.

"Virg, this sure looks to be an open and shut case to me," said Hara, shrugging his shoulders. "As far as I'm concerned, it's strictly under the purview of the local police. Granted, Yost was a problem, but it doesn't appear that he's a problem anymore. Anything else, Virg?"

"Did you have anything to do with this?" asked Wolcott bluntly. "Tell me the truth, Jimmy, skins off."

Hara gazed at him, cracked his knuckles, and said, "Virgil, we've known each other for almost twenty years and I've *never* told you anything but the truth."

Wolcott examined Hara's impassive face; his lack of expression and blank eyes reminded him of when he had first met Hara as a jaded half-breed teenager in the ruins of Hiroshima. He knew that Hara was providing a true and accurate account of Yost's demise, even if he wasn't filling in some of the most pertinent details.

"So, *answer* me, Jimmy, did you have *anything* to do with this?" demanded Wolcott, waving the newspaper.

"If you recall, Virgil, we had a problem and you told me to fix it. That was back in August. It took me a while, but I fixed it."

Wolcott rolled his eyes, grimaced and declared in a low voice, "Tarnations! If Mark Tew ever heard of any of this, or even suspected it, he'd throw both our asses in jail, if he even lived through his initial conniption fit."

Hara nodded solemnly, opened the manila folder on his lap, and slid a sheet of paper across Wolcott's desk.

"What's this?"

"You know the deal," replied Hara, leaning over Wolcott's desk and speaking in a muted voice. "No job is finished until the paperwork is done, and I still have a body to dispose of. I need your signature for that little transaction, Virgil, since you and I both know that there's nothing to be found in the Great Miami but old tires, scrap metal, dead fish, and rotten logs."

Wolcott nervously glanced at the form, quickly endorsed it, turned it face down, and slipped it across the desk to Hara. "No more of this, Jimmy. Please, brother, I'm beggin' you. We can't do this again."

"Oh, I don't think there will be a need again, Virg," replied Hara, returning the form to the folder and standing up. "I blundered *this* time, but I'm back on top of things now. No more slip-ups. Granted, there are still some details to be tied up, but I'll finish those before I go."

"*Go?* Where are you going?"

Hara extracted a medical report from the folder and placed it in front of Wolcott. "That's the results of my lab work," he observed calmly. "I've been diagnosed with acute lymphocytic leukemia. It's a type of cancer. It's very aggressive and almost always fatal. I suppose I was killed by that atomic bomb, but it's just taken a few years to catch up with me."

Stunned, Wolcott muttered, "I'm so sorry, Jimmy. What can we . . ."

"Nothing, Virg. Thanks, though. I want to spend some time with my family before I die, but I'm going to settle this matter first. I really don't like loose ends."

10

HASTY ENCOUNTERS

Woodland Cemetery, Dayton, Ohio
8:19 a.m., Tuesday, October 7, 1969

S eated on a concrete bench under a massive oak, Morozov waited
for Jimmy Hara to appear. If nothing else, it was a pleasant day for
an outdoors excursion. The sun was shining and the temperature
was accommodating. Squirrels scampered close by, chirping birds flitted
through the spreading branches overhead, but a flock of pigeons warily
kept their distance.

He had called Hara at home several times in the past two weeks,
but the American had expressed reluctance to talk. Last night, in the
aftermath of Yost's alleged murder over the weekend, the engineering
technician finally conceded to a meeting.

Morozov shook his head as he reflected on the events of the past
week. While he hoped that this new contact would be forthcoming with
information about the activities within Hangar Three, Morozov had his
doubts. He was furious that Yost had not provided anything of value.

He was due to submit a detailed report to Washington by the end of the week. As it was, he couldn't fathom how to explain why he had no results commensurate with the time and money expended on Yost. And now, Yost had been killed, apparently by criminals.

Even in death, Yost was an irritating thorn in his side. Several weeks ago, he had issued Yost an expensive German-made Minox-B miniature camera and trained him on its operation, but Yost had failed to produce any pictures of merit. Morozov suspected that the American airman had pawned the camera, probably to finance yet another liquor store sortie. Now that Yost was dead, the Minox was yet another loose end that he would be called to account for.

Once holding such great promise, this elusive endeavor had become the most frustrating experience of his career. He felt sure that his return to the Motherland was imminent, since he had been unsuccessful in gathering any useful information about Project Blue Book or the Aerospace Support Project. His GRU bosses clamored for substantial evidence that the Americans were studying alien technologies. Morozov had placated them for several months now, but they were growing increasingly impatient.

He surmised that his safest option was to gradually extricate himself from this Blue Book debacle, ideally handing it over to a GRU officer even more junior than himself. He had tendered a request to be reassigned to the GRU's Hanoi bureau, which had been turned down repeatedly. In truth, to make matters even more aggravating, his appeal for the transfer had not been denied, but rather had just been ignored every time he resubmitted it.

He looked up as he heard someone approach, folded a newspaper and draped it over his lap as a safety signal. The newcomer was wearing the specified colors. "Mind if I sit here?" he asked, pointing towards nearby headstones. "My uncle is buried in that section over there."

Morozov gestured towards the bench and studied the man as he sat down. His features were vaguely Oriental, like he was the progeny of a mixed marriage. Morozov had trained with GRU officers from Kazakhstan, which was formerly part of the Mongol Empire, and this man looked like he could easily be from that region. But at least the Kazakhs were pure-blooded; this man was obviously a despicable mongrel, as were most Americans. "Jimmy Hara?" he asked.

Hara nodded.

"Your identification, please."

Comparing Hara's appearance to the photograph and details on the identification card, Morozov was shocked. It was obviously Hara, but the man looked horrible. His trousers sagged at his waist; it was obvious that he had recently lost a considerable amount of weight. His skin was ashen and the whites of his eyes—at least the parts that weren't blood-shot—were tinged with an unnatural shade of yellow.

"It's really me," noted Hara, as if to apologize for the seeming discrepancy. "I've been very sick recently."

"Sorry," replied Morozov, handing the card back. "Listen, Jimmy, you can call me Anatoly."

"*Anatoly?*" asked Hara. "Isn't that a Russian name? Yost claimed that you were Israeli."

"I am. My family left Russia right after the Revolution. There were a lot of Jews in Russia and Eastern Europe. Of course, not so many after the War."

"Oh. Makes sense," replied Hara. "Before we talk, there's something I need to tell you. We won't meet again after today, so if you have something to ask, you need to ask it today."

Morozov nodded, but said, "I can understand your reluctance, Jimmy, but I'm *sure* that I can offer you *something* to motivate you to overcome your inhibitions . . ."

"That's very tempting, but I really don't think so," interjected Hara. "Like I said, this will be our first and last chat. I'll tell you: initially, I didn't want to come here, but I'm doing it for Eric Yost."

"But he's dead, isn't he?"

"Yeah, but he was still my friend. Look, he used to be a really solid, dependable guy, but he went way overboard with his drinking and gambling, and landed himself in a tight jam. Now, would you care to tell me what you want?"

"I would like you to tell me what you do in Hangar Three," said Morozov bluntly.

Quizzically raising his eyebrows, Hara asked, "Hangar Three? Pardon me, Anatoly, but didn't you say you were from Israel?"

"I did."

Hara sighed and said, "If you're really an Israeli, then you shouldn't be much interested in what we do in Hangar Three."

"Why would that be?" asked Morozov. "Indulge me."

"If you insist. Hangar Three is a reverse-engineering facility. We study foreign aircraft, especially Soviet fighters. We rip them apart, tinker with them, put them back together, and then send them out West for flight-testing in Arizona and Nevada."

"And why wouldn't we Israelis be interested in that?" asked Morozov.

Hara laughed. "Are you kidding? Honestly, you don't *know*? Where do you think we obtain most of our Soviet hardware?"

"Oh. But what about the UFOs? Alien spacecraft?" asked Morozov. "Yost said . . ."

"*Yost* said? If you haven't figured it out yet, towards the end, Yost was an unreliable drunk and a pathological liar. He would probably tell you that the sky was green if you stuck enough money in his pocket, and I suspect that you probably did. He was in a pretty desperate situation, so I'm sure that he was willing to tell you anything you wanted to hear. You obviously wanted to hear that there were UFOs in Hangar Three, so that's what he told you. Honestly, I don't even think he knew that he was lying anymore. Very sad, since he used to be such a great guy."

Morozov nodded in agreement. "I suppose that you're right."

A trickle of blood-tinged saliva oozed down Hara's chin. "Excuse me," he said. "My gums are getting so damned loose that my teeth are falling out." Hara gingerly stuck two fingers in his mouth; grimacing as he tugged and wiggled, he easily extracted a bloody molar. He examined the tooth briefly and then tossed it away. It rebounded off an old tombstone and landed in the neatly preened grass.

"I shouldn't have asked you here," said Morozov. "You're obviously very sick."

Hara smiled; at least half of his teeth were missing. "Yeah, I'm sick all right. I'm dying. It's leukemia, a type of cancer. It's terminal. I won't be around much longer."

Morozov looked at Hara and felt pity for him. In his childhood, when Stalingrad was under siege, he had watched many people die. Most died swiftly, but some—like his aunt and cousins who starved to death because they could not force themselves to eat that which was objectionable—had the unfortunate fate of an agonizingly slow and painful demise. So Morozov was familiar with death, and he saw in Hara the detached demeanor of a man who had reconciled himself to his fate. "I'm very sorry for you," he said solemnly.

"Thanks. I should be on my way," said Hara, slowly standing up. "Goodbye, Anatoly."

"Goodbye, Jimmy." Morozov reached out to shake Hara's hand, but thought better of it. Checking for his keys, he felt a bulge in his pocket. He rushed after Hara, holding out a thick envelope. As he closed the gap, the flimsy stitching of his right shoe fell apart, and the loose sole flapped loudly with his every step, like a prop from a slapstick comedy. He nudged Hara's shoulder and declared, "Wait. This is yours. Your money."

Hara turned slowly and looked at the envelope in Morozov's hand. A weak smile crossed his face. "I have no use for that any more. Why don't you keep it?" He looked down and added, "If nothing else, you could buy yourself some decent shoes."

The Pentagon
9:45 a.m., Wednesday, October 15, 1969

Summoned with Tew and Wolcott to a hurriedly called meeting, it was Heydrich's first visit to the Pentagon. The purpose of the meeting was a mystery, and although the trio discussed theories along the way, Heydrich suspected that it had something to do with the surprise record-setting development in the Soviet space program. At this very moment, three spacecraft of the new *Soyuz* design were orbiting the Earth at the same time, with a total of seven men aboard.

But Tew believed that it had nothing to do with the Soviets, but was more likely some sort of inquiry. "Gunter, it's probably prudent that you leave most of the talking to Virgil and me," he advised. "Certainly, if someone asks you a direct question, speak with candor, answer to the best of your ability, but don't be forthcoming with additional information." Tew wore his dress uniform; three stars now adorned his epaulets and four rows of ribbons decorated his chest.

"That's sage advice, pard," added Wolcott. "I'm sure your buddies at Nuremburg probably heard the same guidance from their lawyers." Wolcott was attired in a rather conservative gray pinstripe business suit; Heydrich was still slightly shocked that the erstwhile cowboy owned any clothes that didn't look like they had been borrowed from John Wayne's closet.

After presenting his credentials to a stern-faced sentinel at the entrance to the secure conference room, Tew adjusted his jacket, brushed an errant dandruff flake from his lapel, and said, "Maybe we'll come out of this intact. Or at least moderately intact."

As soon as they entered, their destiny was revealed. The conference room held the atmosphere of an impending inquisition. Besides General Kittredge and his usual retinue, Tarbox and Russo were present. Both wore smug expressions that telegraphed that they had more than an inkling of what would come next. Heydrich heard the door open behind him and turned to see the arrival of Brigadier General Isaac Fels, commander of the 116th Wing at Eglin.

"Isaac, friend, congratulations on the star," said Tew, extending his hand. "Certainly no one deserves it more than you."

"And congrats on your third," replied Fels. "But *you* earned yours, Mark. I'm sure that someone will eventually realize that mine was the result of some clerical error."

"Let's hope we're both still wearing them at the end of the day," said Tew quietly.

Kittredge motioned for the newcomers to take their seats and opened the meeting. "Gentlemen, our main focus today is to discuss some emerging requirements and the proposed extension for Mark's project. But before we start, Leon has a complaint he wants to air."

With an unnerving sound that resembled an asthmatic cat coughing up a fur ball, Tarbox cleared his throat and then stated, "General Kittredge, I accuse General Tew of blatant disregard of legitimate orders and flagrant dereliction of duty, and I insist the Air Force take appropriate disciplinary actions against him."

With that, Tarbox rattled through a litany of accusations, painstakingly outlining the events of the August mission. He concluded, "And we will show indisputable evidence that he instructed the Mission Three crew to disregard my orders to destroy their target after I had clearly conveyed a command decision to do so, based on time-sensitive strategic concerns. In addition to presenting our evidence, we also want to interview Mr. Heydrich here, since he was probably the best witness to General Tew's actions on that day."

Following Tarbox, in a riveting performance worthy of a television courtroom drama, Russo solemnly recited expert testimony concerning discrepancies in the communications transcripts. His spiel included a five-minute dissertation—replete with color diagrams professionally mounted on poster boards—that provided technical evidence clearly showing that the Gemini-I's onboard voice recorder had been momentarily disabled on or about the time of the Azores contact and that at least twenty seconds of tape had been erased and recorded over.

Russo's arguments were compelling; in another setting, he probably could have convinced the Warren Commission that a battalion of Soviet snipers had been stationed on the Grassy Knoll. After he spoke, the meeting swiftly devolved into a shouting match of accusations and counter-accusations, with the centerpiece being Tarbox and Wolcott aggressively leaning over the walnut conference table, thumping each other's chests with their index fingers.

Abruptly halting the fray, Kittredge commanded, "*Enough!* Leon, since I was under the mistaken impression that we could all handle this matter like well-mannered gentlemen, I granted you a few minutes to state your grievance and ask for an apology. Obviously, you aren't content to do that, so I'll settle this case now, once and for all, and we will not speak of it again."

As Tarbox seethed, Kittredge continued. "Leon, although you wrote the check for the mission, Mark and his guys accomplished your objectives exactly as you defined them prior to launch. The bottom line is that you don't plot a new course in mid-tack. If you wanted that radar platform destroyed while Mark's crew was still on orbit, you should have stated that early on. But you didn't, and while you certainly had a good idea, it was too late in the game to make a deviation. Consequently, since the deed was done per your original instructions, you don't have a valid gripe."

"Yes, General," squeaked Tarbox, his face scarlet with mixed anger and embarrassment.

"Two more bullets, Admiral," said Kittredge. "First, if you're really interested in fostering inter-service cooperation, I would strongly caution you against capriciously accusing one of my senior officers of misconduct. Second, for future reference, in the unfortunate event that we find ourselves in similar circumstances, you might again succeed in buying Air Force hardware, but that doesn't mean you have the right to levy our flight crews into indentured servitude."

Obviously glimpsing an opportunity, Tarbox slid a folder towards Kittredge. "Since this situation might come up again, I have a proposal," he explained. "Assuming that we might have a similar Navy-specific requirement in the future, I want to permanently assign a Navy flight crew to Mark's project, under *my* command, to fly purely Navy missions. We'll provide support personnel as well, to reduce any burden. Lieutenant Colonel Russo here has plenty of experience in Mark's shop, so I'll probably position him to run that operation on my behalf."

"Intriguing concept, Leon," praised Kittredge, donning reading glasses and scanning through the papers in the folder. "It definitely has merit, especially since it would alleviate the command and control issues, and it would also be a fix for our shortcomings in flight crew personnel. *Kudos*. Well done. Well done, indeed."

Heydrich looked at Russo, who was grinning like the Cheshire cat, and knew immediately where this vile idea had hatched. He closed his eyes and audibly groaned; in his thoughts, all he could see was Russo's resurrection at Wright-Patterson, like an insatiable vampire who just could not be slain.

"Mark, obviously I need to consult with you on this matter," said Kittredge, slipping off his glasses and massaging the bridge of his nose. "Are you receptive to some of Admiral Tarbox's guys being permanently assigned to your project?"

"No, sir," replied Tew adamantly. "While I appreciate the offer, I'm confident that our crews can execute any requirement that might surface on the Navy side."

"Well, I suppose that resolves this situation," declared Kittredge abruptly, sliding the folder across the table towards Tarbox. "Thanks but no thanks, Leon."

"We're not done with this," groused Tarbox, handing the folder to Russo.

"Duly noted, Leon," observed Kittredge. "I'm confident that this horse will be beaten until it begs for a swift and merciful death, but don't walk out of here believing that your plan is going to happen without a fight, regardless of *who* you call."

"Yeah, bub," muttered Wolcott, menacingly glaring at Tarbox. "You can rest assured that it'll happen over your dead body."

"Virgil, I think you mean to say 'It will happen over *my* dead body,'" offered Russo.

"That would work as well, youngster," agreed Wolcott. "And if you correct me again, that will occur sooner than later."

"*Enough bickering*," said Kittredge. "Let's focus on more pressing matters, specifically Phase Two of your project, Mark. Our budget guys are tying up the loose ends, but we now have a fairly good perspective on what your future looks like, provided the funds are approved."

Since the discussion had shifted to matters pertinent to the Project, Heydrich wondered why Kittredge did not dismiss Tarbox and Russo from the meeting.

Opening a briefing binder placed before him by one of Kittredge's aides, Tew asked, "When do you expect the Phase Two budget to be approved, Hugh?"

Gazing at the ceiling, Kittredge sniffed. "Mark, you know the story. This budget is driven by potential political outcomes. I don't see it approved any earlier than mid-November of 1972."

"So our wagon is hitched to Dick Nixon being elected again?" stated Wolcott.

"Correct," replied Kittredge, nodding as he looked towards a framed portrait of the president on the wall. "But the close-out of your current phase should roughly coincide with the beginning of the second phase. There shouldn't be much of a gap, so you can plan on twelve more missions after Phase One, with a targeting emphasis on Soviet reconnaissance and communications platforms. The OBS mission will become a secondary priority."

"And if the Democrats win the '72 election?" asked Tew, poring over the new budget.

"If they do, God forbid, then it's highly unlikely that your project will be continued. Regardless of the potential outcome, we still need to discuss some new wrinkles. The Operational Review Committee is tacking on a new requirement that we want you to start looking at right away. We want you to be prepared to execute extravehicular activity operations in Phase Two."

"EVA?" asked Tew. "*Walking in space?* I can't think of any logical reason for our men to leave their vehicle and float around outside. Are you folks not happy with the way we do business right now?"

"Mark, we're thrilled with what you're doing, but the committee wants other attack options in case we are confronted with more sophisticated targets. Also, we want to have a capability to retrieve select components from target vehicles," explained Kittredge.

Attack? thought Heydrich. He had never heard that expression applied to their missions. The operative word had always been *intercept, not attack. Why the change in terminology?*

"I don't like this idea," declared Wolcott. "But I strongly suspect that it's going to be shoved down our gullet regardless. Besides, Hugh, we don't have the facilities for EVA training. We can't exactly saunter down to the base pool at Wright-Patt and practice walkin' in space, can we? That might raise a few eyebrows."

"Hugh, Virgil is right," observed Tew. "We don't have the appropriate facilities, and I don't foresee borrowing any from NASA, considering the security issues."

"Begging your pardon, Mark, but I have a suggestion," uttered Heydrich. "A few months ago, back when the MOL program was still active, I was talking to one of the MOL simulation guys. He said that they had a really nice arrangement at a place called Buck Island, down in the US Virgin Islands. Their crews used to fly down there to do weightless simulations underwater. We could probably use that same place to train our guys." Those words had barely left his lips when Heydrich noticed Russo squirming in his seat as he glanced nervously at Tarbox.

"Buck Island?" retorted Tarbox. "That facility is now *closed*. It is *not* available for your use."

"I just thought . . ."

"You just *thought* it would be a good destination for Caribbean sightseeing junkets disguised as training." Tarbox smirked. "But I'm confident that you'll find a swimming hole closer to home."

"Okay," interjected Tew. "We'll figure out something, but there's a greater problem, Hugh."

"And that is?"

"You've promised us pilots, but there are none coming through the pipeline. I'm down to two flight crews, with nine missions still looming on the horizon. And to be frank, while one of my crews is the cat's meow, I don't exactly have absolute confidence in the other. When are you going to make good on your promises to plus up my flight crews?

Even two more pilots would be an excellent start. That would at least give us another crew."

Kittredge shook his head and answered, "You know it's not that simple, Mark. There are more flight test programs going on right now than any time in history, so everyone graduating at Edwards is immediately spoken for. And if you haven't noticed, we're in the middle of a shooting war that has more than a few pilots gainfully employed. And sadly, a lot our guys are sitting in prison cells in North Vietnam. So things are tough all over. Granted, you only have two flight crews, but you also have plenty of time between missions, so it's not like we're killing them."

But at the rate they're expected to fly, that loose end will eventually be tied by the law of averages, thought Heydrich.

Tarbox snorted. "I just offered you a flight crew, Tew. But you turned them down. Are you sure you don't want to reconsider my offer?"

"We need pilots, Leon," replied Tew, not wavering. "We just don't need *your* pilots."

"Aviators," noted Tarbox. "*Naval aviators.*"

"Okay, to clarify, we don't need your *Naval aviators.*"

"I'm a little confused," said Heydrich.

"How so, Gunter?" asked Kittredge.

"Since the MOL program was cancelled in June, I would have thought we could have picked up some of *those* pilots," stated Heydrich.

"Again, Gunter, there's not a simple answer," explained Kittredge patiently. "Your project is operating under the radar. *Way* under the radar. On the other hand, the names of the MOL flight crew personnel are in the public eye. A lot of the MOL pilots were swapped over to NASA after the program was cancelled. And believe it or not, some of the MOL guys are now flying in regular squadrons, biding their time until the Air Force deems it safe for them to fly in Vietnam. We can't arbitrarily siphon former MOL flight personnel into your project without raising some suspicions."

"General Kittredge, I still have a question about the Can . . . MOL astronauts," interjected Heydrich, adjusting his black-framed glasses.

"I'm aware that a *fourth* group was already selected, because"—Heydrich pointed at Russo—"*he* claimed that he was already assigned to it. The names of *those* men were never released to the press, so if their names aren't public knowledge, why aren't *they* available to us?"

Russo looked nervously at Tarbox, who said, "Good question, Gunter. Other than Russo, the remaining members of the fourth group were *Naval aviators*. Since your skipper has already clearly voiced his disdain for *Naval aviators*, it's a moot issue."

Tarbox cleared his throat and continued. "Since those men were never publically announced, they will matriculate to flight test or fleet aviation assignments. Certainly, we'll afford them the opportunity to transfer to the NASA astronaut program, but they'll have to apply under the normal selection process when NASA makes their next call for applicants."

"I think we've discussed this issue sufficiently," stated Tew, glaring at Heydrich. "I just want to emphasize that we want new pilots as soon as the pipeline becomes unclogged."

"Noted," said Kittredge. "But don't plan on seeing any new faces until your Phase Two funding is approved. Now, Mark, tell me about your new computer. I understand that there's been some recent progress."

"The new Block Two computer? Well, there's good news and bad news. The good news is that the MIT guys delivered. But it took them a year longer than they anticipated."

"In their defense, Mark, the MIT eggheads were sort of busy goin' to the moon," observed Wolcott. "It ain't like they were sitting on their thumbs."

"True," said Tew. "Functionally, the new computer works as advertised. The MIT team adapted most of it from the Apollo Guidance Computer, which is head and shoulders above the older Gemini Block I computer. The Block Two will automate most of the work that the crews are doing manually right now. Anyway, Hugh, that's the *good* news."

"That's excellent," said Kittredge. "But what's the bad news?"

Wolcott chuckled and said, "Well, Hugh, it's sort of like sendin' away for a mail order bride. We expected a sleek little number, but it's a might heftier than anticipated. It's roughly twice as big as the old computer. The first prototype is in St. Louis right now. The airframe honchos are trying to figure out some way to shoehorn it into the spacecraft, but it just ain't lookin' so hot. It's causing some big weight and balance issues. It's really throwing off our center of gravity."

"So we're not going to be flying it anytime soon, right?" asked Kittredge.

"No earlier than Mission Six, and even that's contingent on a lot of technical issues being resolved in time," said Tew. "So it's at least a year out."

"Disappointing," replied Kittredge, pouring a glass of water. "Your next mission is in March?"

"Affirmative, sir. Mission Four," answered Tew. "The hardware is already stacked and waiting in San Diego, ready to fly."

"And the crew?"

"Hugh, I'll have to defer to Mr. Heydrich, who is our chief of operations and training. He has the most contact with the flight crew and can best lend you a feel for their readiness. Gunter?"

Heydrich cleared his throat and answered, "I am relatively optimistic that this crew will be ready to fly in March."

"*Relatively* optimistic?" asked Kittredge. "Heydrich, I appreciate your honesty, but you need to qualify why you're just *relatively* optimistic."

"It's a complicated mission, General," replied Heydrich. "Crew Three—Jackson and Sigler—is assigned to this mission, and while they're good, I'm concerned with their ability to respond to potential changes once they arrive on orbit. That's why I'm *relatively* optimistic, General."

"How about your *other* two guys, your big heroes? Carson and Ourecky?" asked Kittredge. "Can they hack this one? Couldn't you fly them again?"

"We're really trying to give them a break," answered Tew. "Additionally, we don't want to become overly reliant on them."

"Okay, that's understandable. But theoretically, if you determine that the prime crew isn't ready, how long is it going to take to spin up Carson and Ourecky?"

Tew nodded towards Heydrich, who confidently answered, "They could probably launch tomorrow, but I would be a lot more confident if they trained for at least two weeks."

"Two weeks?" Kittredge sipped his water and said, "Mark, Virgil, I don't want to climb into your sandbox, but I will say this, and I want to be as emphatic as possible—you need to make this crew decision in a *timely* manner. If the primary crew can't cut the mustard, then yank them out of the line-up and put your big guns in."

"I'll take that under advisement, Hugh," replied Tew. "But at this juncture, we're sticking with our plan. Crew Three will continue to train. They *will* fly the mission in March."

"Fine. It's your call." Kittredge looked at a clock on the wall and added, "This is an opportune moment to grab a smoke break, so be back in here in fifteen minutes. I want to discuss contingency recovery plans for polar missions. I'll be down in my office, making a secure call, and I'll be back as soon as possible."

As the participants filtered out of the room, Wolcott drew his Zippo, lit a cigarette, and inhaled deeply. "Gunter, are you familiar with the layout of this place?"

"The Pentagon? Not really, Virgil. I've never been here, but I did help draw up plans to destroy it with antipodal rocket bombers." Heydrich sighed plaintively, shrugged, and added, "We just never had time to finish the prototype and move into production."

"That's tragic, hoss. I weep for you," replied Wolcott, loosening his tie and undoing the top button of his shirt. "Look, there's no reason for you to sit in on this next session. There's a little coffee shop located smack in the center of the complex. All you have to do is walk inwards on a spoke corridor. Why don't you go there, relax, and we'll gather you up when we're done?"

"But, Virgil, if you don't mind, I really would like to sit in and listen to the rest of this. No one has ever mentioned the possibility of polar orbit missions before, and I . . ."

"Maybe you're not listening, Gunter. There ain't any reason for you to sit in on this next session. That's a polite way of saying that you need to make yourself scarce. Just go wrangle yourself a cup of hot arbuckle and a couple of fresh sinkers, and we'll herd you up shortly. Fair enough, pard?"

"Fair enough," replied Heydrich.

"And Gunter, the next time we ask you to speak with candor, please try to restrain yourself."

Tri-Border Region, South Vietnam
6:30 a.m., Friday, October 17, 1969

As the sun crept over the horizon, gradually dispelling darkness with light, Nestor Glades watched as the earth slowly took form. At this point, he wanted nothing more than to just *sleep*, but slumber was a luxury he could not afford. He and his five teammates hadn't slept a wink in the past four days, primarily because they were fleeing from the greater part of a North Vietnamese Army battalion decidedly intent on killing them. To the NVAs' credit, their ire was sincere and entirely understandable; Glades and his little crew were responsible for the deaths of over a hundred of their companions.

Crouching in the scant shelter of a bomb crater, he unbuttoned his left chest pocket and extracted his pill kit. He flipped open the cloth cover and quickly selected two APC tablets and two Dexedrine capsules. The APC tablets—a mixture of acetaminophen, phenacetin, and caffeine— would hopefully alleviate the pounding headache that accompanied his severe dehydration. The speed was to keep him awake and focused. He crammed the four pills in his mouth and choked them down, wishing that he still had some water to help him swallow, but their canteens had run dry two days ago.

As an afterthought, he gulped down a tetracycline pill as well. He hoped that the antibiotic would help stave off the infection in his throbbing left foot, the result of minor frag wounds he had sustained from an exploding B-40 rocket three days ago.

The swollen foot has caused his boot to balloon so grotesquely that it looked like something out of a cartoon his kids would watch on

Saturday morning. Thick greenish-yellow pus and blood oozed from the brass vent holes in his instep, as well as from several perforations where jagged shards of shrapnel had ripped through the leather. He sniffed the air; the foot was rank, but at least it didn't have the unmistakably putrid stench of gangrene. Not yet, anyway.

His recon team consisted of four tough little Montganard mercenaries and one other American. He normally ventured out with five more "Yards" and a third American, but the others were incapacitated with severe dysentery when this mission was handed to him. Rather than augment with attachments from other recon MACV-SOG teams, Glades preferred to go with a small and stealthy team of men long conditioned to working with one another.

Inserted by helicopter six days ago, their mission was to locate the headquarters of an NVA regiment. They had successfully pinpointed the headquarters on the second day; in the process, Glades had initiated an impromptu raid that had killed the NVA regimental commander and several key members of his staff.

Their action quickly resonated all the way to Hanoi, resulting in their pursuit by the five-hundred-man NVA battalion ordered to annihilate the recon team at all costs. Now, it seemed as if the battalion had ceased to be an amalgamation of men, but became more like a massive predator, lurking in the wilderness, vindictively eager to snuff them out.

Glades had done this job long enough to clearly know the dire prospects of their survival. By his reckoning, they had less than an hour left; they would either be long gone from here, or they would be dead, left to rot in this stinking jungle.

If they were killed, Glades would prefer that they be abandoned to decay, but he knew full well that was not to be; other SOG recon men would come here to recover their remains. They would come at great risk to their lives, without the slightest reluctance, and the NVA knew it and would be lying in wait. Many more men would die, but the corpses of Glades and his team would not be left in the field. Despite this, he had given his team strict instructions to leave him if recovering his body meant placing themselves at risk, and he had long ago convinced Deirdre to be ready to bury an empty casket.

He looked up to study the faces of the men who jammed into the crater beside him; they shared the blank, callous expression of men who had long since abandoned fear as they had accepted the certainty of death. He pointed at the gray magazine of his CAR-15 carbine and shrugged his shoulders.

Two of the men shook their heads, and two others indicated that they had less than a magazine left. The fifth man—one of his longtime Yard stalwarts—was unconscious, mercifully rendered so by a potent dose of morphine. Missing both arms below the elbows, the warrior was barely alive; if Glades and the others had an hour yet to draw breath, he had less so.

Glades heard a faint drone of a prop-driven aircraft. It was the Covey Rider, an OV-10 Bronco bearing an Air Force pilot and a SOG recon man who would coordinate the effort to extract the team. Cupping a tiny signal mirror in one hand, he carefully lined it up with the sun while using his other thumb as an aiming sight. He wagged the mirror twice.

Immediately, the Covey Rider's SOG man spoke over the radio, "Cottonmouth One-Zero, this is Covey Rider, I have your position, I have an extraction package stooging five minutes to the west, and two F-4s standing by for a run-in. What is your plan? Over."

Glades tucked the mirror back in his shirt and whispered into the microphone of his radio handset, "Covey Rider, there is a small clearing two hundred meters north of my position. Request tac air expend all ordnance between my pos and that clearing. Request STABO extraction immediately after ordnance drop. Over."

There was a long pause before the Covey Rider responded. "That's mighty close, One-Zero. Are you sure? Over."

"Covey, Cottonmouth One-Zero. I am *sure*. Over."

"One-Zero, Covey Rider. Good copy. How many do you have for extraction?"

"Six to string. Request extraction birds drop four and two," whispered Glades, describing his plan for the helicopters to drop ropes and yank out the men by the STABO extraction method.

"Cottonmouth One-Zero, Covey Rider, I copy you have six to string out. Any wounded?"

Glades was tempted to laugh. *Any wounded? Let me see—is anyone NOT wounded? If you're not wounded, raise your hand. Uh, belay that last command, Do Cao—you seem to have lost your hands.*

"Covey Rider, One-Zero, five wounded. Over." Glades didn't include himself in the count; by his standards, his Fearless Fosdick flesh wound didn't merit reporting.

"One-Zero, I copy five wounded. Any KIA? Over."

"Not *yet*. Over," replied Glades.

"One-Zero, good copy. Six to string. Good luck to you, brother. Hold tight. I'm setting up the orchestra now . . . break break . . . Voodoo Child, what do you have?"

"Covey Rider, this is Voodoo Child, bearing gifts of snake and nape," answered an F-4 Phantom driver, miles away and thousands of feet overhead. His cool, disembodied voice sounded more like a radio station DJ's than an Air Force pilot's. "On orbit one minute from communications checkpoint, ninety seconds from target on your call."

"Voodoo Child, copy snake and nape, stand by for high pass and further instructions . . . break break . . . Argyle Flight leader, this is Covey Rider, stand by to run-in on my call. Six to string. One-Zero requests you drop four lines on first stick and two lines on second stick. *Heavy* enemy presence in immediate area. Over."

"Covey Rider, Argyle Flight leader. I am bringing two slicks and two gunships to the fight. Copy six pax for extraction. Will drop four and two. Setting up for run-in. Over."

As Glades and his men burrowed deeper into the crater, striving to bury themselves in the shattered black earth and reeking muck, he heard hushed voices all around them. The wary NVA knew where they were, and were approaching cautiously. He heard the unified roar of two F-4's making a high pass overhead as the Covey Rider oriented them to the terrain and the recon team's location.

He passed the leg straps of his STABO harness between his legs and quietly clicked the snap hooks into the V-rings in the front of the rig. He gestured for the others to do the same, and made sure that his One-One—the American assistant team leader—took care of Do Cao's harness. Integrated into their load-bearing gear, the STABO rigs were like parachute harnesses; with the leg straps fastened, they were awkward to run in, but Glades figured they wouldn't be that much of a hindrance at this point.

There was nothing to do now but wait. Listening to the muted radio handset pressed against his ear, Glades heard that the F-4s were cleared hot for the run. Cradling his CAR-15 in his forearms, he covered his ears and opened his mouth, and the others followed his lead.

Seconds later the earth became flame, smoke, and noise. Hard clods of dirt rained down, and shattered trees pin-wheeled through the air. One flaming trunk landed square on their crater, missing Glades by a fraction of an inch. Each successive detonation jammed him further into the gummy black mud. He half-expected his guts to come spewing out his mouth and nose from the invisible waves of concussive force.

When the debris stopped falling, Glades rose to his knees to quickly take stock of the situation. He yelled for the others to run. He and the One-One grabbed the badly wounded Do Cao by the STABO harness; they half-carried and half-dragged him at a dead sprint through the smoldering underbrush.

Unburdened, the three Yards could have easily run twice as fast as Glades and the One-One, but they obediently stayed close by, protecting the two Americans and their wounded comrade. Incinerated bodies were strewn everywhere. The morning air was saturated with the sweet stench of partially burned napalm and barbecued human flesh. Now more liquid than meat and bone, Glades' festering left foot squished with every painful step.

Two horribly burned but still resistant NVA soldiers blocked the route to the clearing; they screamed in agony as they emptied their

AK-47s at the fleeing recon team. Firing his CAR-15 left-handed from the hip, Glades slayed each with a single bullet to the chest.

Momentarily distracted, he tripped over the charred body of another NVA soldier. The sudden fall twisted his grip from the Do Cao's harness. In a single motion, an almost superhuman gesture for someone so debilitated by dehydration and exhaustion, Glades's One-One swung the wounded Montagnard onto his shoulders and kept running.

Elated, they made it to the clearing. This spot had sustained the brunt of the airstrike. A few deafened and bewildered NVA soldiers were still alive, but they were all but ineffective. Firing carefully aimed single shots, Glades and one of the Yards made short work of any that offered even the slightest resistance.

He heard the Huey slicks approaching. To the west, he saw a clump of NVA fighters running towards them; they had been outside the zone of devastation wrought by the bombs and napalm, and were not the least bit inclined to let the recon team slip out of their grasp.

Glades started to speak into the radio, but saw that the Covey Rider was clearly aware of the NVA's last-ditch assault. The twin-engine OV-10 was working much closer to the ground now, firing white phosphorus rockets to mark the enemy formation for the pair of helicopter gunships that would cover the extraction. Incandescent ropes of green tracers danced and arced up into the air as the NVA desperately tried to knock down the persistent observation plane.

The chattering slicks came on station overhead, hovering just above the treetops. The pilot of the first extraction bird centered it on the clearing. Glades looked up momentarily and saw the four rope bundles fall out of the back of the helicopter.

The ropes, encased in weighted canvas sleeves, deployed cleanly as they tumbled to the ground. As the gunships fell into place, laying down withering fire on the remnants of NVA still trying to fight their way to the pick-up zone, Glades snatched a set of ropes and connected them to the STABO harness of the unconscious Do Cao.

The three less wounded Montagnards clicked their oval-shaped snaplinks onto the other liftlines. With two holding Do Cao upright between them, the fourth stretched his arms out to his sides, signaling that they were ready for pick-up.

Immediately the four men were snatched into the air, in a scene reminiscent of some ghastly surreal version of *Peter Pan*. Jet black with a coating of soot and smudge, dripping blood, they slowly rose through the smoldering remains of what had been a peaceful forest just hours before. The extraction bird lifted them almost free of the trees before transitioning to level flight. Glades hoped that the pilots didn't drag them too far; it wasn't unheard of for recon men to have their spines snapped by branches if the aircrew got too hasty to ascend clear of danger.

The other slick flared to a hover over the clearing. Two more rope bags were kicked out. One caught into the branches about fifty feet above the ground; the men in the back of the helicopter immediately chopped the rope free and hurled out a replacement.

To his dismay, Glades saw that three NVA soldiers had leaked through the gunships' rain of fire and were screaming through the woods. Lacking other options, he and his One-One charged them. Bellowing as if possessed by demons, with flames licking his calves, Glades fired his last round, killing one of the NVA. His One-One tackled another; straddling the man, he bludgeoned him into submission with the flimsy buttstock of his CAR-15. The third NVA elected to turn and run, apparently deciding that he was meant for a longer life and higher purposes.

With the PZ temporarily secure, Glades and the One-One scrambled back to the ropes and swiftly snapped in. Glades gave the signal for pick-up, and they levitated into the air, as if swept up by an invisible hand. The helicopter pilot was kind to them, or at least patient enough not to be rushed; the two recon men weren't pulled through the trees.

As the helicopter accelerated into level flight, catching up to the other slick stringing the four Yards, Glades relaxed his body and spun around momentarily to look to the rear. He glimpsed the black tear-shaped scar in the otherwise lush verdant forest, and watched as the two gunships broke off from the fight and joined the two slicks for the twenty-minute ride back to the launch site. He glanced at his One-One to see if he was still intact. The One-One grinned like a crazed possum and then immediately went limp; drained beyond the point of exhaustion, dangling from a skinny rope hundreds of feet in the air, he simply fell fast asleep.

Glades arched his back and spread his arms; doing so caused his body to turn so that he faced in the direction of flight. He enjoyed this sensation; it felt like free-falling sideways through the sky. He thought of the days ahead. For Glades and his teammates not immediately hospitalized, this afternoon and tomorrow would be taken up with mission debriefings.

Squinting through the rushing wind, he saw the launch site in the distance. Speculating on how long he would continue to beat the odds, he thought of Deirdre and wondered if he would live to see her or their children again.

11

TRANSITIONS

Naval Aerospace Support Office, Los Angeles, California
9:50 a.m., Wednesday, November 19, 1969

E d Russo had never been so frustrated in his entire life. The Air Force's massive MOL program had been cancelled in June, so his assignment was drawing to an end. In fact, his tour would have already concluded if Admiral Tarbox had not temporarily extended the assignment to coordinate September's Blue Gemini mission to intercept and destroy a Soviet maritime surveillance satellite. At present, he was still assigned to Tarbox's small organization, which had received a vague new name—"Naval Aerospace Support Office"— and had physically moved to a government office complex several miles away from the old Air Force MOL complex. Even as the MOL project was being dismantled, Russo's surroundings had changed little. The El Segundo sky was perpetually overcast with a grimy pall of smoke, and the air still reeked from the pungent fumes of nearby oil refineries.

He was alone in an open office bay that was the workspace for the six men—five Naval aviators and himself—who comprised the last group of military astronauts selected before the MOL program folded. A transistor radio played in the background; reading a summary of the morning's news, a broadcaster casually announced that two Apollo 12 astronauts—Charles Conrad and Alan Bean, both former Navy men—were walking on the moon.

Leaning back in his chair, Russo grinned as he tore open a manila envelope containing a long-anticipated message. As he perused the official letter from the Air Force's personnel office, he gritted his teeth and groaned. Several weeks ago, he had applied to attend the Air War College at Maxwell Air Force Base in Alabama. The year-long academic course was effectively a prerequisite for advancement to general rank. But although he met the course's prerequisites, his enrollment application was denied. Cursing, he angrily ripped the dispatch into small scraps.

Russo had to return to the Air Force, but the transition would not be nearly as simple as he had previously anticipated. Although he had been chosen—but not formally announced—as an MOL astronaut, and had participated in extremely significant activities with far-reaching strategic implications, he was now two years behind his contemporaries with very little to show for his efforts. Even though he had never flown in space, he might as well have, since it was as if he had just vanished off the face of the earth while his contemporaries built and reinforced their careers with squadron commands, multiple combat tours, and other challenging assignments.

He now lamented his stint as Tarbox's protégé, as well as many other choices he had made, particularly since it was very obvious that he would not be allowed to quietly slink back to the Air Force. With every day that passed, he became more painfully aware of General Tew's far-reaching influence. This morning's rejection letter was an excellent example of Tew's broad reach. In addition to the Air War College, Russo had submitted paperwork for numerous follow-on assignments—squadron command in Vietnam, squadron command in

Europe, test pilot duty—but his applications were summarily rejected. Try as he might, he was frustrated at every turn; Russo could not shake the clinging stigma that came with his affiliation with Tarbox. His military career was in limbo, on the verge of disintegration, thanks to the vengeful general.

As if to pour salt into his wounds, seven MOL astronauts were afforded an opportunity to transfer to NASA, forming "Astronaut Group Seven" back in August. Even as the new group arrived at Houston, NASA's astronaut bench was filled to overflowing; primary and back-up crews had already been designated for the remaining Apollo flights, so the erstwhile MOL astronauts probably would not fly for years, if ever. Ironically, even though Russo had endured the same selection process as the MOL astronauts, his group was not publically announced, so he was not granted the same opportunity to move to NASA.

Of course, things could eventually change, and in time NASA might need more men, and Russo certainly met or exceeded all of their prerequisites. But still, it might be years before the space agency put out a call for another crop of astronauts. Even then, regardless of his qualifications, it was doubtful that he would ever ride a NASA rocket. Russo had a close friend who worked at Mission Control at the Manned Spaceflight Center at Houston, who solemnly informed him that he should not waste any time or energy applying to fly for NASA, since he would *not* be selected. It was yet another example of Tew's insidious influence.

Even though the Air Force wasn't welcoming him back with open arms, Russo was absolutely sick of the Navy in general, and his five cohorts in particular. He was weary of their insistence that he abide by their culture and speak with their unique language. As far as they were removed from ships and the sea, they couldn't just call a wall a wall, it had to be referred to as a bulkhead. Likewise, a ceiling was an overhead, a window was a porthole and the floor was a deck.

With little else to occupy their time, his Navy counterparts constantly engaged in tireless efforts to make his existence miserable. He had been the butt of countless pranks, which had gradually transitioned from benign harassment to mean-spirited and downright cruel hazing. Lately, it had grown much worse, with his tormentors striking at him not only in the workplace, but also at home. It had come to a head two months ago, when he heard an ominous buzzing sound as he turned the key to unlock the door to his apartment. Cautiously, he had entered to discover his humble abode was occupied by a large and very agitated rattlesnake.

Consequently, Russo was compelled to always stay a step ahead, perpetually moving from one dismal low-rent apartment to the next. He existed like a wandering migrant, living out of suitcases and never fully unpacking his belongings. His social life was nonexistent; he didn't date and had no friends.

Russo was surprised that the Naval aviators had not already moved on to other duties, now that the MOL project was cancelled. Tarbox regularly met with them, apparently to discuss opportunities for follow-on assignments, but Russo was not a party to any of those meetings. Moreover, the Navy men weren't inclined to share any details with him; treating him as an interloper, they remained tightlipped in his presence. Russo suspected that something was afoot, that Tarbox was working on some sort of scheme for the future, but he had no idea what it might be, and at this point he really didn't care.

The phone jangled, interrupting Russo's thoughts. He answered; the pleasant feminine voice on the other end was Tarbox's secretary, summoning him to a meeting with the admiral.

Hanging up the phone, Russo did not rush out of the room, as he had on previous occasions when Tarbox beckoned. He watched the clock on the wall, allowed a full minute to pass, and then called Tarbox's secretary back to ensure that the summons was legitimate and not another prank. It was, so he headed upstairs.

10:20 a.m.

He entered the admiral's office and stood quietly before the expansive metal desk. Papers were neatly arrayed on desktop, like fighter planes arranged on the limited space of a carrier's flight deck, carefully set for efficient arrival and departure. Russo was startled at the volume of paperwork; as the head of an organization that seemingly had no active mission, at least at present, it seemed as if Tarbox's workload had not dwindled in the least. As Russo watched, waiting patiently, Tarbox gradually worked his way through a row of paperwork, scanning each document in turn, signing some and setting others aside in a neat stack, apparently for further consideration or clarification.

Assuming that the admiral had not heard him enter, Russo softly cleared his throat.

Tarbox didn't even look up to acknowledge his arrival, but continued working. Finally, he spoke. "I know that your time here is nearly up, Russo, but I have an offer for you to consider, if you're interested in staying here with us."

Stay here? Surely, the admiral must be kidding, or at least he wasn't aware of how urgently Russo sought to leave this place and get on with his career. Before arbitrarily replying, he quickly considered his other options, which were few and dismal. But as anxious as he was to leave the Navy, he was still curious. "What do you have in mind, sir?" he asked.

"I have received authorization for a long-term project," explained Tarbox tersely. "Most of the key personnel in this office will be assigned to this new effort, to include your five counterparts. As you might imagine, some flying will be involved. I have one additional billet to fill; if you take it, you will remain here for the next five years."

Five years? Five more years of agonizing harassment? Ugh. "That sounds intriguing, sir, but . . ."

"I've been allocated three slots for the Navy's nuclear propulsion school in Maryland," said Tarbox abruptly. "I've already identified two officers to attend, you'll be the third if you accept my offer, and your follow-on assignment is contingent on satisfactory completion of the nuke course. Moreover, when and if you complete the nuke course, which is six months in length, you will matriculate to a six-month temporary assignment aboard a naval vessel or at an on-shore research facility."

"Nuclear propulsion school?" asked Russo, trying to conceal his skepticism. "Did I hear you correctly, sir?"

Nodding once, the admiral did not look up, but continued to scan and sign documents. "I've reviewed your college records, and you meet the basic academic requirements to attend the nuke course," he said. "Bear in mind that you will still have to undergo a personal interview with Admiral Rickover before you can be enrolled, but I am confident that we can adequately prepare you for that hurdle."

"And the follow-on assignment, sir?" asked Russo. "What would that entail?"

"One step at a time, Russo. If I didn't make myself clear, there will be *no* follow-on assignment if you do not complete the nuclear propulsion course. Was that not clear? In any event, we will not discuss the specifics of the follow-up assignment until *after* you return from your nuclear training."

Russo considered the admiral's offer. He was already over two years behind his contemporaries, and now Tarbox expected him to essentially mark time for an *entire* year without any clear objective on the horizon? And *nuclear* training? *To what end?* The admiral's offer was puzzling at best.

Tarbox cleared his throat and asked, "Well?"

"I appreciate your offer, Admiral, but since I intend to continue my flying career, I don't see how this nuclear training could possibly be relevant to—"

"*Relevant? Ha! I* decide what's relevant," hissed Tarbox, finally looking up from his paperwork to fix Russo with a baleful gaze. "You can accept my offer, or you can go back to the Air Force. As for flying, from what I've gleaned of your prospects, you would be lucky to fly *anything* with wings."

Russo's heart sank; the admiral was absolutely correct. "Can you at least grant me some time to think about this, sir?" he asked. "Please? Five years is a very significant commitment."

"You want time to think, Russo? Certainly. My offer expires in two minutes, so you now have 120 seconds to make up your mind."

Waffle 'n' Egg Diner, Dayton, Ohio
9:30 a.m., Friday, December 19, 1969

Morozov had grown accustomed to eating breakfast in this diner. Usually bustling, it was presently deserted, save for an elderly man at the counter, reading the morning newspaper, and three middle-aged women occupying a booth near the entrance.

The trio of women plied themselves with hot chocolate as they prepared for a last-ditch shopping foray, comparing objectives and strategies like field marshals before a major campaign. The place was festooned with chintzy decorations. Strands of paper garland were draped between the light fixtures over the Formica counter. From the jukebox, Bing Crosby crooned "White Christmas."

The short-order cook, a ponderous bald man in his fifties, warily kept an eye on the door to the stockroom. A ragged scrap of mistletoe hung down from the doorframe and he obviously stood vigilant to pounce on any waitress dim-witted enough to linger in the portal for more than a fleeting second. By Morozov's observation, none of the women appeared likely to oblige.

Morozov dumped sugar into his coffee and stirred it with his fork. He was disgusted with the sweet tooth he had developed since being posted to Ohio. Before coming here, he had maintained a fairly ascetic existence, depriving himself of all but the most basic staples.

Savoring the syrupy concoction, he thought it was just as well he spoil himself while he still had the opportunity, since this was likely the last time he would enjoy sugar in such plentiful supply. It was all but unavailable back home in the Soviet Union, despite their cozy relationship with Cuba. Perhaps the sweet stuff was a luxury reserved solely for the *nomenklatura,* much like the rumored shiploads of rum and cigars that arrived weekly from the Caribbean bastion of Communism.

He sighed as he looked at this morning's headline: "AIR FORCE SHELVES BLUE BOOK." A smaller headline stated "UFO Research Project to Shut Down." He had already received official word that his mission was to be curtailed, now that the Blue Book project was defunct.

Tomorrow, he would pack up his belongings, move out of his rented room, and board a Trailways bus for the long ride back to Washington.

Since this mission had been such an abject failure, his future prospects were woefully dim. His dreams of a vaunted posting to Hanoi were quickly fading. In fact, he now dreaded a hasty summons back to the Aquarium—the GRU headquarters at Khodinka Airfield near Moscow—with demands for a complete accounting of the money and resources poured into this dry hole.

Despite Hara's assurances that Hangar Three was home to a facility that reverse-engineered Soviet aircraft, Morozov still harbored thoughts that perhaps Yost had been telling the truth all along and the Americans were so protective of their UFO secrets that they had been willing to sacrifice him.

He even considered Jimmy Hara's death to be peculiar. According to Carr, his information source within the base hospital at Wright-Patterson, Hara had been earnest when he claimed that he was dying of leukemia. In fact, he had succumbed in late October, not long after they had rendezvoused in the cemetery. After reviewing Hara's medical reports, Morozov learned that Hara's particular medical condition was often caused by exposure to intense radiation, which caused to him to question what it was that Hara worked with in Hangar Three, and whether it was really hand-me-down MIGs and *Sukhois* as Hara claimed. Another odd twist was that Carr was hastily reassigned to Alaska not long after he had provided the information about Hara's records. *Anyway, there is just no way of knowing,* thought Morozov. The only absolutely sure things were that this aggravating mission was finally over, despite his lingering questions, and he had a bus to catch tomorrow.

Auxiliary Field Ten, Eglin Air Force Base, Florida
2:45 p.m., Saturday, January 10, 1970

Sitting cross-legged at the base of a large pine tree, Glades had positioned himself on a slight rise, at a spot where he could observe the kill zone and all the inner workings of the ambush. The kill zone was a short stretch of gravel road in an isolated area of Aux One-Oh. The road transected a sparsely wooded area dominated by small groves of scraggly pines, sumac shrubs, and various scrub oaks.

It was a dreary afternoon. A slight but steady wind blew from the south, bringing cold air from the Gulf of Mexico. Shivering slightly, Glades buttoned the two buttons of the green acetate "sleeping shirt" that he wore under his jungle fatigue jacket for extra warmth. He had been back at Eglin for less than a month, returning just in time for Christmas. Over the years that he had cycled between Eglin and Southeast Asia, he discovered that while he was able to quickly shift his mental gears to adapt to the different pace and circumstances, his body did not acclimatize as swiftly. His internal thermostat seemed permanently set for the moist heat of Vietnam, and he didn't adjust well to the damp cool winters of the Florida Panhandle.

Leaning back against the pine's scaly bark, he looked at the tiny metal calendar clipped to his watchband. He had been in the field for eighteen days out of the past three weeks. He looked forward to tomorrow, when he could stretch out on the couch, drink a cold beer or two, watch the Super Bowl to see if the AFL Kansas City Chiefs were all they were cracked up to be, and spend time with Deirdre and the kids.

Today, he was working an additional duty assignment, one that he didn't especially enjoy. Because of his pre-existing arrangement with the 116th Aerospace Operations Support Wing at Aux One-Oh, Glades had recently been pressed into service as a technical advisor and mentor to a special unit currently being formed within the Wing. The new unit, led by a captain, was a "Rapid Response Flight" of eighteen handpicked men that would deploy immediately in time-sensitive situations. All would attend Army Ranger school, where Glades worked as an instructor between stints in Vietnam, and all eighteen would eventually go to the Army's HALO free fall school at Fort Bragg.

Today's exercise was a daytime rehearsal of a night ambush. More specifically, the drill was one of a series of exercises to refine tactics and techniques for a "silent ambush," in which the ambush force would use silenced weapons and special procedures to quickly assault an enemy vehicle or small convoy that was carrying captured US personnel.

In the scenario, which would be rehearsed twice more in daylight before being run twice again tonight, two US pilots were grabbed after ejecting over a remote area in a Third World country. The Rapid Response Flight would stop the convoy, take back the pilots, and then immediately prepare them to be extracted by a Fulton STAR—Surface to Air Recovery—"Skyhook" system flown by an MC-130 Combat Talon aircraft.

The ambush was arranged in a classic "L," with the bulk of the airmen forming an assault line in the long axis of the "L" and a support element, consisting of a sniper element and a two-man M60 machine gun team, in the short leg of the "L." The support element was situated so that its fires were oriented down the long axis of the kill zone.

In broad daylight rather than darkness, the ambush force looked rather odd. Attired in tigerstripe pattern camouflage uniforms, the men's faces were daubed with jagged slashes of black and green grease-paint, and they wore heavily tinted goggles to simulate night conditions. The targets of the ambush, two medium-sized cargo trucks recovered from a scrap yard, were already situated in the kill zone. Several dummies—including two representing the captured pilots—were seated inside the vehicles. As with virtually all of the 116th's training exercises, the ambushers' weapons were loaded with live ammunition, so a post mortem of the drill would unequivocally show which dummies—enemy or friendly—were hit or not.

The nascent detachment was under the command of Captain Ed Lewis, who until recently had been charged with leading the Training Flight that assessed and trained new candidates for the 116th. Glades had previous experience with Lewis. As a "lane grader" for the Florida Phase of Ranger School, Glades had evaluated Lewis during a night raid patrol roughly a year ago. Later, as a training advisor to the 116th Wing, Glades had observed Lewis as he evaluated several candidate-led search and rescue exercises; in particular, he remembered the arrogant captain levying a failing grade on a black candidate after the airman had conducted one of the best patrols that Glades had ever witnessed.

In Glades's opinion, Lewis had a relatively good grasp of the tactical basics, but like more than a few of his comrades at MACV-SOG, Lewis was unduly prone to rely on the latest gadgets or gimmicks. A prime example was his weapons choice for the assault force. Except for the sniper and machine-gunner, the men were armed with Ingram M-10 9mm submachine guns, all equipped with Sionics silencers. Last year, Glades had evaluated the stubby M-10 for his MACV-SOG recon team in Vietnam; his assessment was that it was a weapon ideally suited for murdering someone in a phone booth, but little else. The short-barreled little gun—which looked like a slightly oversized pistol—climbed rapidly when fired on full automatic. Fitting the gun with a silencer—more accurately speaking, a suppressor—alleviated this rising tendency to some degree, but also made the weapon awkwardly front-heavy.

In the discussions leading up to Lewis's choice, Glades had argued for the utilitarian M3A1 .45 caliber "Grease Gun," a WWII-era submachine gun which could also be outfitted with a suppressor. The M3A1 was clunky and a bit bulky, not sleek and sexy like the alluring M-10, but it was exceptionally simple, dependable, and surprisingly accurate. Moreover, it fired the tried and true .45 caliber round, a proven man-stopper, rather than the 9mm round selected by Lewis. Lewis asserted that because the 9mm round was smaller and lighter, his men could carry considerably more ammunition. Perhaps this was true, but killing men was not a theoretical exercise for Glades, as it was for Lewis; Glades didn't revel in taking lives nor did he keep count, but it was a safe bet that he had killed at least enough people to populate a small village. In his vast experience of separating men from their souls, he had found that it didn't matter how many rounds you carried or fired, since one bullet at a time, judiciously administered, usually did the trick. And besides, compared to the lighter 9mm bullet travelling at a higher velocity, the .45 caliber projectile—a 230-grain chunk of copper-jacketed lead moving at a relative snail's pace of 880 feet per second—packed the kinetic energy of a freight train when it impacted flesh.

Glades had no affinity for tricks and gadgets; he was a firm believer in fundamentals in all endeavors. Two years ago, when the local high school football coach had been suddenly felled by a heart attack, Glades—with the blessing of the Ranger Camp commander—had stepped in as a volunteer coach until a replacement could be found. His entire playbook fit neatly on a single index card, with room to spare, and he spent every available minute of practice hammering home the essential skills of blocking, tackling, running, passing, and catching. Parents complained about the strenuous work-outs and the seemingly shallow repertoire of plays, but the young men learned to be a team unlike any seen previously or since. They were unbeaten in their regular season, and came within a whisper of winning the State championship, and probably would have, had Glades not been called back to Vietnam and a newly hired coach hadn't shown up with a thick notebook brimming over with "sure fire" plays.

Shifting his attention back to the matter at hand, Glades watched the proceedings. Well disciplined, the ambushers waited absolutely motionless and silent for the past thirty minutes since quietly occupying their assault line. Glades saw Lewis whisper a code word into the radio handset, a signal for the sniper to be ready to fire. Diverting his attention to the sniper, Glades could see that the airman was carefully aiming his suppressor-equipped weapon—an M21 7.62mm sniper rifle—at the cab of the lead truck waiting in the kill zone.

The sniper was using his rucksack as a makeshift rest; a good idea, thought Glades, since tonight the LART—Leatherwood Automatic Ranging Telescope—scope currently mounted on his weapon would be replaced with a heavier PVS-2 night vision sight. Fifteen feet to the sniper's left, the machine gun team was poised to "close the back door" by isolating the far side of the ambush with a continuous stream of automatic fire, should anyone attempt to escape. The machine gun team's M60 was the only "loud" weapon in the ambushers' arsenal, and they wouldn't fire unless they had a definite target in their sights.

Things looked good, mused Glades, and then suddenly they didn't. With the apparent intent of illuminating the kill zone, Lewis commenced the ambush by firing a hand-held parachute flare. While it was admirable that Lewis remembered to include the flare in the daytime rehearsal, since he obviously planned to fire it at night, he just shattered a tactical commandment.

Even as Glades was logging the violation into memory, the air was shattered by a loud report. As the sniper pulled the trigger and broke his round, his suppressor, apparently not adequately tightened onto the M21, blasted off the threaded coupling of the rifle's barrel. Tumbling end over end, the black-painted cylinder whistled through the air, thumped against the grill of the first truck, and then ricocheted off at an abrupt angle, very nearly hitting one of the ambushers lying in wait. There was no way of knowing where the bullet went, except that it almost certainly did not strike the intended target in the cab of the truck.

Much to their credit, the other ambushers were unfazed by the suppressor's malfunction. Operating in pairs, half rose and closed the gap to the vehicles while the others fired at the few hostile targets that were immediately visible. The assaulters split neatly into teams that covered the cabs and cargo areas of each vehicle, and then quickly and quietly dispatched the remaining hostile targets.

Rising from his location, Glades fell in behind Lewis as the captain orchestrated the remaining actions on the kill zone. Speaking quietly into the radio, Lewis dispatched reinforcements to the security teams positioned a few hundred yards on either side of the kill zone. The security teams functioned as mini-ambushes to provide early warning and to interdict any enemy reaction forces.

"Send up the blimp," whispered Lewis. Four men walked back in the direction of the ambush position, retrieved a hefty bundle of equipment, moved back to the kill zone, and then began inflating a large finned balloon from a bulky cylinder of helium. As the blimp took shape, hissing and billowing, Lewis's radio operator made contact with an MC-130E "Combat Talon" aircraft orbiting a few miles to the west. The Combat Talon was a turboprop C-130 Hercules uniquely configured for special operations.

Two of the ambushers had volunteered to stand in as the pilots for the Fulton STAR pick-up. Assisting each other, they donned hooded one-piece suits that were fitted with integral self-adjusting harnesses. As the balloon rose into the air, trailing a thick braided nylon "lift-line" connected to the harnesses, the two men sat down, shoulder to shoulder.

Glades checked the connections that linked the airmen's harnesses to the bridle that attached them to the lift-line. He looked at the two men, who were now more than a little bit apprehensive, and asked, "Are you boys absolutely sure you want to do this? It ain't too late to bow out. We can send the dummies in your place. No one would think any less of you for not riding this elevator." The men clenched their teeth and shook their heads. Glades heard the MC-130E droning in the distance, lining up on the balloon as it approached.

He had ridden a Fulton Skyhook rig once and was not particularly fond of the contraption, particularly since it had almost killed him. A few years ago, he had been dispatched to locate a downed U-2 spy plane in the southern mountains of Uzbekistan. After locating the crash site, Glades spent two days smashing the sensitive components of the U-2 before he and the pilot's corpse were extracted by an MC-130E Combat Talon that raced in from nearby Iran.

The pickup had not gone without incident. Normally, it takes less than ten minutes to snatch someone from the ground and reel them into the warm safety of the cargo compartment, but a jammed capstan had caused Glades and his dead companion to be dragged behind the MC-130E for over an hour, slowly pin-wheeling in the plane's turbulent wake.

By the time the crew finally managed to pull the pair aboard, Glades was deeply unconscious, so stiff and cold that the airmen could not immediately differentiate between him and the cadaver. So, Glades was none too enthused about ever riding a Fulton rig again, nor did he see the wisdom of any other sane person riding it, except perhaps in life or death situations where there were no other options for escape.

Now, obviously pondering their decision, the two airmen waited for their moment of truth. In just a few minutes, the MC-130E roared low overhead, aiming for red streamers tied to the lift-line. An odd-looking "pick-up yoke," protruding like giant spindly whiskers from the nose of the MC-130E, snagged the lift-line and the two men were abruptly snatched up into the air. Moments later, the transport disappeared in the distance, trailing the pair as they were gradually winched aboard.

With the "pilots" successfully rescued, Lewis blew a whistle, calling an end to the exercise. "Okay, gents," he said. "We're burning daylight, and I want to get in another run before we do this tonight, so we're not going to practice our withdrawal to the pick-up zone. Gather in. We'll have a quick powwow, go over lessons learned, and then we'll run it again." With that, the ambushers compared notes, mostly congratulating themselves on how smoothly the operation had gone.

Eavesdropping on their comments, Glades opened an extra ammunition pouch on his pistol belt and extracted a small can of C-ration sliced peaches. Deftly wielding a tiny P-38 opener, he ripped into the can, held it to his nose, and inhaled deeply. He spooned the sweet fruit from the green can, one dripping slice at a time, and savored each delectable bite.

The peaches reminded him of his West Virginia upbringing, when his father splurged every year at Christmas, walking down to the company store and plunking down just enough money to buy a gallon-sized can of peaches. That was the sole recurring treat in his childhood; the rest of his young life was mostly misery and deprivation, as he waited for his turn to descend into the mines or to find his way out of the Appalachians. Glades found his way out.

Hearing a slight noise, he looked up and saw a small deer well off in the distance, on the other side of the road. Watching the deer as it skittishly navigated its way through a tangle of greenbrier vines, he chewed slowly on the last slice, swallowed it, and then turned up the can to drink the syrup. Sticking the empty can in his spare ammo pouch, he listened to Captain Lewis's final comments on the ambush.

As Lewis's critique drew to a close, the team's sniper, an airman just barely familiar with the concepts of long distance shooting, laughed as he threaded the silencer back onto the barrel of the M-21. "I'll cinch it down this time, Captain," he avowed, tugging a small pair of Vise-grip pliers from a cloth Claymore mine bag hanging at his side. "It'll stay on. That won't happen again."

"It had better not," noted Lewis sternly. "Everyone, let's set up for the next run."

"Are you nuts, boy?" asked Glades, walking over to the sniper and snatching the rifle out of his hands. "You mean to tell me that you're dumb enough to fire a round through this can after it's already blown off your barrel, sailed over a hundred yards through the air, and smacked into the front end of a truck? You can rest assured its guts and baffles are knocked out of alignment, and if you pull that trigger on it again, you're going to need that rifle pried out of your face, if you're even still alive."

"Uh, sorry, Sergeant," said the hapless sniper. "I guess I wasn't thinking."

"You'll be thinking a lot less after that bolt flies out the back plate of your skull, son," noted Glades.

"Point taken," said Lewis. "Thanks, Sergeant Glades. Do you have any more observations for us?"

"Well, as a matter of fact, sir, I do," declared Glades, using his thumbs to shift the weight of his load-bearing H-harness back onto his shoulders. He pointed at the lowest ranking airman present, and asked, "What did we do wrong?"

"Uh, sir, I don't think we did anything wrong," muttered the young airman, apparently awed that Glades would call upon him for his opinion.

"It's sergeant, not sir," said Glades. "For starters, son, you always—always—always initiate any contact with a casualty-producing device. If you don't, then you're just granting the enemy the split-second of reaction time that he needs to dive out of that truck or duck down below the dashboard where you can't land a clean hit on him."

"I don't understand, sergeant," stated the perplexed airman. "I thought we did use a casualty-producing device."

"Well, let me spell it out to you, son. A casualty-producing device produces casualties. That means it kills or wounds folks, preferably as many as possible at the same instant. A flare is not a casualty-producing device. I'll grant you that it's not a bad idea to put one up after you initiate contact, so you can illuminate targets, but if it ain't killing people, it's not a casualty-producing device." Glades noticed that Lewis was frowning; it was clearly obvious that the captain was annoyed at being criticized in front of his men.

"And while we're on the subject of flares," observed Glades. "You can aim those hand-held flares so they pop just about anywhere you want. Now, ideally, if your assault line is back there" —he gestured at where they men had lain in wait—"like yours was, and it's dark out, where would you want that parachute flare to go?"

"Over yonder," said one of the men, a gaunt NCO from Mississippi, pointing out towards the stunted oaks and sawgrass on the distant side of the kill zone, opposite from the assault line.

"Why?" implored Glades.

"If the flare is behind the kill zone, in relation to us, it's going to backlight the targets, even after it hits the ground," answered the NCO.

"Correct. But your flare went one-hundred-eighty degrees in the opposite direction," noted Glades, pointing into the pine forest behind the assault line. "Is that a problem?"

"It is," answered the NCO. "Because even though it would illuminate the targets, it would backlight us."

"Correct. Would there be another problem?"

"Can't think of one."

Glades leaned his weight on his right foot and frowned. "Well, there is *another* problem. Like you said, that parachute flare is likely to continue burning for quite a while after it hits the ground." He knelt down, picked up a clump of pine needles, and held them in an outstretched hand. "This stuff is dry as a bone, and that parachute flare could catch it alight in an instant if you're not lucky. And by the time you search the vehicles, put up a balloon, execute your Fulton pickup and complete the rest of your actions on the objective, chances are pretty good that there will be a fairly substantial scrub fire burning. And that's a problem because . . ."

"Our pick-up zone is in the direction," answered the lowest ranking airman. "Six hundred meters past where the fire would be. We would have to run through the fire to get to the helicopters."

"Correct."

Lewis, who had been tapping his feet and fidgeting, chose to quibble. "All good points, Sergeant Glades, but since this was a silent ambush and there were friendlies on the trucks, I didn't have the latitude of using a Claymore mine or firing indiscriminately into the trucks, so I didn't have the option of using a casualty-producing device to initiate."

"You didn't?" asked Glades, arching his eyebrows. "How about your sniper, sir?"

"That's why I fired the flare," argued Lewis. "To light up the trucks so he could pick up his target."

"Okay, sir, but let's remember that he'll be using a PVS-2 scope tonight. Even without the PVS-2, are you telling me that he couldn't line up low center of mass on the driver's side of that windshield and put a bullet through it? Maybe he won't hit that driver square on, but he's sure going to do sufficient damage to distract him from driving that truck, don't you think?"

Lewis nodded. "I'll concede that."

"Okay, Captain. So while we're on the subject of your sniper, you positioned him so he was firing smack down the long axis of the target, so if he was successful in placing that first bullet through the windshield, it would have smacked through the glass, then drilled through the driver's head, then cruised right on through the flimsy canvas separating the driver's compartment from the cargo area, where it could . . ."

"Possibly hit one of the pilots," interjected the youngest airman, grinning. "Our precious cargo."

"Correct. So on the next go-round, you might consider positioning your shooter at an angle to where it's less likely that he'll accidently kill one of the folks you're supposed to be rescuing. Last but not least, speaking to that suppressor that your sniper launched at the truck, you always check and double-check your equipment before you leave for a mission, and then you check and double-check it again before you move out to occupy your assault positions. And if it's a piece of equipment that's mission-critical, you check it, double-check it, triple-check it, and quadruple-check it. Understood?"

The airmen nodded.

"Anything else, Sergeant Glades?" asked Lewis.

"No, Captain. I think that just about covers most of my observations."

"Okay. Thank you, Sergeant," said Lewis. "Gentlemen, head back into the woods and fall in on your gear. I'll be there in a minute, and then we'll go through the whole drill from the top."

The airmen quickly got to their feet, gathered up their kit, and walked back into the woods.

Twisting his M-10's suppressor to ensure that it was sufficiently tightened, Lewis waited until the men were out of earshot before he defiantly asked, "Is there any particular reason that you need to be so critical of my actions in front of my men?"

Glades took a sip of water from his canteen, screwed the lid back on, and replaced it in the canvas carrier at his hip. "Yes, Captain, there's a reason. In combat, things happen quickly and situations tend to deteriorate rapidly. People die. When and if you execute an actual mission, any one of those men, including that young airman, has to be ready to immediately step into your shoes. Now is your opportunity to pound the basics into them, to train them to develop simple and effective plans, and then to execute those plans with audacity. So that's why I'm critical. And with all due respect, Captain, General Fels told me to pull no punches. So I'm not."

"General Fels?"

Glades nodded.

"Fair enough. I'm sorry for being thin-skinned, Sergeant Glades. I earned this Ranger tab, which is a fairly big accomplishment of an Air Force officer," said Lewis, pointing at the black and gold emblem on his right shoulder. "But there's still much to learn."

Glades nodded. "But it's all fundamentals, sir. Everyone wants to think that Rangers are some kind of super-soldiers, but in fact they're just well-disciplined troops who execute basic soldier skills exceptionally well. That simple, Captain. Once you have a good grasp on the basics, and you consistently use good judgment and common sense, your troops will follow you anywhere. You don't want to be the leader whose men follow you out of sheer curiosity."

Taking in Glades's words, Lewis was silent for a moment. Then he spoke. "Mind if I ask you a question? Totally off the subject."

"Yes, sir?"

"I heard you'll be headed back to Nam soon, back to MACV-SOG. I guess you know that General Fels encourages us to go over for combat time. Is there any chance that . . ."

Glades shook his head. "No, Captain. And please don't ever ask again."

Rehabilitation Hospital #6
Novomoskovsky Administrative Okrug, Moscow, USSR
4:15 p.m., Friday, January 16, 1970

Finishing his reports early, General Yohzin unexpectedly found himself with the luxury of an hour of free time and decided to honor a favor. One his colonels had asked him to check in on his son. The eighteen-year-old had recently incurred extensive head injuries and was staying indefinitely—perhaps permanently—at a Moscow infirmary.

Although he conscientiously kept at his duties at Kapustin Yar, the colonel was inconsolable. He and his wife had doted on the boy, and they were beside themselves with grief. To make matters even worse, the despondent colonel was convinced that he had created the circumstances that led to his son's hospitalization.

While exceptionally intelligent and studious, the boy wasn't a good test-taker; while still in high school, he had bungled the crucial standard examinations that determined his potential opportunities for post-secondary education. As such, he wasn't even considered for a college or university that would have actually challenged his considerable intellect.

Just as the boy was on the verge of being drafted into military service, the colonel had struck a deal that landed him at a "commissioning school" on the outskirts of Moscow. The school, one of roughly a hundred such institutions in the Soviet Union, offered a four-year curriculum that culminated with an engineering officer's commission in the Soviet Army. It certainly wasn't on par with some of the more premier commissioning schools, like the renowned M.V. Frunze Higher Naval School in Leningrad, but it provided an adequate college education and a solid path to the future as a military officer.

Slightly less than a week after he walked onto the school's grounds, he departed—unconscious and barely alive—in the back of an ambulance. As was often the case, there was an official account of the incident, and then there was the truth; in this situation, the circumstances of the boy's injuries were not just tragic, but bordered on the criminal. Although the official report stated that he tripped down a flight of stairs, the truth was that he had suffered his devastating injuries at the hands of his classmates.

Even upon arrival, the boy was immediately targeted by upperclassmen, particularly because he was perceived as a physically weak bookworm. Moreover, since there were usually at least three applicants for every vacancy at the school, and since the colonel's son had not been present at any of the screening examinations, it was obvious that he had been granted a precious spot because of political influence.

Late one night, he was subjected to ritual hazing that had quickly escalated to a severe beating that had left him with a cerebral hemorrhage. Not appreciating the true extent of his injuries, hoping that he might sleep it off and quickly recover, the school's administrators delayed bringing the young man to medical care. By the time a doctor saw him and diagnosed his injuries, too much valuable time was lost. The physician administered powerful drugs to alleviate the swelling, but it was too late and the damage was irreparable; the edema had squeezed out years of knowledge and snuffed out his spark of curiosity.

Sorely tempted to hold his nose, Yohzin followed an orderly who escorted him to the small ward that held the colonel's son. Poorly lit and dirty, the dreary sanatorium more resembled a warehouse than a medical facility.

Pausing at the door, he removed his overcoat as he took in the tableau. He handed the bulky garment to the orderly, who hung it on a shoddy wooden coatrack in the hallway. Yohzin was aghast at the squalid conditions. Ten other beds—also occupied by invalids—were jammed into the same room, which probably should have accommodated four patients at most.

The orderly gestured at a bed, but Yohzin had to almost strain his eyes to recognize the occupant. The boy's scalp was shaven as clean as an egg. A transparent tube, protruding through a small gauze-packed opening in his skull, drained away yellow-tinged clear fluid. His head lolled to the side, and a string of drool trickled from the corner of his chapped lips. Clearly oblivious to his circumstances and surroundings, he wore a blank expression. While a few of the other patients occasionally groaned and emitted incoherent mumbles, the colonel's son was entirely silent.

A young student nurse sat at his bedside, patiently spooning some sort of weak gruel into the boy's mouth.

"May I speak with him?" asked Yohzin.

"You can talk to him, Comrade General, but don't expect him to reply. He doesn't talk," replied the white-clad nurse. "He's entirely unresponsive."

"Will he know who I am?"

"*Nyet*, Comrade General. But mercifully, I don't think he knows who he is, either."

Yohzin seethed with anger. The boy wasn't dead, but he wasn't entirely alive. He had once demonstrated immense potential, but all that energy was now dashed, like a charging locomotive derailed, once spewing steam and fury, but now inert and useless. His mind wasn't merely addled; it was *gone*. Instead of merely pounding him senseless with their fists, his young assailants might as well have decapitated their classmate with an ax.

Yohzin sat in a chair offered by the orderly and stared at the colonel's son lying motionless in the steel-framed bed. His ire was soon dispelled by an overwhelming sense of dread. Fear gnawed at his guts as he realized that one of his own sons could readily suffer a similar sordid fate. Such institutionalized brutality was a normal aspect of Soviet life and was absolutely rife in the military, where the strong routinely preyed upon the weak. The long-established systematic abuse of conscript soldiers was known as *Dedovshchina*. It was nearly as prevalent in academic institutions like the commissioning school, but the consequences were usually not as dire.

Yohzin himself had escaped such ritual abuse because he had attended the *Technische Hochschule* in Berlin, under a student exchange program in the thirties, long before Germany and the Soviet Union became sworn enemies. There, he had been granted the gift of focusing his youthful energies on his studies rather than perpetually scrambling for survival.

He was obsessed over providing similar security for his sons but knew that there was no failsafe mechanism to shield them. While he hoped that his sons would matriculate into one of the more prestigious engineering universities, they could literally land anywhere, even drafted into the military. Certainly, he felt confident that he could wangle an appointment at a good school, given his rank and position, but there were no guarantees. Moreover, the cloak of his fatherly protection would fall away abruptly at the schoolhouse door; once inside the institution, they were subject to the same sort of hazing that befell the colonel's son.

While Yohzin was sure that his older son, a self-assured and sturdy youngster, could hold his own in a scuffle, he wasn't nearly as confident about his younger son's prospects. Although not frail, the fourteen-year-old was a small boy who constantly had to fend off bullies.

He would sacrifice literally *anything* to protect his sons. He agonized over the potential alternatives, and as much as he deplored the notion, he knew that some desired outcomes often called for desperate measures. As he contemplated one particularly repugnant option, he was almost consumed by guilt. But the truth was that he would do *anything* to safeguard his children, and if that entailed wallowing in guilt for the rest of his life, then so be it.

Filyovsky Park, Western Administrative Okrug, Moscow, USSR
10:01 a.m., Sunday, January 18, 1970

With Magnus trotting alongside, Yohzin strolled down the path that ran by the Moscow River, looking for Smith. Keeping a brisk pace, he jammed his hands in his pockets for warmth as he walked and contemplated the choices that had brought him back to this place. Certainly, he was looking out for his family's welfare. He wanted his sons to be able to go to college anywhere they wanted and not have to constantly fear for their safety instead of being able to focus their energies on their studies. He also wanted them to be free to aspire to be whatever they desired after college, without fear of being jammed into an uncomfortable niche by unknowing bureaucrats.

Of course, he felt shame for betraying his Motherland, but he also felt an intense desire for vengeance, to settle his longstanding grudge with the senior officers and asinine Party officials who had mismanaged his career. He felt that the military had grossly squandered his talents and stymied his true ambitions; he should have been released from his duties to join his contemporaries in the exploration of outer space. It wasn't that his wishes were unknown; for years, he had voiced them incessantly, to anyone who seemed even virtually receptive to listening.

Snow crunched under his feet as he spotted Smith in the designated meeting place. The American greeted him warmly, as if they were long-lost friends. "So, General, I assume that you have changed your mind about working with us?" he asked.

"Yes." Yohzin answered in a tone that was as frigid as the air. He gestured for Magnus to sit.

"Well, I'm glad that you've overcome your reluctance," said Smith, tugging back his woolen mitten to glance at his wristwatch. "Now, we should talk some specifics. As I indicated before, the last time we met, we're willing to compensate you handsomely for any information you're willing to provide, commensurate with the information's value. Because of what you described the last time that we met, the money will accumulate in a special account, gathering interest, and will be made available to you once we extract you and your . . ."

"Listen to me, Mephistopheles," snarled Yohzin, abruptly cutting him off. "Don't bother dangling the usual array of incentives. If I'm compelled to sell my soul to you, the least I can ask is to set the terms. Are you ready to hear my conditions?"

Exhaling a pall of steam, Smith nodded.

"My youngest son is fourteen years old. If we arrive at an agreement, I will collect information for you until he is ready for college, and then I want my family to go to the United States. If it can be arranged, my wife and I will be content with a modest retirement in a quiet place, ideally somewhere in the American Southwest. We don't need a lot of extravagant amenities, just the basic necessities, but I do want both my sons to have the benefit of a good education."

"Wait," interjected Smith. "If your son is fourteen years old now, and you want to be out in time for him to go to college, then we're only talking about roughly four years."

"That's correct. And I want my *entire* family to come out, including my older son."

And then the bargaining process began. "So you're only willing to remain in place at Kapustin Yar for four years?" asked Smith. He shook his head and quietly added, "Honestly, Comrade General, I seriously doubt that I could convince my superiors of the value in such an arrangement. After all, you're asking for a retirement in the United States, as well as college for your sons, in exchange for a mere four years' worth of gathering information. That's an awful lot to accommodate."

"We're wasting time, Smith," growled Yohzin. "Don't try to manipulate me. I know the steps of this dance. Usually, you might struggle for months or even years to recruit some underling clerk. You'll cultivate him for *decades*, winnowing your way through the marginal drivel that he delivers, hoping against hope that he will ascend to become a powerful bureaucrat where he *might* have access to information that is actually of value."

Yohzin continued. "But I'm at the pinnacle of my career *now* and have routine access to volumes of priceless information. I can provide you with technical specifications, test schedules, launch schedules, communications frequencies . . . *everything*. Face it: you could exploit thousands of sources for thousands of years, but you will never acquire the same quality and quantity of information that I could provide in a single *day*."

Striving to stay warm, Smith lightly stamped his feet in the snow. "You present a strong case, General. I'll bring your terms back to my superiors. If my bosses determine that the information that you provide is of the magnitude that you claim, then we'll compensate you accordingly. But in the meantime, as we await their decision, I'll make you an offer that it is within the scope of my authority to grant."

"And what could *you* grant, Smith?"

"This. We will still bring you out after four years, as you wish. You and your family will be granted asylum and given new identities, but you will not immediately retire. Instead, you continue to work for us, in some capacity, in the United States. As for your sons' education, for the next four years, the sums you are paid will accrue in a special account. Granted, it might not necessarily be sufficient for them to attend Harvard or MIT but should be at least enough for a very respectable state university."

The two men were silent for well over a minute, as Yohzin considered the arrangement. Finally, Smith spoke. "Are my terms acceptable, General, at least as an interim arrangement?"

"*Da*," replied Yohzin, in a small voice that was almost like a croak. "Yes. Your terms are acceptable." In the distance, an unseen hawk emitted a bloodcurdling screech.

"Good," answered Smith. He removed his black-framed spectacles and wiped condensation from the lenses.

"Now that the die is struck, what's next?" asked Yohzin quietly.

"You always stay at the same hotel when you visit, correct?"

"I do," replied Yohzin.

"The next time you visit, call us to let us know you're in Moscow. The next day, at a time that the operator gives you, call the hotel's front desk and report that you smell a faint gas leak. A gas fitter will come to fix it. He works for us. He'll give you a camera and some other items, and show you how to use them. He'll also show you the procedures that you use for placing materials at a site where we will retrieve them."

"A dead drop?" asked Yohzin, summoning one of the few cloak-and-dagger terms that he knew.

"Correct," answered Smith. "It's imperative that you follow the instructions to the letter. From this point on, that's the only means by which we will communicate, unless your circumstances change, or unless there's an emergency. If that's the case, you call the number and we will set up a meeting. Just be aware that it normally takes at least twenty-four hours to make all the necessary arrangements."

"And you will be able to spirit my family out of the Soviet Union, to the United States, when the time comes?"

"We will," affirmed Smith. "Assuming that you will be at Kapustin Yar for another four years, there is more than enough time for us to make those plans. Additionally, we will develop contingency plans to extract you swiftly if you suspect that your activities have been compromised or that you're otherwise in danger."

Nodding, Yohzin was silent. He suddenly arrived at the brutal realization that there was no deviation from the course he had chosen. Even if he was suddenly overcome by a change of conscience, it was far too late to change his mind. There was no opting out; his life could never revert back to what it had been before.

"Do you have any other questions?" asked Smith, interrupting his thoughts.

A stiff wind gusted down the path, blowing loose snow off of tree branches. Yohzin shivered, almost uncontrollably, and shook his head.

"Then we'll be expecting your next call, Comrade General."

Cap-Haïtien, Haiti
1:10 p.m., Monday, February 9, 1970

This was his third sortie into Haiti since August, and Henson had grown accustomed to the clamor of the airport. He had grown accustomed to the people as well; spotting familiar faces in the crowd, he waved and smiled. A wizened old vendor shoved a ripe avocado into Henson's hand. *"Pou nou manje midi,"* she declared, displaying a toothless grin. *"For your dinner."*

"Mèsi," he replied, smiling. *"Mèsi anpil."* In his earlier visits, he had come to learn that his first impressions were not entirely accurate. While most Haitians lived in desperate poverty, they were an inherently industrious people who made the best of their dire circumstances. Most were distrustful of outsiders; they still referred to Henson as a *blan* even though he was much darker than the average Haitian and now spoke a very passable *Kreyòl.*

Certainly, he still didn't understand the Haitians' tenacious attachment to voodoo culture, and he also had no theories about how a single island—Hispaniola—could be divided into two countries of effectively equal natural resources, and yet one nation prospered while the other seemed doomed to wallow in eternal wretchedness. And more than anything else—he could not fathom how these people could allow themselves to be constantly subjugated and exploited by one cruel dictator after another.

His tenure in Haiti had proven exceptionally lucrative. He had a sizeable amount of cash in reserve, primarily because he had been so successful in obtaining the requisite goods and services at bargain basement prices. He now suspected that his natural charm and practiced negotiating skills had little to do with it, but his success probably could be more attributed to the fact that he was often seen in the company of Colonel Roberto, and everyone was eager to please Roberto.

In any event, he had a rubber-lined canvas bag stuffed with cash— over $10,000 in US and other currencies—carefully buried near the shed in Morne Bossa where he lived and worked. Since he wasn't sure when and if he would return to Haiti after this mission, he planned to excavate the bag and sneak the cash back into the States at the end of this four-week mission.

He heard a familiar voice call his name and turned to see Roberto approaching. As usual, his uniform was impeccable and not the tiniest droplet of sweat dared make an appearance on his brow. "Matthew," he said. "*Bonswa, zanmi mwen.* Still searching for bauxite ore?"

"*Bonswa*, Roberto. But of course." They greeted one other like best friends who hadn't seen each other in decades. They chatted, casually alternating between Haitian Creole and French, and then Henson dug into his gym bag. "Here, I brought you something," he announced, handing the *Fad'H* officer a small parcel neatly wrapped in brown paper.

Roberto excitedly tore at the wrapping like a child on Christmas morning, revealing a brand new copy of *The Andromeda Strain* by Michael Crichton. "*Mèsi anpil*, Matthew. I've been extremely anxious to read this. You're very kind."

Henson laughed. "I just wanted to spare you the effort of ransacking tourists' luggage until you found a copy." Strolling up to the Customs desk, he casually flashed his passport and smiled. The woman smiled back and waved him through the gate.

"So you want to deprive me of my simple joys? I only have so many distractions here."

"Actually, I figured that if I gave you that, you might not feel so inclined to rummage through *my* luggage. "

Roberto grinned. "Oh, I'll probably do that anyway, just to stay in practice. Matthew, do you have any plans for the evening?"

"Nothing significant. Do you have something in mind?"

"My wife wants to try a new recipe and insisted that I invite you."

"A Haitian specialty?"

"Oh, no. Something from your hometown: shrimp *etouffee*."

"Please tell your wife I would be honored to join you for dinner. And I'll bring the wine."

12

CONTINGENCY PLANS

Aerospace Support Project
4:50 p.m. Wednesday, February 18, 1970

"Gentlemen, we are *done* for the day," announced Heydrich, staggering into the office as he unzipped his parka and removed his Bavarian alpine hat. Dog-tired, he slumped into a chair at the table and removed his condensation-fogged glasses. "I would have a nervous breakdown right now, but I just don't have the time or the energy."

Tew looked up from a document and asked, "Care to render us a progress report?"

"Yeah, Gunter. Tell us," asked Wolcott. "Did Jackson and Sigler put the horse in the barn?"

"*Nein.* Not even remotely close," replied Heydrich, solemnly shaking his head as he toyed with the ivory edelweiss pin on his felt hat. "We flew five intercept profiles in the past two days, and they were zero for five. They really flubbed the last one. They were just about dead tired, so I sent them to the showers instead of running another profile. They should be headed for home just about now."

"So, pard, who's the problem child?" asked Wolcott. "Or is it both of them?"

"Both, but mostly Sigler. He just can't keep pace, especially when the hours stretch out and he's tired. He tries really hard, but he's almost entirely dependent on what the computer spits out."

"I thought that's why the computer was in the cockpit," observed Wolcott. "To relieve most of the calculating workload. Ain't that the whole point, pard?"

Heydrich shrugged. "In a normal situation, if we were intercepting a target that remained constant, working off the computer would be fine, but this *verdammt* target is giving us fits."

"How so?" asked Tew, not looking up as he scrawled his signature on several budget expenditure forms.

"This target's maneuvering, much more so than what we've observed before. We think the guidance is shot up in increments, so they're only making a slight deviation in any given orbit, and a whole change is typically executed over the course of six to eight orbits. It's probably taking them roughly twelve hours to reposition to cover another target or swathe. So if our guys are unfortunate to intercept it during a shift, then they have to think far enough ahead to know where it's going to be at least four orbits in advance."

"The computer can't do that?"

Heydrich shook his head. "The Block I computer was built for NASA's requirements. It relies on tracking data fed up from the ground. If we were flying with NASA's worldwide tracking and relay system, then we might be able to pull it off. But with our intermittent communications, we're lucky if the timing is such that we tell our guys that the target is maneuvering and give them initial data on the increments and intervals of the shift. If there's any saving grace, the target's orbit stays almost uniformly circular, so at least we don't have to factor in altitude."

"Gunter, are you telling me that it just can't be done?" asked Tew.

"No, Mark, I'm not. Carson and Ourecky pulled it off two weeks ago. In fact, they were successful three times out of four and probably would have pulled off the fourth intercept in another two to three orbits if we had left them in the Box."

"*Why* are we having these problems, Gunter?" asked Tew, shaking his head in exasperation. "Why is it that Carson and Ourecky can do this, but Jackson and Sigler can't?"

"Ourecky can work magic with just a few shreds of information, but Sigler is not in the same realm. It's that simple. Like I said, Sigler relies on what the computer churns out. Ourecky stays *ahead* of the computer. *Way* ahead of it. And if the computer fails, Sigler crumbles faster than an oatmeal cookie. A computer failure doesn't even faze Ourecky. He's relentless; he just puts his head down and persists until the problem is resolved."

Heydrich continued. "Beyond that, Carson and Ourecky are just an excellent team. They complement each other perfectly. Carson is an exceptional pilot, and Ourecky is . . . well, Ourecky. I know that you initially brought him in as another whiz kid to work on the guidance system, but the fact is that he *is* the guidance system."

"So what are our options?" asked Wolcott. "Short of sending the Dynamic Duo back up?"

"Honestly? If you still insist on not flying Carson and Ourecky?" asked Heydrich. "We should hand this target back and request another one."

"So you want me to crawl to Kittredge and admit that this target's too hard? That's not a viable option," replied Tew. "That would delay this mission at least a month, perhaps longer."

"Mark's right, Gunter. There are plenty of folks just waiting for us to stumble. They would swoop down on us and pick our bones clean, like starvin' buzzards gnawin' on a cow carcass."

"Can I be frank, Mark?" asked Heydrich.

"Of course."

"I know that this is a terrible thing to say, but in retrospect, our current situation would be a lot less precarious if you had assigned Jackson and

Sigler to the first launch. After all, it was a simple practice mission and well within their level of competence," stated Heydrich coldly. "At least right now, we would have two crews on relatively equal footing, instead of one varsity crew and one that's barely marginal."

"Gunter, you're probably right, but I'm going to ask you to never air that thought again."

"Ditto, pardner," added Wolcott, wringing his hands together.

"So is there anything to add?" asked Tew.

"There is. If there's no recourse but to stick with this target, at least yank Jackson and Sigler off this mission. Crew One can continue to train with the old Block I hardware, since Ourecky doesn't require the new computer, and Crew Three can work exclusively with the Block Two training machine when it arrives. But in the meantime, we have to fix what's broken."

"Gunter's right, as usual," noted Wolcott. "Mark, pardner, I think it's high time you realize that we're obliged to bite the bullet. It's time to gnaw lead and saw some bone."

Examining his desk calendar, Tew grimaced and nodded. "Drag Carson and Ourecky in here first thing in the morning."

Wolcott shook his head. "I don't know if you remember, but Ourecky's taking a couple of days off, pard. Him and the missus are pulling up stakes and moving into their new domicile."

"It slipped my mind. Gunter, if you lock Carson and Ourecky in the Box on Monday morning, will they be ready to launch on the tenth?"

Heydrich nodded. "Launch on the tenth? Without a doubt."

"Okay. Call Carson here in the morning," said Tew. "He can quietly pass the word to Ourecky. And let's all pray that Bea doesn't fly into a hormonal rage and kill us all."

Dayton, Ohio
3 p.m., Friday, February 20, 1970

Wearing Scott's old turtleneck sweater, Bea lounged on a bare mattress laid on the hardwood floor of the living room. With her sock-clad toes almost brushing the metal grate of the electric space heater, she mused on impending events.

Sears was scheduled to deliver their new furniture—a couch, chair, bed, and dining room set—on Tuesday, the heating oil truck would come around on Monday, and the telephone company was supposed to install their phone sometime during the week. Gently patting her swollen belly through the stretched wool cable knit, she kept time with the song—"*Eli's Coming*" by Three Dog Night—pouring from a transistor radio perched on the window frame.

She liked their new place, a comfortably quaint two-bedroom house, especially after her cramped apartment. The neighborhood was a mix of well-established families and young couples, so she was confident that there would be an ample supply of playmates and babysitters for . . . *who would the baby be*? They had yet to settle on names, and that was an issue Scott wanted to resolve long before the baby arrived. She didn't understand why he was so insistent, and the more they discussed the subject, the further they were from resolution. Besides, there was still plenty of time; the baby wasn't due until the middle of next month.

The door swung open, accompanied by a gust of cold air, as Ourecky lugged the last cardboard box from the borrowed truck outside. "Where?" he asked, nudging the door shut with his hip. "Front bedroom?"

"That's pictures mostly. Just set it down right there." She pushed herself up and padded into the kitchenette. "Here. You've earned it, baby," she said, handing him a cold Schlitz. "Wow. At least we know the icebox works. We don't have heat, but at least our food will be refrigerated."

"And how," Ourecky said, levering off the cap with shiny can opener. White foam spewed from the bottle's neck; the beer was more icy slush than liquid. They plopped down together on the mattress and basked in the radiant orange warmth of the gently buzzing space heater.

"I'm anxious to start on the nursery. We can pick out some colors and then go by the hardware store to buy paint and brushes tomorrow afternoon. We can start painting next week."

Swallowing, he shook his head. "Bea, I'm sorry, but I'm not going to be able to paint next week. I'm going to be busy. Something came up. We have a high priority flight test in the middle of next month. There's all the prep work and late hours, and I might not be here when . . ."

"The baby comes? Is *that* why Drew came by yesterday?" She felt like she had been punched in the stomach. Her lower lip quivered and tears started to well in her eyes.

Ourecky nodded again. Setting the beer on the floor, he put his arms around her shoulders. "Really, I should be back in plenty of time before the baby . . ."

"Don't!" she snapped, pushing him away. "You *promised*."

"I didn't promise, Bea. I said I would do my utmost to be here, and I will. You know damned well that I can't make any promises."

"Obviously. Do you suppose you can at least be in town to give her away at her wedding?" Watching his face, she saw his jaw tighten and his brow furrow and knew that he was just as frustrated as she was angry. As much as she loved him, she could not comprehend what it was that he could not tell her. *What could be so damned important that he couldn't offer even the slightest clue? Why must so much of his life remain a mystery to her?*

"Give her away? You're still assuming that it's going to be a girl."

"I'm desperately *hoping* that we'll have a girl. Surely you know that by now. A girl would suit me just fine. Not an engineer and *definitely* not a pilot."

"Okay, if we're discussing what this child will be, then let's talk about something else, another topic we've been avoiding lately."

Here we go again, she thought, *whirling round and round the mulberry bush, and never making any progress towards a decision.* She just couldn't fathom what was to be gained by revisiting this subject until the conversation deteriorated into yet another quarrel. Not wanting the first night in their new home to be spent in angry silence on distant sides of this mattress, she calmed herself and said, "Baby, there's still time. Why do we have to be in such a rush?"

"Because we'll be coming into the window soon. I want us to be prepared."

She laughed softly, hoping to dispel the lingering tension between them. "Coming into the *window*? You make it sound like I'm launching a rocket to the moon instead of having a baby. Okay, honey, do you have any new ideas? Clearly, we've exhausted just about all of the possibilities, and it's obvious that voting isn't an option since there's two of us."

"Then how about your plan that you name her if it's a girl and I name him if it's a boy?"

"I thought you were still opposed to that idea. What's caused you to change your mind?"

"It's grown on me," he replied. "So it is a deal? Speak now or forever hold your peace."

"Deal." She extended her hand out of the sweater's enveloping sleeve, and they linked pinkies. "I've already got mine picked out. How about you?"

"Yeah," he replied. He sipped from the beer and drew a folded-up index card from his shirt pocket. "Ladies first, dear. Who will our little girl be?"

"I want to name her after my grandmother, on my mother's side: Anna Katherine."

"Anna Katherine? That's pretty formal, isn't it?"

"We'll call her Anna Kate," she replied, grinning. "Won't that be sweet? Can't you just picture her in ribbons and bows, in a dainty little First Communion dress?"

"That's good. I like it. Anna Katherine it is. Anna Kate she'll be, if . . ."

"Okay, Scott Ourecky, spill the beans. Who will our little boy be?"

He handed her the folded card. She opened it, read the name, groaned, and rolled her eyes. "Oh, please, please, please, *anything* but this. Wouldn't you be happier with a little Scott Ernst Junior romping around the house? I'll even let you raise him to be an engineer. We could buy him his own pint-sized slide rule. Please, Scott, wouldn't that be better than *this*?"

"No. I want this, Bea. It's important to me."

"Well, since you've obviously taken leave of your senses, let's explore some other possibilities. How about naming the baby after your father?"

"My older brother has already done that. It would be awkward to do it again. Anyway, I thought we had a deal, Bea. I'll concede to Anna Katherine if it's a girl, but if it's a boy . . ."

"Fine. You're right; we agreed. But I wish you would reconsider."

"Why are you so opposed? It's a good name."

She drew in a deep breath, let it out slowly, and answered calmly, "Scott, I'm in this for the long haul. I want us to grow old together." She held out the card to him. "You're right; it's a good name, but no matter what happens, it will always remind me of *these* days, waiting for you and not knowing where you are or what on earth you're doing. I know that someday we'll grow past this phase, and things will eventually become more normal, but when that time comes, I just don't want to be reminded of these days. Does that make sense?"

He nodded, and said hesitatingly, "Bea, you're right. If you want to change . . ."

"No," she said, interrupting him as she put her finger to his lips. "A deal's a deal. But, Scott, there's something that I need to tell you."

"Is it about the baby?" he asked anxiously. "Is something wrong? I thought Doctor Blakely told you that everything was okay."

"He did. I'm sure the baby will be just fine," she replied. "Here's what I want to tell you: Scott, I love you dearly, but I *deplore* your job, even though I don't have a clue what you do, and I *hate* the Air Force. I just wish that you were willing to walk away and leave all of it behind you."

"Bea, I've told you. I can't just walk away. It's not that simple."

"Well then, for my part, I'll make it as simple as I possibly can for you: I will do *anything* you want. As much as I like my job, I would go to Nebraska if you wanted to take over your family's farm. If you want to go back to school, then we'll scrimp and pinch pennies, but if that's what makes you happy, then we'll do what it takes. Then you can go to work in the space program, if that's still a dream of yours. We'll find a way. I just want this baby to live a normal life, to grow up in a stable environment, to have the same friends from year to year to year, and not move from one corner of the globe to the next every time the wind blows."

He held her tightly, kissed her, gently stroked her hair, and said, "You're right. I'm sure that things will settle down someday, and our lives will be normal. Bea, please be patient with me."

"I hope you're right, Scott, because more than anything else, I want this baby to grow up knowing *you*, and not a picture on the wall."

13

LABOR PAINS

On Orbit
8:18 p.m. Eastern, Thursday, March 12, 1970
(Rev 45 / GET: 66:52:35)

Tick . . . tock . . . tick . . . tock. Ourecky gradually opened one eye and glanced up at the wind-up alarm clock fastened to his hatch with Velcro. Because there was an ever-present danger that they could oversleep, they had brought up the clock. For all the amazing technology crammed into this tiny space, there was nothing to ensure that they would be conscious at crucial moments, so they had to rely on a venerable old Westclox Baby Ben.

When they slept in weightlessness, their relaxed arms naturally floated out in front of them. That wasn't good, since their extended hands might inadvertently brush switches and circuit breakers. To prevent this, Carson had hooked his thumbs into his harness shoulder straps.

Snoring loudly, Carson was obviously getting the rest he sorely needed and deserved. This had been a particularly nerve-wracking mission; they had intercepted the maneuvering Soviet satellite, but only barely so. A small glob of saliva formed at the right corner of his lips and slowly drifted away, a glistening orb hanging in the air before his unshaven face.

On his side of the darkened cockpit, hands tucked under his thighs, Ourecky could not will himself to fall asleep. He had been in orbit for almost three days, and in just a few hours, his feet would be back on solid ground. His fitful mind was filled with thoughts of Bea, and he hoped that their baby would be patient just a little while longer so he could be there for the delivery.

For whatever reason, Tew was going to extraordinary lengths to ensure that he would make it back to Ohio if there was even the slightest chance that the baby would not wait. Parch Jackson was standing by at White Sands—their planned touchdown site—with a T-38, fully fueled and primed for immediate takeoff, with no other mission than to deliver the expectant father. Sigler was on hand in New Mexico as well; he would assist Carson with the post-flight chores of off-loading exposed film and other mission materials from the spacecraft.

The alarm sounded with the familiar jangle that had jarred many millions from their slumber. Ourecky pushed in the stop plunger, stowed the clock in a side pouch, and nudged Carson. "Time to get busy," he announced, removing the cover from his window. "Rise and shine."

As usual, Carson's eyes opened slowly, almost furtively so, as if he wasn't certain whether he was awake or still dreaming. "Oh man, that snooze hit the spot," he declared, unhooking his thumbs and carefully stretching. He smeared his dry lips with a Chapstick and grabbed a quick swig of water from the dispenser. "Boy, do I feel completely refreshed now. How about you, Scott? Log any Z's? It seemed like you were just tossing and turning over there."

"I got plenty," fibbed Ourecky. "Ready for power-up? We have a contact in fifteen minutes."

"Then let's get cracking." Consulting a checklist, Carson said, "Okay. Top breaker panel first. Maneuver drivers set to Primary. Set maneuver thruster breakers to closed."

"Maneuver drivers to Primary. Maneuver thruster breakers to closed," replied Ourecky, throwing a series of switches over his head.

On Orbit
8:33 p.m. Eastern, Thursday, March 12, 1970
(Rev 45 / GET: 67:08:05)

Carson finished a cheese sandwich on tortilla bread, chased it with water, and asked, "Crypto?"

"It's in," answered Ourecky. "I verified the settings while you were stuffing your face."

"Thanks. Aren't you going to eat something? Surely you have to be hungry right now."

"I don't think I could eat anything if I tried," replied Ourecky. "My gut's full of butterflies. I'm not too fond of reentries, and at this point, I just want to make it home in one piece."

"I understand." Carson adjusted his headset and turned up the volume. "It's time. Go hot."

Ourecky switched on the cryptographic device, and the two men waited for the transmission to come up from the EC-135E tracking aircraft flying over a hundred miles below them. Several seconds passed before they heard the distorted voice in their headsets: "Scepter Four, Scepter Four, this is Indian Ocean Sentry on Channel Two . . . Do you read?"

"Indian Ocean Sentry, this is Scepter Four," replied Carson. "I'm reading you five-by on Channel Two. We're standing by to copy reentry guidance."

"Scepter Four, here are your next three reentry shots. Your primary remains Zero-One on Rev 46, GET 68:49:04, 20 plus 18, 25 plus 12, roll left 50, roll right 45, more details to follow," chanted the controller aboard the distant aircraft. "Zero-Two on Rev 47, 70:19:16, 20 plus 47, 26 plus 31, roll left 55, roll right 45. CRZ Three-Six on Rev 48, 71:49:12, 20 plus 18, 25 plus 56, same bank angles. How copy?"

Ourecky jotted down the series of numbers and then read them back. The numbers represented the basic information they would need in the event they had to reenter without additional guidance. Each string of numbers told them the contingency site to steer for and the Ground Elapsed Time—in hours, minutes and seconds—at which they would fire the retros. The Gemini reentry vehicle had an offset center of gravity to generate aerodynamic lift during reentry. Controlling the lift enabled them to more precisely control where they would eventually land, so the other numbers told them when to apply bank angle and how much bank to apply.

"Good copy, Scepter Four, stand by for DCS data upload."

Watching the DCS light blink on, Carson stated, "Uploading data now." At this point, their onboard computer was automatically receiving detailed reentry data from the tracking aircraft.

"We have the load," observed Ourecky a few seconds later.

In the next few moments, they verified some specifics concerning the reentry and got a weather update for the touchdown site at White Sands. Closing out the session, the Airborne Mission Controller said, "Good luck, fellas. Have a safe trip home. Indian Ocean Sentry Out."

10:15 p.m. (Rev 46 / GET: 68:49:18)

"Thirty seconds to retrofire. Ready to head home?" asked Carson.

"More than ready," replied Ourecky.

"Hey, brother, I told you we would get you back to the house before the baby came."

"I greatly appreciate it, Drew."

"Anything for you. Okay, retro squibs to Arm," stated Carson. "Arm Auto Retro is amber."

"Arm Auto Retro is amber," confirmed Ourecky.

Carson pressed three telelights in turn, watching them change from amber to green. "SEP OAMS is green," he stated. "SEP ELEC is now green. *Bye, bye, big guy.* Nice knowing you."

Several feet behind them, mechanical guillotines sheared electrical cables and explosive shaped charges detonated, separating the adapter section from the spacecraft. The big adapter was essentially their life support module, containing the critical elements for an extended mission. Now, they were reduced to only that required for a safe return to Earth.

"And SEP ADAP is now green. Adapter is jettisoned," noted Carson. "Retros in fifteen."

"Adapter jettisoned. Fifteen seconds to retros," replied Ourecky.

"Count-down to retros, ten, nine, eight, seven, six, five, four, three, two, one—*Mark*."

Holding his breath, Ourecky waited for that inevitable solid thump of the first retro firing, but it didn't come. For whatever reason, the retros had failed to ignite.

"No joy," stated Carson bluntly. The retro rockets were set to fire automatically at a pre-set time. Not wasting a moment, he punched the manual fire button, but there was no response from behind the spacecraft's ablative heat shield. He tried again, and then a third time. "Manual activation is not working, either. Man, this isn't good. This is not good at all."

Exhaling, Ourecky scanned the instruments, looking for some clue to the retros' failure, but there was nothing out of the ordinary.

Both men were quiet for several seconds, and then Carson broke the uncomfortable silence. "Okay. Obviously, we've suffered a setback. Let's try to resolve the problem before we come up on our next reentry window. What's our next shot? Patrick Air Force Base?"

Ourecky cleared his throat, then answered, "Right. Zero-Two. Patrick on the next rev."

"Okay. Go ahead and load the data for Patrick."

"Got it," replied Ourecky.

"Scott, I don't have to tell you that with the adapter gone, our consumables are at a premium. We hope for the best but plan for the worst. Hopefully, we'll fix this situation and scrape the runway in Florida shortly, but regardless of what happens, we'll keep calm, keep working the issues, and keep flying the machine until the bitter end."

Bitter end, thought Ourecky. Only minutes ago he had been concerned about making it home on *time;* now it was a question of making it home at *all.* He was painfully aware of their limited supplies. With the adapter gone, they were entirely reliant on their secondary oxygen system—just two 6.5-pound cylinders—which was intended to sustain them through reentry.

Since the average human consumed eight pounds of oxygen a day, the two cylinders equated to roughly nineteen hours, or approximately twelve and a half orbits. Most of their contingency recovery zones were scattered around the Equator; on average, they had one reentry window per orbit, two at most, and rarely were they located in prime real estate.

To make the prospects even more ominous, they had reentry windows on the next two orbits, and then they were stuck for three orbits—four and a half hours—until the next window, which would take them into Clark Air Force Base in the Philippines in broad daylight.

A bright light flashed in Ourecky's face; startled, he looked out his window and realized that it was sunlight reflected off the white paint of the adapter. The adapter flew alongside them now, a few hundred yards away, like a giant Dixie cup hovering nearby. Seeing it reminded him that not only were they disconnected from a plentiful supply of spaceflight essentials—oxygen, water, maneuvering fuel and electrical power—but they were also separated from another key piece of equipment. The DCS—Digital Command System—was located in the adapter.

The DCS was the electronic conduit that received updates from the ground and loaded them automatically into the computer. From this point on, Ourecky had to cross-reference his notes and painstakingly type in their reentry data into the MDIU—Manual Data Insertion Unit—keyboard. Since the computer had little tolerance for errors, plugging in the data would be trying in even the best of circumstances, but now, operating under a severe time crunch and groggy from sleep deprivation, the crucial task would require every last bit of his attention.

Dayton, Ohio
11:04 p.m., Thursday, March 12, 1970

With Scott out of town, and the prospects of his timely return gloomy, Bea was staying with her friend, Jill, and her mother. Unfortunately, Jill's infant daughter was suffering from colic, so no one was getting much restful sleep. Only an hour ago, the little girl had quit screaming.

Wide awake, watching the luminous hands of the alarm clock on the nightstand, Bea realized that everything was going to happen much sooner than expected. She finally decided that there was no sense trying to wish away what was obviously inevitable. She shoved aside the heavy quilt, sat up in the bed, and pulled on her flannel nightgown.

She stood up slowly and groaned as another strong contraction hit. Striving to keep her balance, she let it pass and then navigated her way into the living room. The television was still on, displaying a flickering test pattern. Bea switched it off, turned on the lamp, and nudged Jill, who was sleeping soundly on the couch. "It's time," she quietly announced.

Rubbing sleep from her eyes, Jill sat up quickly. "The baby's coming? I thought you still had another week," she said excitedly. "Okay. I need to brush my hair and I'll—"

"There's no time," urged Bea. "We have to leave *now*. Stay calm and just do what we planned. Take my suitcase down and start the car. I'll call the doctor, and then I'll wake your mother to let her know to take care of Rebecca until you return. Then I'll come down to the car."

"Okay," said Jill, slipping into the jeans and sweatshirt she had laid by the couch. She slid into her clogs, grabbed her hairbrush, and headed for the door.

Bea woke Jill's mother and then called the hospital. She was sure she had time for one more call, so she dialed the number. When he finally came to the phone, she said, "Hello? Yes, I know you're probably busy, but it's time. . . . Yes, *right* now. Please tell him if you can. Bye."

She gasped as another contraction gripped her. Panting, she sat down until it passed and then went to the crib. Rebecca had been roused by the commotion; she was awake but barely so. "Wait here, little girl," said Bea softly, bending over to adjust the baby's plush covers. "And I'll bring you a friend to play with. Maybe you can grow up together."

On Orbit
11:44 p.m. Eastern, Thursday, March 12, 1970
(Rev 47 / GET: 70:19:01)

With the data loaded in the computer and retrofire just a few minutes away, Ourecky and Carson checked the switches and breakers yet again. Everything was set properly, but there was always the possibility that something unseen was amiss, maybe a miniscule glob of solder that had broken free behind an instrument panel, or a fickle relay that just refused to relay.

Ourecky thought about reentry. The process of leaving orbit was an incredibly complex undertaking. It was as if their lives depended on an enormous and complicated slot machine, where hundreds of cherries, lemons and oranges had to spin into exact alignment when the handle was thrown. The one potential jackpot garnered a terrifying ride earthwards in a flaming meteoric chariot; the infinite multitude of losing chances would result in a slow death by asphyxiation, crammed shoulder-to-shoulder in a dark and cold capsule, staring out into the harsh void.

At a minimum, a retro failure meant waiting for the next available window; in this case, that would entail sweating through another agonizing ninety minutes, fixated on the gauges that displayed the slow but steady depletion of their oxygen supply.

"Thirty seconds to auto retro fire," stated Carson. "Time to head for those beautiful Florida beaches to eyeball all of those hot bikinis. Hey, I know a great little Tiki bar over in Cocoa Beach. Maybe if we're down there . . ."

Frowning, Ourecky double-checked the computer as he said, "Good for you, Drew. Personally, I would rather get home as quickly as possible."

"Oh, man, I forgot about Bea!" exclaimed Carson, slapping his forehead with his palm. "Look, buddy, I'm sure that Parch is in the air, already halfway there from White Sands."

"I hope so."

"Auto retro in ten," noted Carson. "Five, four, three, two, one, Mark . . . retrofire!"

There was silence. "Manually firing retros," said Carson, stabbing a button with his finger. Still nothing. "Okay, Scott. Dump the computer and load for Three-Six. Haiti, right?"

"Yeah."

"Keep your chin up. We're going home on this next go-around."

On Orbit
12:30 a.m. Eastern, Friday, March 13, 1970 (Rev 48 / GET: 71:04:23)

They had a brief commo window as the spacecraft passed over Patrick Air Force Base. There was little said, mostly because there had been virtually no changes in the situation. It was obvious that the mission controller was painfully aware of their bleak outlook, and did his best to remain upbeat yet professional.

At least they didn't have the nuisance of the cryptographic equipment; the mission rules prescribed that once the adapter was separated and battery power was at a premium, all communications were conducted in the clear.

"Everything is still on track for CRZ Three-Six," stated the controller succinctly. "Retrofire at 71:49:12, 20 plus 18, 25 plus 56, roll left 55, roll right 45. All other guidance remains in effect."

"Roger," said Carson. "We'll shoot for Three-Six. Let them know to expect us."

"I'll make sure that they leave a light in the window and the welcome mat on the front porch. Good luck, guys. Fifteen seconds to loss-of-signal . . . Hey! I almost forgot, Scepter. I have a message for your right-seater. Task 99 is in progress. I say again, Task 99 is in . . ." As the signal swiftly faded, the controller's voice faded into a nonsensical warble and then died altogether.

"*Task 99?*" asked Carson. "Scott, is that what I think it is?"

Ourecky glumly replied, "Yeah. It means Bea went into labor. I guess she's at the hospital."

"That's great news!" exclaimed Carson, lightly punching the right-seater's shoulder. "You're going to be a father!"

"Great news? She could be having the baby right now. As a matter of fact, as slow as the news gets to us up here, she could have delivered hours ago. And if our afterburners don't light, I may never see that baby at all."

"Have some faith, Scott. We'll get you home. In the meantime, let's walk through the power-up sequence, step-by-step, and make sure we didn't miss anything. While we're at it, we'll do a wiggle test on every switch and breaker to see if we can catch the one that's bad."

"Sure," replied Ourecky. They had already gone through this process, and both men were confident that the instruments were good. He was virtually positive that the fault lay in one of the relays within the sequencer system, in which case there was nothing they could do about it.

They couldn't exactly pull this jalopy over on the shoulder to pop the hood and ferret out a loose sparkplug wire. The all-important sequencer relays were inaccessible to them, so if the sequencer didn't light the retros at the opportune moment, and they weren't able to light them manually, there wasn't much that could be done.

"On this next lap, I'm going to manually fire the retros in advance of the sequencer," said Carson. "Maybe that will circumvent the problem."

"I sure don't think it could hurt. Look, Drew, I've got another idea, but it's sort of out there."

"Let's hear it," said Carson.

Ourecky explained, "I'm fairly sure that the problem's in the reentry sequencer. We may not be able to work around that, but if we set the computer to Ascent mode and then reset the switches, we might trick the platform into thinking we're lifting off instead of reentering."

"Interesting notion," replied Carson, scratching his forehead. "But I'm missing the point."

"Drew, in order to reenter, we have to light the retros. This is an ugly fix for the problem, but if the platform thinks we're deep into the boost phase and then you throw the abort lever . . ."

"It will start the sequence to salvo-fire the retros for a Mode III abort. Hmmm. Neat approach. You're right. Prior to retrofire, we can have our angles set to at least come out of orbit. Do you think you can reset the computer in time to fly the rest of the reentry?"

"Probably not. But once the retros burn, gravity will pretty much take care of everything else. All we have to do is dump the drogue and deploy the paraglider."

"Okay. Let me ponder this a while," noted Carson. "You sketch it out and start tinkering with the numbers. I think it's a good concept, but I see one glaring problem."

"What's that?"

"Just about all of our remaining reentry windows take us into an island of some sort. If we execute this plan, we can forget about any precision landing options. Going down in the drink is effectively committing suicide, so I wouldn't want to try your plan unless we have a window that takes us into a major land mass, preferably the United States."

"Our next window into the States is Edwards, on Rev 59," stated Ourecky.

"Rev 59?" Carson was silent for a while, and then said, "Scott, let me ask you a couple of hypothetical questions. First, are you confident that your scheme could work?"

"Not *absolutely*, but if nothing else works in the meantime, I think we should try it."

"Fair enough. Second question, if you had to, do you think you could set this up and fly it by yourself? No help from me?"

"I'm sure I could," answered Ourecky. "Why would you ask that, Drew?"

"Because we'll run out of air about an hour before we hit the retro-fire mark on Rev 59. I would just like to be confident that you can drive yourself home, *by yourself*, if need be."

A chill passed through Ourecky when he realized Carson's implication. "We're *both* going home," he said emphatically.

Carson chuckled and said, "Brother, we'll see. We'll take this an orbit at a time, and we'll make our best effort on every pass. But mark my words, Scott, my priority is to get *you* home safe. Bea and that new baby are waiting for you downstairs. There's no one waiting for me."

14

DESCENT

On Orbit
1:13 a.m. Eastern, Friday, March 13, 1970 (Rev 48 / GET: 71:47:05)

"Okay, here we go again, except this roll's going to be the big winner," declared Carson. "Like I told you, I'll try the manual switch first. How do the squib batteries look?"

Ourecky verified the power settings. "All three squib batteries are reading rock solid. Hey, Drew, you have this, right? You can handle this next part by yourself?"

"Sure. What's on your mind?"

"Well, I know you're not the religious sort, but if you don't mind, I'm going to say a prayer."

"Sure. At this point, any and all options are welcome." In the center console, an indicator flashed yellow. "There's the ARM AUTO RETRO light. Thirty seconds. Scott, if you're going to chuck up a prayer to the Almighty, you had better make haste."

Ourecky closed his eyes. He didn't opt for a Hail Mary or any of the ritualistic recitations he'd learned during his childhood of catechism at Saint Wenceslaus, but instead silently said a simple and heartfelt prayer: *God, please let us return safely to Earth so that I can see Bea and my child.* Then he sucked in his breath, clenched his teeth, opened his eyes, and waited.

"I will manually fire retros," stated Carson, almost officiously. "Auto retro fire in five, four, three . . . manually firing retros." He calmly and firmly pressed the MAN RETRO FIRE button.

Ourecky exhaled sharply as the first retrorocket roared to life.

"And there's One," exclaimed Carson. "I think you can quit praying now, Scott. We're going home. We *are* going home!" Separated by intervals of five and a half seconds, the remaining three retros joined the welcome chorus. After the retros had served their purpose and the retrograde section was jettisoned, the spacecraft began its long plunge into the atmosphere. It didn't take long for the G's to build and the vibrations to commence.

Although he might eventually become accustomed to the tense moments of takeoff, Ourecky wasn't sure that he would ever become entirely comfortable with reentry, even though it meant that they were returning to Earth. They entrusted their lives to the heat shield, a relatively thin slab of ablative materials that was designed to literally burn and flake away to dissipate the heat that would otherwise incinerate them in an instant.

Outside his window, an undulating orange glow gradually built up, eventually replaced by a raging inferno, like they were being held at the sharp end of the Devil's poker and dipped into a roaring blast furnace. The cabin interior heated up quickly, and soon Ourecky was drenched in sweat, listening to the rattle and hum as he tried to focus on his instruments.

This must be what it feels like to descend into Hell, mused Ourecky. He could not know just how right he was, and as he fell into the unknown, he also could not know that almost half a world away, at the very same moment, Bea was delivering their child into the world.

Morne Bossa, Haiti
1:25 a.m. Eastern, Friday, March 13, 1970

Ten minutes prior to his scheduled contact time, Henson clamped the earphones over his head and listened to the shortwave radio. His instructions were simple. At pre-set times, he was to monitor the radio and have his recovery equipment prepared to function at a moment's notice. He had positioned railroad flares to outline the designated touchdown area. Outside his simple shed, the portable generator was fueled and running.

As he waited for his appointed window, Henson dialed the shortwave to the BBC World Service and listened to the news of the day. There wasn't much good news to be heard. A series of bombs had been detonated in schools, shopping centers and office buildings in New York City, Pittsburgh and Appleton, Wisconsin. In Laos, Communist Pathet Lao forces were steadily advancing on the capital city of Vientiane.

Speaking in a clipped British accent, the news announcer droned on, relating the events of the day. The Cambodian government ordered North Vietnamese and Viet Cong forces to immediately leave their country. Angry Cambodian mobs burned and pillaged Vietnamese neighborhoods in Phnom Penh. On a bright note, Expo '70, the latest incarnation of the World's Exposition, opened in Osaka, Japan. Henson chuckled at one of the last news items, a quick tidbit from America. Although North Carolina schools had been ordered to desegregate, the state could not afford the buses required by the new integration plan.

Checking his watch, he saw that his contact window opened in one minute. He spun the shortwave's channel knob to the prescribed frequency and waited; exactly on time, the message came through. It was a seemingly illogical string of numbers, but Henson knew that each sequence of numbers held a specific meaning. He listened intently for his station number: Three-Six; repeated three times, it would precede the sequence meant especially for him.

And then he heard it—*Three-Six Three-Six Three-Six*—and he rapidly jotted down the series of numbers and letters that followed: *Zero Seven Zero Four Zero Six Four Six Tango One Two Echo Romeo Yankee Lima.* The sequence was repeated three times before the distant operator requested that Henson acknowledge the message by repeating it back.

Confirming receipt, Henson swiftly tapped out the Morse code authentication, and then examined the message. The first part of the message was two sets of four numbers; the first set represented the time, stated in "Zulu" Greenwich Mean Time, that a "vehicle" was theoretically supposed to touch down on his dirt strip, the second time was when he was most likely to establish radio contact with the new arrivals. The next segment, preceded by the letter "Tango" told him to set his portable TACAN beacon to transmit on channel twelve.

The last segment caused his heart to skip a beat. He had been through this routine innumerable times, but this was the first indication that this was not just another inane drill, but the real thing. "Echo Romeo Yankee Lima" was shorthand for "En route Your Location." Whatever it was that was due to fall out of the sky, it was going to fall out of the sky *here*.

Henson tugged off his headphones, set aside his Morse key, and reflected on what was to ensue. At long last, his curiosity would be satisfied; he would finally know what all this fuss was about. Calmly and deliberately, he made the last preparations. He switched on a UHF band radio, went outside, checked the generator's fuel level, and powered up the TACAN beacon.

He checked the winds with a handheld anemometer/wind vane and verified the current pressure with a lightweight field barometer. Now he would wait for the crew to announce their approach; then all that remained was to light the flares that marked the touchdown pattern and talk the crew in. Then he would help them conceal the vehicle, whatever it was, and hide them while waiting for a recovery team to arrive.

It surprised him that they were coming here. Haiti was such a low priority site that he hadn't even been assigned an assistant to help him with the task. He hoped that everything was all right, and that the men—whoever they were—would arrive safely.

1:35 a.m., Friday, March 13, 1970 (GET: 72:10:05)

The sky was pitch dark as they plummeted to Earth. "Sixty thousand feet," announced Carson. He pushed the HI ALT DROGUE switch. "Manually deploying drogue . . . Drogue's out. 60K light is lit also. Man, it really stinks in here!"

Just a few moments later, the paraglider opened flawlessly. "Oh man," exclaimed Carson. "That was the slickest, smoothest opening ever. This is stacking up to be a cakewalk. We'll have you home in no time. Okay, Scott, hang on tight. I'm going to two-point suspension."

Ourecky checked his harness, made sure he was square in his seat, and braced his hands on either side of his window. "Ready for two-point."

Both men were jolted forward sharply. "Ouch!" declared Carson. "Sorry. That could have been a lot smoother. I'm going to do my control checks. You stay with the post-retro checklist."

"Okay. I'm still working the checklist. RCS Control circuit breakers open. RCS Control is off. ACME Control One through Six breakers off." Ourecky went through the rest of the checklist, then asked, "I'm going to raise the recovery crew on the UHF. Sound good to you?"

"Go for it."

Ourecky keyed the mike and spoke. "Three-Six, Three-Six, this is Scepter Four, transmitting in the blind. We are en route to your location. Over."

There was no reply. About a minute later, he tried again. "Three-Six, Three-Six, this is Scepter Four, en route to your location. Over."

The voice on the other end was reassuringly calm. "Scepter Four, I have you loud and clear on UHF Two. TACAN is transmitting on channel twelve."

Ourecky selected the TACAN channel; immediately, the indicator bulb lit. In the DME—Distance Measuring Equipment—portion of the TACAN panel, two arrays of digital numbers—like matching odometers from a car's dashboard—started clicking into place to indicate the direction and distance from the landing site. Laughing, he couldn't contain his sense of relief.

"Hey, let's maintain a sense of decorum over there," said Carson, punching Ourecky lightly on the shoulder. He grinned, keyed the radio, and transmitted, "Three-Six, we have a *solid* TACAN on channel twelve. DME is showing slant-range eight point three miles out Zero One Zero degrees from your location. Please state field conditions for arrival."

"Scepter Four this is Three-Six. Field is packed earth. Land heading one-eight-zero. Winds are currently eight knots out of the southeast, gusting to twelve. Current altimeter reading is 29.90. You have about a quarter moon out there. Read back, please."

"Three-Six, I copy that field is packed earth. Will land heading one-eight-zero. Winds are eight out of the southeast, gusts to twelve. Current altimeter is 29.90. Quarter moon. Are your marker lights on? I see the beacon for Cap-Haïtien Airport to my right front, but I don't see you."

"Scepter Four, field is currently lit. You should see red flares shortly. I am located four and a half miles south-southeast from Cap-Haïtien airport. Also, be advised that there's a big storm brewing to the north, back behind you, but you should arrive well in advance of the rain."

"Good copy," answered Carson, peering through the window. He dimmed the cabin lights to see better. "Thanks for the warning, Three-Six. We'll put on our galoshes and rain slickers."

"Roger. Be advised that I'm running this station by myself, so I will be off the air for about a minute. I need to call my headquarters on my other radio."

"You're there by *yourself?*" answered Carson, almost incredulously. "Roger. We'll keep an eye out for your lights while you tend to your other chores."

As they crossed the coastline, Ourecky noted, "We're at six miles straight-line. We should be over land right now."

"Feet dry," observed Carson, breathing a sigh of relief as he recalled their experiences training in the Gulf of Mexico. He reached out and threw a series of switches. "Opening skid bay doors and lowering skids." He heard a hissing noise just below their feet; the landing skids' extension was powered by compressed gas. The hissing ceased, and three green lights glowed on his instrument panel. "All three skids are down and locked."

Humming in exuberant glee, Carson kept a disciplined scan between the dimly lit instruments and the window. In the distance, slightly off to the right, he could see the twinkling lights of Cap-Haïtien. Far to the left of the port city, slightly beyond and to the left of the airport's rotating marker beacon, he picked out a pattern of red pinpoints arranged in a box configuration that marked the landing site. He made slight adjustments to steer towards it and then held a steady course, reminding himself that they were mere minutes from being safely back on Earth.

Then his humming abruptly stopped as his hackles went up. He realized that the red pinpoints were coming *much* too fast. Years of flying had taught Carson to trust his gut instincts, and right now his gut was *screaming* to him that something was considerably out of kilter.

Since it was designed to fly in the vacuum of space, the Gemini-I was not equipped with an airspeed indicator. Carson might well have been born with accelerometers in his butt, because he sensed that they were moving fast—*very* fast—and even though the paraglider lent them some considerable forward speed, he could not account for why they were moving so swiftly.

"Something's wrong," stated Carson, nudging Ourecky. "Really wrong. Time that DME and give me a rough estimate of our ground speed."

Watching the sweep second hand of his watch, Ourecky timed the digital numbers rapidly clicking away on the TACAN DME display. After thirty seconds, he declared, "This can't be right!" According to this, our ground speed is over fifty miles an hour!"

Carson was perplexed. It was as if a giant hand had grabbed them and was shoving them inland at a prodigious rate of speed. Suddenly, the dark countryside was illuminated by a tremendous flash of lightning; he momentarily glimpsed a castle—the Citadelle—in the mountains overlooking Cap-Haïtien and the northern coast. They were in *big* trouble; at the rate they were going, they would quickly overshoot the forgiving plains of the coast and soon be over the mountainous terrain of the central highlands.

As a second brilliant flash lit up the land, Carson realized the source of their dilemma. The controller had warned them about impending bad weather. Obviously, a powerful thunderstorm was arriving from the north, and it wasn't going to let them down easy. Remembering the story of a Marine F-8 pilot who bailed out over a thunderstorm—ejecting over Virginia and eventually landing in North Carolina roughly forty minutes later—the question was whether the storm was going to let them down at all. When and if it did spit them out, it was doubtful that they would come to rest in safe terrain. For all of the carefully deliberated preparations, it was obvious that Mother Nature had not been consulted, and now she was demonstrating her intense discontent.

In his experience, Carson had learned that roughly ninety-nine percent of flying was excruciatingly routine, almost like driving a milquetoast family sedan down a painfully boring stretch of straight and level highway to Grandma's house. That said, the vast majority of pilot's training—approximately ninety percent of the ground work and practice flying—was focused on the remaining *one* percent of flight, when things go horribly wrong in the blink of an eye, when instruments fail on a complex night approach, or when a plane swerves into a blistering spin or when the weather turns in an instant, altering the world from blissful calm to liquid shit. And that one percent of flight is *why* Carson existed. That one percent was why Carson *was* Carson. He was born to become one with his machine, to complete the mission and return safely to Earth.

Besides having too much airspeed, they were also too high. In all of their paraglider training, they had never once contemplated a scenario where they would have too *much* altitude. Normally, their dilemma was not having sufficient altitude, which caused them to come up short of the desired landing site. And even tonight, returning from orbit, coming up short would not be a horrific problem, had they been coming down at White Sands or Edwards or any of the other normal landing sites where there was an abundance of flat unoccupied desert.

In his head, he pictured the situation. Although Three-Six's weather information was obviously correct, it only provided a static snapshot of the conditions. Carson knew that weather could be incredibly dynamic, not static. Because it was three-dimensional, simple reports and meteorological charts could not effectively convey its invisible complexities. Weather front boundaries could not be accurately depicted with neatly drawn lines, because those lines were anything but neat as they projected off the surface of the Earth and into a churning atmosphere.

Carson surmised that a moisture-laden mass of turbulent warm air—bearing the powerful thunderstorm—was literally bearing down on and riding over the top of a mass of calmer, cooler air. The warm front was like a huge inverted wedge, like a wave on the verge of cresting, miles ahead of torrential rains and gusting winds.

"We are *way* too high," he announced. He threw the hand controller to the left, and the paraglider responded in turn, spiraling the vehicle down in corkscrew turns to bleed off excess altitude. "Scott, we're just going to have to make the best of it. Keep your eye on the TACAN DME. About every thirty seconds, I want you to call our direction and distance to the controller down there, so he has some idea of where we're coming to roost. And cinch your harness good and tight. I'll try my best to slide us in somewhere safe, but I doubt that this will be pretty."

"Three-Six, this is Scepter Four. Three-Six, this is Scepter Four," stated Ourecky. "My pilot advises that we have too much wind at altitude and will likely land well to your south. I will call TACAN DME until we land, so you'll have a reference. How copy?"

"Scepter Four, good copy. Where are you now?"

"We are two miles out, twenty degrees *from* you. We should pass overhead shortly."

"Scepter Four, this is Three-Six, I copy two miles out, twenty degrees from my location. Will continue to monitor and record. Good luck to you."

"Thanks," replied Ourecky. He turned to Carson and added, "I swear that I know that voice from somewhere, but I can't pin it down."

"Probably from a practice drop. Scott, go ahead and switch on the camera," said Carson. "I know that we're not going to make the field, but I want to know what's below us. I'll let you know when to switch on the landing light."

"Got it," said Ourecky. He threw a pair of switches on his panel, then turned on a small television screen in the center panel. A blob appeared in the screen; the monitor, connected to a fiber optics camera mounted alongside the forward strut skid strut, gradually warmed up, eventually displaying a dim image of the landscape below.

Carson surmised that spiraling down was counterproductive. When they came around the circle so that their back was into the wind, the paraglider was thrust sharply forward, so that they really weren't flying a corkscrew, but a series of tight buttonhook turns, still drawing ever closer to the dark mountains. Pulling out of the spiral, he pointed the paraglider's nose into the wind to hold a steady course. Now, even with the airspeed generated by the paraglider, they were effectively flying rearwards. Ourecky continued to talk to the controller over the UHF radio, maintaining a running commentary of their distance and direction to the now distant landing site.

Used to zooming straight ahead at supersonic speeds, flying backwards was entirely counterintuitive to Carson. He still wanted to lose altitude and stay as far north as possible, away from the mountains. As he watched the distant lights of Cap-Haïtien gradually climbing up in the window, he realized that they were going into the mountains regardless. A few moments later, he estimated that he was roughly five hundred feet up. "Switch on the landing light, Scott."

Ourecky turned on the landing light, which projected a focused beam like a car's headlight. While it was an essential to landing, it would consume their remaining battery power at an enormous rate. Even at this point, Carson wanted to conserve their batteries to have some residual juice to power the radios, so that they could summon the cavalry.

Scanning between the television monitor, the window, and his instruments, he identified a lighter shaded area below them and to the left, surrounded by darker terrain. He guessed that it was a cultivated field. Instead of gradually flaring the paraglider, which would be the normal procedure to lose speed and hold their position, Carson did the opposite: he dipped the paraglider's nose so that the wing's forward acceleration would cancel out their rearward flight.

It was a drastic and risky move, but it apparently was working. They were still moving backwards, but the craft seemed to be dropping—rather quickly—into the field that Carson had selected for their landing site. As they descended, he made slight adjustments to the sluggish paraglider, feverishly tweaking their trajectory to stay clear of the darker ground. It was tremendously awkward, because his steering inputs were backwards from normal flight.

Suddenly there was an unexpected and unwelcome change. Their rearward movement abruptly slowed, as if they were going into some sort of reverse stall, and then they surged forward. Carson realized that they must have descended through the "floor" of the storm front, and were now being overtaken by the surface winds that Three-Six had warned about earlier.

Instead of flying backwards, they were now accelerating forward. Trying to avoid any hasty corrections, Carson gradually tugged back on the hand controller, bringing the tail of the paraglider down. He glanced at the monitor, saw that they were still over the light-shaded area that he selected for touchdown, and then shifted his attention to the contact light that would blink on when a sensor, dangling fifteen feet below on a wire, touched the ground. That was his signal to "hard flare" the paraglider, just prior to the skids making contact, when Ourecky would fire the pyrotechnics to discard the fabric wing. Gripping the hand controller, he calmly stated, "Scott, call DME to Three-Six and stand by to jettison the paraglider."

"Three-Six, preparing to land, we are at 8.8 miles, 20 degrees *to* your location!" blurted Ourecky. There was no reply from Three-Six.

"Watch that contact light, Scott. Eyes on that light."

"*Contact*," announced Ourecky, as the tiny light flickered green.

"*Flaring*," stated Carson, tugging the hand controller fully back. What happened next was puzzling, but only momentarily. Instinctively waiting for the usual scraping noise as their skids bit into the ground, he heard nothing but an odd rustling sound. What he didn't know was that they were landing in a field of sugarcane, and that the tall stalks had brushed the contact sensor, triggering the contact light. So immediately after Carson hard-flared, the craft plummeted approximately twelve feet to the ground, still moving forward at roughly ten miles an hour.

Not engineered for sharp falls, the left skid strut snapped like a dry twig. The little craft lurched sharply to the left, plowing up earth as it slid to a stop. Despite their restraints, both men flailed forward. Ourecky smashed face-first into his instrument panel, shattering the clear visor of his flight helmet. With his instincts conditioned by years of boxing, Carson's right forearm snapped out, blocking him from a similar fate. As a final insult, the television monitor broke free of its bracket, crunching into Ourecky's midsection.

Carson took stock of the situation. The craft was laid over hard to the left, almost upside down, so that he was literally suspended by his shoulder straps. He couldn't see anything through his windows and realized that his side of the vehicle was probably lying on the ground.

Looking to his right, he saw that Ourecky was obviously injured, bleeding, and apparently unconscious. He hoped that Ourecky's hatch was clear of the ground, so that it could be opened. Then he glimpsed the flickering orange light through the right window, realized that the hatch was probably not impeded, and breathed a sigh of relief.

With a start, he suddenly realized the source of the orange glow. Since they landed in some sort of cultivated field, the still sizzling heat shield likely had ignited crops of some sort. *Not good.* Carson reached down and yanked a toggle that "safed" the ejection seats and assorted pyrotechnic charges intended to blow the hatches open. Outside the spacecraft, the world was quickly becoming a maelstrom of raging fire.

Carson reacted quickly. He undid a series of connectors and squeezed free of his harness. He reached over, swiftly unfastened Ourecky's restraints, and wriggled until he was squeezed crossways in the cabin. Jamming his shoulder against the hatch, he unlatched it; pushing with all his strength, he swung it open as far as he could. The hatch opened barely enough to facilitate their escape. Aided by gravity, Carson extricated Ourecky from his seat, and both of them plopped headlong onto the hard ground below.

Apparently awakened by the abrupt fall, Ourecky came to life. With blood spurting from a gash in his forehead, he was obviously dazed and disoriented. "Drew?" he muttered. "Are we down? What's burning? What's that awful smell?"

Even with the billowing smoke of the burning sugarcane, the spent spacecraft's collective fumes were almost overwhelming. The air was heavy with the scorched metal smell of the heat shield and the acrid fumes of residual propellant slowly venting from thrusters.

Without thinking, Carson raised into a crouch and placed his hand against the edge of the hatch to regain his balance; the insanely hot metal instantly seared through the thin leather palm of his Nomex flight glove and fused to the skin underneath. Howling from the intense pain, he quickly snatched his hand away, ripping away a layer of flesh.

Frantically assessing the situation, Carson considered grabbing the survival kits, but realized that their *immediate* survival hinged on escaping the rapidly growing flames. Hopefully, once the fire died down, they would be able to salvage some gear from the spacecraft.

"Let's go!" yelled Carson, wrapping his arm under Ourecky's. "Come with me!"

They barely escaped through burning sugarcane. In the flames, Carson saw a tree line, about two hundred yards away, and ran for its asylum, thinking that the fire would likely burn out when the dense cane was consumed.

They reached the safety of tree line. In the light of the flames, he tended to Ourecky's wound, dislodging a thumb-sized shard of clear

Plexiglas. The singed right sleeve of his flight suit was just barely hanging on by tattered remnants of fabric; he ripped it the rest of the way off and used it to fashion a makeshift bandage to stanch Ourecky's bleeding.

He heard yelling, and saw two ominous figures wielding machetes, backlit by the pyre, about a hundred yards distant. They appeared extremely angry, probably seeking vengeance from whomever it was that dropped from the sky to set their crops ablaze.

Half-carrying Ourecky, he retreated further into the woods. Suddenly they were spattered with huge drops of rain. The cooling moisture was welcome for a moment, but the rain quickly worsened into a torrential downpour. The thunderstorm that had shoved them into the mountains was now here, and it was drenching the landscape. The ground sloped sharply; Carson realized that they were now descending down the west side of a steep ridgeline. There was so much rain that making any forward progress was almost like swimming through ink.

The next hour was a cat-and-mouse game; straining to keep his balance in the steep terrain, Carson fled deeper into the woods, maybe a hundred yards or so, and then looked back to see if they were still being pursued. Periodic flashes of lightning revealed that they were still being stalked by the two fearsome machete-armed men, so he went still further downhill, sometimes rolling head over heels when he lost his footing on the slick ground.

Finally, as the rain started to taper off, it appeared that the men had abandoned the chase. Carson leaned the barely conscious Ourecky against a tree and then plopped onto the ground next to him. Cupping his burned hand to his chest, he cringed at the horrific pain and the smell of scorched flesh. He had hoped that they could stay close to the spacecraft, but decided it was ill advised to linger in the area. It wasn't the best of circumstances, and it would obviously be hours if not days before they were entirely safe, but at least they weren't still trapped in orbit. He decided to wait here until dawn and then assess their situation in the light of day.

Milton, Florida
4:35 a.m., Friday, March 13, 1970

The phone jangled on the nightstand; Nestor Glades woke up, slipped out from Deirdre's sleeping embrace and answered it. It was the duty officer at the Ranger Camp. He listened to the brief instructions, mumbled assent, hung up the receiver, and switched on the lamp.

Deirdre awoke. "Trouble?" she asked, rubbing sleep from her eyes.

"Somewhere," he replied, standing up and stepping towards the closet. He quickly pulled on blue jeans and a denim shirt. "I have to go."

"Now?" she asked. "How long?"

"Don't know," he replied, stuffing his shaving kit, boots, and some folded field uniforms into a canvas kit bag. "Could be days, could be weeks. Deirdre, I'm really sorry about this."

"Don't be," she answered, standing up and slipping into her nightgown. "It's your job, Nestor. I understand that. Just try not to get shot. And try really hard not to come home with lice again. That was really quite a nuisance."

"Yeah, it was, wasn't it?"

"I'll make coffee," she whispered, padding towards the door.

"No time." He picked up the kit bag and walked towards her. Holding the bag in one hand, he hugged her with the other. "I'll think about you."

"*No*," she answered, hugging him back. "Don't you *dare* think about me. I know that they're not calling you in the middle of the night to go play in a tiddlywinks tournament. Focus on what you have to do, get it done, and then come back to me, Ness. There'll be plenty of time to catch up later. We'll have the rest of our lives. Someday, you'll finally be done, and then we can be together always."

Mission Control Facility, Aerospace Support Project
5:05 a.m., Friday, March 13, 1970

Wolcott had experienced some bad nights, but none quite so bad as this in recent memory. Seated in the glassed-in sanctuary at the rear of

Mission Control, he hoped that they could resolve this sordid situation swiftly. He looked at an aeronautical chart of Haiti spread out over his desktop. He had never been there but had spent a week in the neighboring Dominican Republic a decade ago and couldn't imagine that the countries could be that much different.

Since he could do virtually nothing in this trying situation, Mark Tew had gone off-base an hour ago, to see Bea and her new baby at the hospital. While Wolcott couldn't comprehend how Tew could leave at such as critical moment, perhaps it was just as well. The night had been a horrible roller coaster for everyone in Mission Control, and probably no one had been more shaken by it than Tew. Besides, of the two, Wolcott was much better suited for the sort of seat-of-the-pants decision-making that would surely ensue in the coming hours and possibly days.

As they grew up together in the service, even back during the War, Tew had always gravitated to the operations side of the business at hand, focusing his considerable intellect on forging precise and careful plans, while Wolcott was the one who executed those intricate plans with decisive aplomb, modifying them on the fly—literally—as required, doing what was necessary to execute the mission and bring the boys home when possible.

There was a furtive knock at the door. Wolcott gulped down the dregs of his coffee and bellowed, "Enter!" The Recovery Operations Liaison, a major assigned from Isaac Fels's Wing at Eglin, entered the room and spread a detailed topographic map on the conference table.

"*Update*," ordered Wolcott brusquely.

"We know they landed close to here," said the major, obviously cowed to be in the presence of Wolcott, gesturing at the map with a mechanical pencil's eraser. "That's the last position they reported to our man on the ground. They're in the mountains, near a town called *Dondon*."

"So you've had radio contact with them?" asked Wolcott.

"No, sir. We put an aircraft overhead about an hour ago, but he's heard nothing on the radio and hasn't detected their rescue beacon, so we don't know if they landed safely, or if . . ."

"We will assume they're safe," interjected Wolcott, snapping the filter from a cigarette before lighting it with his Zippo. "And until we know absolutely for sure, I will caution you to not ever imply otherwise, unless you want a size ten Tony Lama planted permanently up your ass."

"Yes, sir," replied the major, swallowing.

"Okay, so they haven't phoned home yet, pardner. So where do we go from here?"

"Our Rapid Response Flight has been on stand-by at Eglin. They'll establish a staging site at Homestead, near Miami. The RRF has sophisticated search and communications gear, and they can execute a tactical rescue if the circumstances warrant. An RRF advance team will go into Haiti tonight and link up with our man on the ground."

Wolcott took a long draw on his cigarette, exhaled a heavy pall of smoke, and said, "Wait, pard. You keep saying *our man on the ground* like you only have one sole hombre out there in Indian country. Please tell me that ain't the case."

"It is, sir. This was planned as a Class Four austere site with two men, but the second man was hurt on a training jump at Eglin last week. We were short-handed, and Haiti was a low priority site, so General Fels elected not to send anyone else down there because he was confident our man could handle any contingency."

"Hmmph," sniffed Wolcott. "I s'pose Isaac and I will chat after the dust settles. Continue."

"The RRF advance team will go in tonight. In the meantime, we're coordinating with the Navy to shift an intelligence-gathering vessel from Cuba, to monitor radio traffic in this area. In addition to the aircraft that flew over last night, we'll have a U-2 overfly around noon to pinpoint the crash . . . uh, landing site, sir. Once it's positively identified, we'll put the RRF on it."

"Noon? You can't move any faster than that, pard?"

"The U-2 will be flying out of Davis-Monthan, sir, in Arizona. Besides, in those mountains, the sun angle at noon will be a lot better for overhead reconnaissance and photography."

"Okay, pard, what else?"

"Sir, with your permission, we want to authorize our advance man in Haiti to execute a hasty search to see what he can find out before the first string players get into the game."

"So we have two men and a multi-million-dollar spacecraft somewhere in the mountains of Haiti, and we really ain't sure if they're dead, alive, or seriously injured, and you're tellin' me that we're going to turn over the whole goldanged search effort to *one* guy on the ground?"

The major nodded. "Sir, Three-Six was always intended to be a clandestine operation. Until we receive presidential authorization to ratchet up an overt search operation, and the State Department notifies the Haitian government, it has to remain clandestine. We have to tread very lightly. Our operational ceiling for aircraft is ten thousand feet, so until that U-2 gets on station, our advance man has the best set of eyes in the country."

"So, pard, you need *my* permission for him to start poking around?" The major nodded.

"Done. Now let me ask you something. I don't have much time to follow the news very closely, but if I recall my current events, Haiti is just plumb ate up with voodoo and crazy people, and the dictator is that nutcase, uh, Puppy Doc Something. Right?"

"Papa Doc Duvalier, sir. But Papa Doc is anything but a nutcase. He's a dictator, true, but he's very smart and very ruthless. The last thing we want is a confrontation down there, especially in the same neighborhood with Cuba."

"What happens if our two boys end up in the hands of the Haitian military?" asked Wolcott.

"That would not be good," asserted the major.

"Clearly," noted Wolcott. "So how about your hombre down there right now? Can he handle this situation until the cavalry arrives? And what's his name?"

Smiling, the major answered, "He can. His name's Matthew Henson. And he's our best."

Morne Bossa, Haiti
5:55 a.m., Friday, March 13, 1970

As he waited for his scheduled radio contact, Henson poured his blue enamelware cup full of coffee and took a sip. Of the few things he had grown to enjoy in this strange country, he had really taken a liking to their uniquely potent coffee.

Hours ago, he had been ordered to stay put and wait for further instructions. Now, after a long and sleepless night, he wanted to get out in the countryside and *do* something. Of course, he also suspected that helicopters had already swooped in to rescue the crew. All that would be left would be the grunt work of picking up the remaining pieces, such as ensuring that the "vehicle," whatever it was, was adequately secured and quietly transported out of the country.

As the second hand swept the top of the clock's face, the short-wave immediately came to life. The message—sent three times in Morse code—was terse but precise and to the point. Henson was to receive a special recovery team tonight and was directed to assist them in any manner that they requested. More to the point, the message indicated that two US personnel had crash-landed near Dondon. They were currently unaccounted for, and Henson was directed to begin a hasty search to locate them or at least determine their circumstances.

Tapping the Morse key fastened to his leg, Henson acknowledged the message, switched off the radio, zeroed the dials, and gulped down the rest of his strong coffee. He stuffed his binoculars, a Panama straw hat, some maps, a canteen, a first aid kit, a short bolo machete, and a few other key items into a shoulder bag and then went outside to gas up his motorcycle. Minutes later, Henson donned his mirrored sunglasses and kick-started the old Motoguzzi to life. While he wasn't intimately familiar with the steep country near Dondon, he knew the area well enough to know that stumbling around the mountains would be futile. He needed assistance, and although it would require incurring some risk, he set his mind to do what had to be done. Gunning the engine, he shifted into gear and roared off in the direction of Cap-Haïtien.

6:25 a.m.

Dawn was breaking as Carson emerged from the undergrowth and onto a narrow roadway composed of loose dirt and coarse gravel. Consulting the rising sun for his bearings, he saw that they should head to the right, roughly north, which would take them towards Cap-Haïtien. In the distance, a rooster crowed, heralding the new day.

If the pair of machete-wielding goons prevented them from safely approaching their vehicle, he thought that they could eventually pick their way to the CRZ site where they should have landed last night. They would have to be careful, putting into practice all the evasion skills that they had learned in the SERE course at Aux One-Oh. Carson suspected that they might have to wait until nightfall before they risked travelling. He also was beginning to realize that he might be forced to make Ourecky as comfortable as possible, hide him, and then go for help.

Terribly thirsty, he knew that finding water was a high priority. He also suspected that recon aircraft would be flying overhead soon, if not already, attempting to locate the crash site. He wanted to find an open area nearby, possibly a meadow, and lay out signal letters with whatever materials he could scrounge. Just a few simple signals would inform the recon aircraft that they were alive, required immediate medical attention, and were on foot headed to the north. Now he was grateful that Ourecky had compelled him to pay attention during their SERE training.

As he went back into the woods to gather his companion, he heard an odd sound from the top of the hill; muted by the dense woods, it sounded like metal rhythmically striking metal, like someone pounding in nails with a heavy claw hammer.

Whiffing the humid air, he gagged. Besides their reeking body odors, he detected the distinct smell of burnt hair; rubbing his face with his uninjured hand, he realized that his moustache and three-day growth of beard had been singed away.

As he helped Ourecky to his feet, shadowy figures swarmed from the lush vegetation on the opposite side of the road, and he was confronted by a surly mob of Haitians bearing machetes and sticks. They had probably been lying in wait, listening to Carson and Ourecky crashing through the woods in the dark.

Even as he trembled in pain and exhaustion, Carson was confident that he could outrun the mob by himself, but there was no way that he would abandon Ourecky to face their wrath alone. He lowered his friend to the ground and then crouched over Ourecky's prostrate form as he brought his guard up. *I might be outnumbered*, he thought, *but I'm not going down without a good fight.*

The mob encircled them; as they drew closer, Carson could see that most of the black men looked terrified, even as they menacingly waved their machetes and clubs. In their ragged flight suits, black with soot, scorched and bloody, Carson imagined that he and Ourecky were probably were a grisly sight to behold.

There was no way to know what was in store, but whoever these men were, they certainly weren't the neighborhood Welcome Wagon offering cordial greetings. As Carson fended off the attackers to his front, a man lunged from behind and whacked the back of his head with a large stick. Stunned, Carson crumpled to the ground next to Ourecky. Just before he blacked out, he heard a single voice, distinctly authoritative in its tone, shouting what were clearly instructions to the others. And then Carson and his consciousness parted company.

15

LOUP-GAROU

Cap-Haïtien Airport, Haiti
6:45 a.m., Friday, March 13, 1970

The airport seemed deserted when Henson arrived. He was concerned that Taylor might be away on one of his frequent jaunts to Port-au-Prince. He cruised on to the south end of the airport property and parked his motorcycle beside Taylor's place. The ramshackle structure of plywood and corrugated steel was divided into a shop area and adjoining living quarters. On the side of the building, hand-painted blue letters declared "Missionary Air Services."

He was relieved to discover Taylor's Maule M-4 in a grassy area on the far side of the big shed. The sturdy little aircraft was parked with care, so that the two landing gear tires and small tail wheel were positioned precisely on small squares of gravel. A faded canvas tarp was draped over the windshield, apparently to stave off crazing and other damage from prolonged exposure to the tropical sun. Two rangy-looking brown dogs hunkered under the fuselage, menacingly growling at Henson as he approached.

Henson removed his sunglasses and knocked on the plywood door. A fetching young Haitian woman cracked open the door and peered out. Scarcely out of her teens, she wasn't wearing a stitch of clothing, but didn't seem the least bit bashful about her nakedness. "*Kisa, blan? Sa ki fè ou vle a?*" she asked, stifling a yawn. "*What, white man? What do you want?*"

"*Mesye Taylor, souple.*" Henson fished Taylor's card from his wallet and held it out to her.

Smiling coyly, she opened the door slightly wider. Henson couldn't tell if it was so she could examine the card or to offer him a less obstructed view of her lithe body. "*Li dòmi. Retounen nan pita,*" she said. "*He is asleep. Come back later.*"

Henson heard Taylor's voice within the shed. "Lydie! *Ki yes sa?*" he implored. "*Who is it?*"

"*Blan,*" she replied, handing back the business card.

"*Mete rad. Fè kafe,*" ordered Taylor, coming to the door and nudging her to the side. "*Get dressed and make us coffee.*"

"Am I interrupting something?" asked Henson.

"No. That's Lydie. She cooks, cleans up and does laundry. And other things."

"Other things? Obviously," replied Henson. "Can we talk?"

"Sure, man. Come in. Henson, right?"

Henson nodded. "Yeah. I need to rent your airplane." It took a moment for his eyes to adjust to the dark interior. The single room was surprisingly tidy, with a well-swept concrete floor and simple furnishings. Lydie disappeared behind a gauzy curtain that concealed a sleeping area.

"Hire my plane? Sure. The going rate is twenty bucks an hour, plus fuel," replied Taylor. He took a clipboard from a rusty nail hammered in a wall post. "The Methodists are swapping out a medical team in Gonaives, so that'll keep me busy the first part of the week, but I can probably slip you into the schedule on Thursday. How's that sound? Thursday morning?"

"Today. *Now.*"

"Nope. No can do, babe. Today's my maintenance day, and I don't fly on the weekends." Taylor wore only white boxer shorts. A sizeable portion of his abdomen and chest was covered by shiny pink scar tissue, apparently a souvenir of a fiery crash. He gestured for Henson to take a seat at a small table, and then bellowed: "*Kafe*, Lydie! *Kafe! Jodia! Jodia, souple!*"

"I need to fly *today*," stated Henson, swatting a mosquito.

"Today? Why the rush?"

"I can explain, but I'll have to reveal some trade secrets in the process. Can you keep a lid on it? I'm willing to pay extra."

"Oh, so you're worried about tipping off the competition? You know, Henson, I've heard that there were some other mining operations poking around in the mountains. Now this makes sense. Anyway, man, I've flown for people who have much bigger secrets than you, so yeah, I can keep my mouth shut. So where do you need to go? And why the rush?"

Henson tugged a topographic map from his shoulder bag and spread it out on the table. With his index finger, he traced a broad oval on the map, delineating an area in the mountains between Grande-Rivière-du-Nord and Milot, and explained, "I need to do some aerial scouting. Just so you know, I'm required to send my company a daily summary of weather conditions. When they found out we had a big thunderstorm last night, they told me to look for some particular mineral deposits here. The rain would have exposed them. They want me to get in the air as quickly as possible this morning, to take advantage of the low sun angle."

Taylor's eyebrows rose as he smiled. "You claimed you were searching for bauxite. I'm guessing that you're on a quest for something a little more valuable than ore for aluminum."

"That's very astute of you," replied Henson, grinning. The two men studied the map for several minutes, with Taylor recommending search patterns to efficiently assess the area.

Interrupting them, Lydie brought two cups of coffee on a serving tray. She now wore a man's white T-shirt as a dress, and her short-cropped hair was covered by a blue kerchief. She handed Henson a ceramic mug, accompanied by a lascivious gaze that made him extremely uncomfortable. "*Sik, blan?*" she asked, offering a bowl brimming over with raw sugar. He nodded, and she dumped a spoonful of the coarse brown crystals in his cup.

"So let's talk money," said Taylor, holding out his mug so that Lydie could add sugar. He frowned at her blatant flirting; with a subtle nod, he banished her to the cooking area at the opposite corner of the room. "So you insist on going up now? A hundred bucks an hour, plus fuel. And an extra hundred bucks buys my silence. Permanently."

Ever frugal, Henson mentally inventoried his wallet. Technically, since his Apex contract specified that he could keep any surplus funds remaining after the recovery site's initial logistical coordination was complete, a task that had been completed weeks ago, all of the money in his wallet was his; consequently, he would be underwriting this portion of the rescue operation out of his own pocket.

Now, he was operating in an uncomfortable gray area, because he would be shelling out his own money without any guarantee that he would be reimbursed. Then he recalled the frantic voice on the radio last night, remembered that there were two guys out there who needed his help, and reconciled himself to do what had to be done.

"A hundred bucks an hour plus a hundred bucks for you not to talk to the competition?" asked Henson. "That's a bit steep, isn't it?"

Taylor chuckled. "You forgot fuel. And my amnesia doesn't come cheap, but you can rest assured that when I forget something, it's damned sure forgotten."

"Okay. It's a deal. When can we go?"

"Let me finish my coffee. Want a biscuit? Molasses?"

Henson smiled and nodded. "Sounds yummy. What's breakfast going to set me back, you scoundrel? Another hundred bucks?"

"For you? On the house, soul brother." Taylor noisily slurped his coffee, looked over his shoulder, and said, "Lydie, *biskwit e melas pou de, souple.*"

7:47 a.m.

Following a fastidious pre-flight inspection, bolstered by coffee and molasses-sopped biscuits, they were airborne. Henson had convinced Taylor to remove the right door. With his feet dangling in the cool slipstream, he sat on the aircraft's aluminum floor with a seatbelt looped around his waist. Taylor held the Maule at roughly a thousand feet, periodically adjusting their altitude to remain clear of rising ground, flying a grid search with back-and-forth parallel legs.

Using the Citadelle fortress and the Cap-Haïtien Airport as reference points, Henson kept his map oriented to the terrain scrolling slowly below. Scanning the landscape, Henson periodically feigned interest in various rock outcroppings and asked Taylor to circle as he marked the locations on his map and studied them through his binoculars. As they passed over roads and villages, curious Haitians looked up and waved; Henson casually waved back.

Roughly thirty minutes into the flight, Henson's heart beat faster. According to his analysis of the terrain, they were passing over the area where the pilots likely came to Earth. As the Maule buzzed about five hundred feet over a ridgeline, Henson saw an odd burned area in what appeared to be a harvested field. He consulted his map; the oval of scorched ground was almost exactly where the pilot had last reported his position. He didn't want to inadvertently call Taylor's attention to the location, but the pilot clearly noticed his interest.

"Lightning strike," noted Taylor authoritatively, yelling over his right shoulder. "That's why it's burned down there. I'm no mining expert, but if I was searching for ore deposits, that's exactly what I would be looking for. You gotta figure that the metal would draw the lightning."

"Well, if you ever decide to give up flying and swindling missionaries," bellowed Henson, "you've got a future in minerals exploration. Hey, is that a crop growing down there?"

"*Kann.* Sugarcane. Do you want me to wheel it around for a closer look?"

"Yeah," answered Henson, almost casually. "Why not? Give me one circle on it." As Taylor banked the plane into a right-hand pivot turn, Henson studied the scene intently. As much as he wanted to capture every single aspect into his memory, he also did not want the pilot to realize the vital importance of the discovery.

Initially, he focused rapt attention on the black oval. Directly in the middle of the charred blotch, he saw an even darker spot; staring through his binoculars, he saw that it was an object of some sort, possibly an aircraft. And then he spotted two figures—apparently two men—in the vicinity of the object; Henson gasped and his heart pounded furiously when he realized that the pair were moving, apparently examining the crash and removing items from the object.

Striving to remain calm, he shifted his focus to soak in other details. The burned spot was almost perfectly centered in an irregular patch of cleared ground, about four acres in size, on the top of a ridgeline that ran roughly north to south.

At the north end of the field was a hut with a thatched roof, with an adjacent cleared area that was probably a vegetable garden. He had studied the map sufficiently to know that the nearest village was Menard, roughly a mile and a half to the west. A larger town, Dondon, was located just south of Menard, along Highway Three, a heavily used gravel road that meandered southwards to Saint Raphael.

For an instant, he considered asking Taylor if he could land the Maule in the sugarcane field, but realized that it was too risky, for various reasons. He yelled for Taylor to fly another orbit around the site, looking for clues about how to gain access on the ground.

Expecting to see a trail or route from Menard, he realized that the west side of the ridgeline was probably too steep to cut a road. Then he discerned a footpath running from the harvested area towards the northeast and realized that it merged with a narrow road, perhaps suitable for a jeep or small truck, that eventually ran towards the large town of Grande-Rivière-du-Nord.

Completing the second circle, Taylor yelled, "See enough? Want to resume the pattern?"

"Sounds good," answered Henson. He decided that he would let Taylor stick with the search grid for another thirty minutes—fifty bucks worth—before he would ask to return to Cap-Haïtien. After all, if those were the pilots back there, they were obviously alive and kicking, so they should be willing to wait just a little while longer for Henson to make contact.

Relieved, he took a swig of water from his canteen and formulated his plan: After Taylor landed him back at the airport, he would return to Morne Bossa, radio in a report with the good news, and then continue on to the crash site. If all went well, he would be talking to the pilots before noon.

Mission Control Facility, Aerospace Support Project
9:50 a.m., Friday, March 13, 1970

Both exhausted, Tew and Wolcott were sound asleep at their desks when the Recovery Operations Liaison Officer pounded at the door. Marginally conscious, Wolcott looked up to see Gunter Heydrich standing behind the major. A broad grin on Heydrich's face telegraphed that there was obviously good news.

"Henson located the crash site!" declared the major. "He also thinks he spotted your men!"

"What?" exclaimed Tew. "That's excellent news!"

"Thank God," muttered Wolcott, leaning back to stretch in his chair.

"So what's their condition?" asked Tew. "When can we bring them out of there?"

The major handed Tew a transcript of Henson's transmission. "Regrettably, he hasn't yet made direct contact with your men, but he's on his way now. He spotted them from the air."

"From the air?" asked Wolcott. "Didn't you say that our ceiling was ten thousand feet?"

"He apparently hired a local plane to look for the crash site," explained the major. "He found the crash site and saw two personnel—alive and moving—in the immediate vicinity."

"He wangled a plane?" snorted Wolcott incredulously. "Henson *rented* a danged airplane?"

"I don't care how he pulled it off," interjected Tew. "He did it. That's all that matters. Now, Major, how about the rest of this operation? When is your team going in to retrieve our boys?"

"Sir, the RRF has moved to the staging site at Homestead. Ideally, we'll drop our advance team directly onto the crash site tonight and have everyone out of there by the morning."

"Excellent," said Tew.

"There's one loose end," said the major. "Since we've pinpointed the crash site, General Fels asked if we can scrub the U-2 reconnaissance overflight. The aircraft has already launched from Arizona, but it can be recalled. After all, there's no sense . . ."

"Good point," noted Tew. "Let's shelve the U-2 pass. I don't see the need to—"

"*No*," interjected Wolcott, reading through the message transcript.

"*No?*" asked Tew, the major, and Heydrich simultaneously.

"That's right, pard. *No.* As much as I would like to trust your guy, until we actually place our hands on the vehicle, it's merely a coincidence that this burned spot coincides with what our guys called in last night. I like to be as hopeful as the next guy, but I still want firsthand proof."

Wolcott added, "And just because Henson spotted two hombres at this alleged crash site doesn't necessarily mean that they're *our* hombres. So before we jump to conclusions and turn back this platform, let's put it overhead to fetch detailed imagery. And let's make damned sure that imagery lands in the hands of your boys at Homestead before they traipse into Hell's Kitchen tonight."

"As much as I hate to admit it, Virgil," declared Tew, "you're exactly right. Major, the U-2 overflight will proceed as planned. Anything else?"

"No, sir."

"Then keep us apprised. Carry on."

Southwest of Grande-Rivière-du-Nord, Haiti
10:52 a.m.

Although it was clearly marked on his topographic map, the road from Grande-Rivière-du-Nord was marginal at best. It was deeply rutted, and large areas were washed out from last night's heavy rains. Henson picked along with his Motoguzzi, frequently climbing off to walk the bike through the more difficult stretches. He finally came to the point where the meandering road petered out to a footpath. A squat hut stood close by the junction. The ground was too soft to trust the kickstand, so Henson leaned the Italian motorcycle against a sturdy tree.

"Bonswa," grunted an elderly man emerging from the hut. He was carrying a machete, which didn't alarm Henson in the least; outside the cities of Haiti, the chopping tools were so ubiquitous that they might as well be articles of clothing. Wearing no shirt, the man was dressed in ragged bib overalls that were more tatters than whole cloth. Seeing Henson, his eyes flew open wide. His machete clattered to the ground, scarcely missing his bare feet.

"Bonswa," replied Henson politely, puzzled at the man's apprehensive reaction.

They talked. In a short exchange, the man timidly explained that he was the caretaker for a nearby grove of banana trees. Henson pointed up the hill and asked who owned the land on the ridgeline. The man said that it was the property of two reclusive brothers who grew sugarcane that they sold to a local rum distiller.

He said that one of the brothers came down from the hill perhaps once a month to venture into town. According to him, the other brother never came down, except to haul their crops at harvest time, when the siblings toted the ripe cane down the trail to make a pile at the roadhead. Otherwise, he remained on the hill.

"*Poukisa?*" asked Henson, pulling off his sunglasses to wipe sweat from his brow. "*Why?*"

"*Lèp,*" muttered the man, gesturing at his face and grimacing. "*Leprosy.*"

"*Lèp?*"

"*Lapenn anpil,*" said the man, nodding. "*Very sad.*"

Now it was Henson's turn to be apprehensive. Virtually all he knew of leprosy was what he had recently read in *Papillon,* where Charrière described receiving assistance at a leper colony on Pigeon Island. Henson also knew that while leprosy was largely eradicated throughout the world, it still persisted in isolated pockets.

He wasn't quite sure how to handle this awkward situation but felt that if he kept his distance and avoided physical contact, he should be safe. He knelt down and snugged his bootlaces for the walk ahead. Reaching into his wallet, he offered the banana farmer a few Haitian *gourde* to watch the motorcycle until he returned.

Fanning his wizened hands in front of him, the farmer nervously refused Henson's offer, claiming that it would be an honor just to assist him. Henson noticed that the elderly man's eyes never met his, but remained intently focused on the satchel hanging at his side.

Chink . . . chink . . . chink . . . chink. As Henson ascended the steep footpath, he heard hammering sounds, and he wondered if the brothers might be building something. The trail was a series of switchbacks gradually climbing the east flank of the ridgeline. It took him almost an hour to navigate his way up, all the while searching his memory to dredge up any facts about leprosy.

At several points, the narrow track was blocked by thick vines bearing harsh thorns, so he dug his short machete from his canvas shoulder bag. Fashioned after a Filipino bolo knife, it was ideally suited for lopping vines and hacking through dense undergrowth.

Finally, he crested the ridge, arriving at the north end of the brothers' land. Their humble dwelling was a one-room shack with wattle walls and a roof of thatch. Beside the austere shack was a meticulously tended vegetable garden enclosed by cacti woven into a dense lattice to exclude animals. Outside the barrier, clucking white chickens pecked at the bare brown earth, while wandering goats foraged in nearby scrub. In a ring of stones, this morning's cooking fire still smoldered.

Chink . . . chink . . . chink . . . chink. The brothers weren't present, so Henson went towards the hammering. He followed a path through still intact sugarcane until he reached the burned area that he had seen from the air. In the middle of the burned area, he saw two black men, obviously the brothers, standing beside an odd cone-shaped object lying on its side. Expecting to find an F-111 ejection pod or something similar, he recalled his "safing" lessons at Aux One-Oh. Startled, he realized that he was looking at a partially dismantled Gemini spacecraft.

Intent with their labors, the two men were oblivious to Henson as he paused to stare in sheer wonderment. Using a large rock as a hammer and a rusty leaf spring as a chisel, they were systematically loosening and prying off the heat-resistant shingles that covered the spacecraft's exterior.

They had obviously been chipping away for quite a while, since virtually all that remained was the heat shield, titanium frame assembly and pressure vessel. The corrugated metal shingles were stacked neatly in three nearby piles. It was an odd sight; the ravaged spacecraft looked like some otherworldly insect undergoing a metamorphosis, shedding off metal scales as it molted.

Henson continued his stealthy approach. Even as he drew close enough to reach out and touch the two men, they didn't sense his presence. Thin and muscular, they looked to be in their late twenties. Shirtless and barefoot, both wore old khaki work trousers, probably gleaned from a bundle of donated clothes from the States.

Leaning into their work, they chattered away at each other while they diligently hammered and tugged. They didn't know that he was there until one dislodged a shingle and turned away from the spacecraft to carry it to a pile. Seeing Henson, the man dropped the black metal tile and screamed at the top of his lungs. Staring at Henson's machete, the man was obviously petrified with fear. Gasping, the other brother threw away his makeshift hammer and chisel. Both men approached Henson and fell to their knees in ashes, bending their heads forward as if to offer themselves for sacrifice. Wailing, they shook with abject fear as they awaited their fate.

Tapping his machete on his thigh, Henson wasn't sure what to make of this surrealistic scene. As the men wept and trembled, he removed his Panama hat and mirrored sunglasses, and then wiped sweat from his face with a red bandana. Squatting on his haunches, he examined their faces.

One brother's face appeared entirely normal, if not handsome, while the other brother had an enormous growth protruding from his forehead above his left eye and a similar tumor jutting from back of his head. The repulsive growth on his face—about the size and shape of a mango—was so large that it prevented him from opening his eye completely. Other than the two abnormal growths, the man looked to be entirely normal; Henson saw that his fingers were intact and that there were no sores or boils on the rest of his body.

Finally, the normal-looking brother spoke. Begging for mercy, he offered his life in exchange for his brother's safety. Hearing that, the leper brother chimed in, likewise offering *his* life if Henson would see fit to spare his sibling. Like the elderly man at the base of the hill, both men would not look him in the face, but instead stared at the canvas bag hanging from his shoulder. Heavy with the weight of his canteen, binoculars, and some other oddments, the bag's strap dug painfully into his neck, so he shifted it slightly. He couldn't help but notice that both men cringed when he did so. As the two men continued to plead, Henson lost his patience with them.

"*Ase! Pa fè bri!*" bellowed Henson, standing erect to examine the spacecraft. "*Enough! Be quiet!*" He saw that one of the two hatches was open. Half-expecting to glimpse the mangled remains of the two pilots, he peered into the vehicle and found it vacant. He quizzed the brothers, asking if any men had arrived with the vehicle, and what happened to them.

Clearly relieved that Henson wasn't going to summarily execute them, the two men enthusiastically recounted last night's events. They said that they had been awakened by the crash. They witnessed two white men climbing out of the vehicle as the fire was just catching, and were initially very angry with them for setting their sugarcane ablaze. Furious, they chased the *blan* strangers into the woods, but after they realized that the men might be injured, they gallantly followed them in hopes of rendering assistance, but they lost them shortly after the rains began.

Henson asked the brothers to show him where the two men went, and they brought him to the edge of the wood line, where he found a damaged flight helmet. He ventured into the woods for about a hundred yards, following a fairly discernible trail of broken vegetation, boot prints, scuff marks and diluted blood spatters. Realizing that the two pilots were probably long gone from this locale and likely holed up somewhere during daylight hours, he decided that the best course of action was to report these new wrinkles as swiftly as possible.

As they walked back uphill, with the brothers becoming progressively calmer and more communicative, he learned that they were twins. The normal twin was named Jean, and the leper twin was Henri. Their mother had hemorrhaged to death after their birth. Their father was killed when they were twelve, and Henri's tumors appeared shortly afterwards. Seeking medical attention, Jean had brought Henri to Grande-Rivière-du-Nord; not wanting a leper in their midst, the townspeople had exiled them to their hilltop farm. Jean was convinced that he was immune to leprosy and took it upon himself to care for his stricken twin. It was only a few years ago that Jean was allowed to return to the town, and even then most of the residents shunned him.

Henson asked the brothers why they were stripping off the shingles from the vehicle. Chagrined, staring at the ground, Jean tearfully apologized for damaging something that wasn't their property. After Henson succeeded in calming him, Jean explained that after they had returned to their plot, the rain had quenched the fire and they realized that they hadn't lost their entire crop of sugarcane. When they examined the object and found that it was covered with some sort of metal shingles, they assumed that it must have be a gift from the Almighty.

"*De Bondye? Poukisa?*" asked Henson incredulously. "*From God? Why?*"

"*Yon do kay!*" blurted Jean, pointing towards their simple hut. "*A roof!*" He apologized again for stripping the shingles from the vehicle, but stated that their sole intent was to build a new roof for their shanty. He explained that with a good metal roof over their heads, he might be able to entice a woman into marriage. And since having a sturdy roof was perhaps his only chance at love and marriage, surely then the shingles had been delivered by God's divine hand.

Covering a grin and stifling a laugh, Henson considered the absurdity of the situation. Here were these two secluded and destitute brothers, ripping apart a sophisticated spacecraft that probably cost millions of dollars, just so that one of them could woo a woman. And that *God* sent it here? Only an *idiot* would believe such a ridiculous thing. But looking at the three heaps of black shingles, Henson mused that the truly faithful might believe it also.

Struggling to maintain a straight face, Henson sternly declared that the object was his property. He scolded the twins, but also told them that he would be lenient if they assisted him. He told them that that he had to leave but should return within a few hours.

He offered to compensate them if they concealed the object, guard it until he could return with other men, and not ever speak of it with anyone. He instructed them to immediately bathe and wash their hands thoroughly, and to dare not touch the object again.

There was a practical reason for his last admonition. Besides all the unexploded pyrotechnic charges and other hazardous materials aboard the spacecraft, Henson knew that some of its metal components, made from special alloys and beryllium, could be toxic to handle, and he hoped that the brothers hadn't already made themselves sick with their impromptu home renovation project.

The twins swore their vigilance. Henson donned his hat and turned to leave. As he walked away, Henri called out after him, asking him how much he would pay them to do these things. Henson thought about it for a moment, and replied that if they did everything precisely as he asked, he would give them at least enough *gourde* to purchase a new roof of corrugated tin.

Peristyle de Beasujour, Dondon, Haiti
1:35 p.m., Friday, March 13, 1970

Blinking his eyes, Carson slowly came awake. Groggy, he surveyed his bizarre surroundings. He guessed that they were being held in some form of occult temple. The orange-painted walls were decorated with pictures of grinning skulls and demon-like figures. The room was roughly thirty feet square; a thick wooden pole, decorated with bizarre carvings, was the centerpiece of the space. It stank with an indescribably foul stench, as if years of sweat, smoke, blood, and grime had permeated the stucco walls and were now slowly leaching out into the air.

His right hand throbbed with indescribable pain, and his head ached like it was being crushed between two locomotives. Running his left hand along the back of his skull, he discovered that his hair was matted with blood, probably from where someone had bashed him.

Beside him, Ourecky was tucked in a fetal position, hands jammed between his thighs, snoring and gurgling quietly. Both men were clothed in only the long underwear that they wore under their flight suits. Carson's white cotton garment was sweat-soaked and smelled heavily of smoke and noxious chemicals. His expensive new Hamilton pilot's chronograph was missing.

He sat up to examine the gash on Ourecky's forehead. The bleeding had stopped, but he was concerned that his friend had suffered a severe head injury, possibly a concussion. Listless in the oppressive heat, Ourecky tried to speak, but his speech was slurred and delirious.

He obviously had no idea of where he was, but then again, neither did Carson. Last night's events seemed like a distant dream—a nightmare—and today was shaping up to be a continuation of the same. Ourecky smiled weakly and then went limp as he lapsed back into blissful unconsciousness. Carson started to stand, but two large black men made it clear that he was not free to leave the premises.

All the while, Haitians came and went, gawking at the two white men like they were on display at a zoo. For whatever reason, something that Carson could not quite fathom, most of the Haitians—even the pair of hulking machete-armed men guarding them—appeared frightened to be in their presence.

Few visitors lingered long, and most seemed reluctant to make eye contact. One Haitian had apparently been pressed into service as a tour guide. As a newcomer entered the room, the guide pulled them off to the side and quietly spoke to them, perhaps offering an explanation of how they came to be here. Carson noticed that the guide frequently gestured with his hands, often using them to mimic a bird's flight.

Bearing two wooden bowls and a burlap sack, two ancient black women squatted on the dirt floor next to Carson. Jabbering in a strange language, they prodded his face and body with their fingers, as if they were subjecting him to some crude sort of doctor's examination.

Mumbling some incantation, one of the women dipped her hand in a bowl and swished it around. She spoke to the guards, who snatched Carson by the wrists, and then she slathered his burned hand with a rancid-smelling salve. The glowering guards held him firmly as he struggled and flinched, but the crude ointment slowly took effect. As disgusting as the goo smelled, Carson had to admit that it was soothing, and gestured his sincere thanks.

Toothlessly grinning, the second woman offered him a mango from the bag. He furtively bit into the ripe fruit and found that its juices were like the sweetest ambrosia. Ravenous, he consumed the mango in short order and gestured for another.

Chuckling, the women kept pace with his demands, lavishing him with mangos and bananas, each one tasting more delectable than the last. He pointed at his lips and made a swallowing noise, and they gave him a cup of cool water, and then another. He tried to wake Ourecky so that he could share in the offerings, but was unable to rouse his snoring companion.

After gorging himself and slaking his thirst, he realized that he had to answer the call of nature. Carson pointed at his crotch and then towards the entrance, attempting to convey that he needed to go outside to relieve himself. As the crinkled old crones cackled at his discomfort, the guards debated the situation before jerking Carson to his feet and escorting him to the door.

Staggering outside into the glaring sunlight, he saw that the building was constructed of concrete cinderblock with a roof of corrugated metal. The exterior was garishly painted, mostly in pastel shades of lime green and pink. In front of the building, several ornately colored flags snapped in the wind.

A throng of children gathered to stare at him. He saw a gravel road nearby, possibly the one that he had seen earlier today. There were several wattle huts in the vicinity; perhaps this was a worship center that served a neighborhood.

Walking back inside, Carson tried to make sense of it all. Were these civilians holding them as they waited for government or military authorities to arrive? Was there some form of reward involved? As he listened to the women's nonsensical rant, he was confident that it was only a matter of time before rescuers were sent for them. In the meantime, he accepted that matters could be much worse; they weren't being beaten or tortured, they had a roof over their heads, and were provided with food and water. Still, he was distraught over Ourecky's condition, and desperately hoped that the rescuers would come before it was too late.

Morne Bossa, Haiti
2:25 p.m., Friday, March 13, 1970

Anxious to make radio contact with Ohio, Henson mentally composed his message in his head as he raced north up Highway Three. Turning off the main road onto the dirt track that ran to his shed, he saw a *Fad'H* jeep and truck waiting. As he braked to a stop and killed the motorcycle's engine, he saw Colonel Roberto and a squad of *Fad'H* soldiers. All were dressed in cotton Army fatigues; unlike the others, Roberto's uniform was heavily starched and accented by a brown leather Sam Brown belt that bore a Colt .45 pistol in a polished leather holster.

Roberto had obviously just finished his lunch and handed his lunch pail to one of the soldiers. Carrying a Thermos bottle, he greeted Henson. "Matthew, you're a difficult man to find. Have you been busy this morning, by chance?"

"Just the usual work. Poking around, raking up ore samples, the normal routine," replied Henson casually, closing the motorcycle's fuel valve as he swung out the kickstand. "And you?"

"It's been an *unusual* morning," said Roberto, taking off his cap. "Let's go inside and chat."

"It's hot in there in the afternoon. Why don't we just stay out here in the shade?"

"No. Let's go inside, Matthew."

As Henson walked up, he realized that the brass padlock on the door was undone, but distinctly remembered closing it this morning when he left to go into the mountains.

Entering the shed, Roberto gestured towards a chair and said, "Have a seat." A *Fad'H* sergeant stood beside the door. Roberto unscrewed the cup from the Thermos and filled it with steaming hot coffee. Henson could not conceive of drinking coffee in this heat. In moments, his clothes started to dampen with perspiration. In contrast, Roberto might as well have been seated on an iceberg, because he remained cool, dry, and unflappable as ever.

"Interesting radio you have," noted Roberto, using a pocketknife to snip the end from a Dominican cigar. "I hope you're not offended, but I took the liberty of examining it while I was waiting for you. I'm somewhat of a radio aficionado, much like those ham radio operators you have in the States. That's a top-of-the-line Heathkit shortwave, isn't it? Very nice. Do you use that for business or are you a ham radio operator yourself?

"Business. I'm sure you're aware that the telephone cables aren't very reliable down here, but my company—Apex—wants daily reports, so we use shortwave radio to communicate."

"Interesting," said Roberto, lighting the cigar. Puffing at it, he added, "And that UHF radio over there? What's that for? It seems like it would only be good over relatively short ranges."

Henson nodded. "We have that so that when and if more specialized teams come here to do more precise survey work, we can talk to them on the land and in the air."

"Specialized teams," mumbled Roberto. "And that gadget outside in the tent? What's that? It looks like an aerial beacon of some sort."

"Precisely. We plan to eventually use airplanes to do most of the spotting work. That beacon will allow us to determine an exact location when they report a finding."

"Fascinating," said Roberto. "I am so impressed with the scientific nature of your business."

Anxious to report the latest information, Henson stole a quick glance at the shortwave radio.

"Am I keeping you from something, Matthew?" asked Roberto.

"Nothing too important. Nothing pressing, anyway."

"Good. I was going to ask your assistance, since you have the latitude to travel around so freely in the outlying areas," said Roberto. "Here's my dilemma. During their routine patrols this morning, my soldiers took several reports in villages in the mountains south of here. It seems that late last night, immediately before the big thunderstorm, there was some sort of strange apparition in the sky."

"An apparition?" Henson smirked.

"*Wi*," replied Roberto. He puffed on his cigar and added, "The villagers insisted that they had witnessed a giant bird carrying another animal clutched in its talons. They said that they saw it clearly because it was illuminated by the lightning flashes."

"But, Roberto, you're an educated man. Surely you don't believe their claims."

Roberto clicked his tongue and said, "Matthew, it's easy for you to just wave aside these sorts of things, but this isn't an incident I can afford to dismiss so readily." He turned towards the sergeant and gruffly ordered, "*Pote m' kat la.*"

The grim-faced sergeant nodded, left the shed and returned with a map.

"If there had been merely one sighting of this giant bird, perhaps I could disregard it," explained Roberto, tapping on the map with the earpiece of his gold-framed Foster Grant sunglasses. "But there were *many* sightings. The details were very consistent. Granted, it was after midnight and most of these people were probably returning from voodoo rituals and were fairly well soused, but it's difficult to discount the uniformity of their descriptions."

"You're right," commented Henson, nodding. "Interesting." He felt sweat pouring down his spine, like someone had stuck a fire hose down the back of his shirt. He couldn't understand how Roberto could remain so incessantly cool.

Roberto laughed. "Oh, yes, *very* interesting. Anyway, most of these people swore that they had witnessed a *loup-garou.*"

"A shape-shifter?" asked Henson. "A werewolf?"

"Werewolf? Not every *loup-garou* assumes the visage of a wolf. They can be birds, snakes, dogs, anything. A true *loup-garou* can become any creature that he wants to be. They can change from human form to animal form and back again, at will. So the legend goes."

"But all this is just superstition isn't it?"

Roberto frowned. "Personally, I don't think that there are any *loups-garous* around here, at least not any real *loups-garous*. In many places, when a child comes up missing, the ignorant and superstitious are quick to attribute the disappearance to a *loup-garou*. I suppose that blaming a mythical monster is easier than accepting the knowledge that a child molester lives in their midst, especially someone so evil that he's willing to murder a child to cover his tracks."

Roberto finished his coffee and threaded the plastic cup onto the Thermos. "So, Matthew, have any of *your* contacts mentioned seeing a giant bird carrying a wounded animal? Perhaps you could lend me some insight into this apparition."

Henson shook his head. "Sorry, but no, Roberto. I wish I could be of more help."

"I see. One more question." Roberto pointed at a town—Dondon—on the map. "Here's something odd. I had no such reports from Dondon, even though some of those people should have seen it as well. It struck me as so odd that I sent some of my soldiers there to canvass the inhabitants, but they all insisted that they saw nothing unusual last night. Strange, isn't it?"

Examining the map, Henson nodded.

"Have you been to Dondon?" asked Roberto. "Is there any chance of you passing through there in the near future? Perhaps you could let me know if you hear anything."

"I've not been to Dondon," said Henson truthfully. "I usually stay east of the river and Grande-Rivière-du-Nord. Besides, from what I hear of Dondon, it sounds like a place to avoid. There's apparently a lot of bad voodoo and criminal activity there: trucks hijacked on the road and people found murdered. So I stay out of there. I don't need any unnecessary confrontation."

"Good idea." Smiling, Roberto stood up. "Please accept my apologies for interrupting your day. It's obvious that you have much to do, so I'll leave you to your chores."

Henson stood up and shook Roberto's hand. "I hope you find your answers," he said. "I'll keep my ears to the ground while I'm out in the countryside. Sounds like quite a mystery."

"Quite. Oh, Matthew, one other thing . . ." Roberto snapped his fingers. The gruff sergeant stepped forward and handed him a thick rubberized packet, something that Henson recognized immediately. Roberto opened it, and yanked out a fistful of Haitian currency, as well as a rubber-banded bundle of large denomination US bills.

"My men found this buried out back," explained Roberto, thumbing through the bills. "I personally know the man who owns this property. Most of his residual earnings end up going to rum, whores, and gambling, so I suspect that this is yours. Am I correct?"

"Yes," replied Henson quietly. His mind spun as he tried to invent a plausible explanation, but there was none. He swallowed; his throat felt like it was on the verge of swelling shut.

Roberto tucked the money back into the pouch and handed it to Henson. "There's quite a sum in there, Matthew. I suppose mining exploration is quite an expensive endeavor."

Henson nodded solemnly.

Roberto slipped on his sunglasses and gestured at the door. "Care to walk me out?"

The two men walked outside. "I suppose we should be on our way," said Roberto, smoothing the tips of his perfect black moustache. "I consider you a dear friend, Matthew, but let me reiterate something. So long as you insist that you're doing nothing but innocent work and I have no reason to suspect otherwise, my soldiers will not interfere with you. But if you're up to anything else, it would behoove you to tell me immediately. Am I making myself *clear?*"

"Yes."

Roberto nodded, then grinned and asked, "So, are you free tomorrow morning?

"It's Saturday. Roberto, you know that I *never* work on Saturday."

"I know," said Roberto, dropping his half-smoked cigar before crushing it under the heel of a custom-made combat boot. "Julienne and I would love to have you for breakfast. We just received a package of peppered bacon from Santo Domingo. Is seven too early?"

"Seven? Oh, no. Seven is fine. I'll see you then."

It took Henson several minutes to calm down. As his heart rate gradually quieted, he switched on the shortwave and dialed its transmitter to the emergency frequency. Keeping a watchful eye on the door, he tapped out a message in Morse code. The gist of the dispatch was to let them know that he had physically confirmed the location of the vehicle, that it was damaged but secure—at least for the near future—but that he had still not made contact with the two pilots.

Although it took only a few minutes, it seemed like he waited an eternity for the reply. The distant operator acknowledged the update and told him the rescue team was still arriving tonight to take over the mission. The rest of the brief message gave the specifics of the team's infiltration plan, particularly focusing—in detail—on the role that Henson would play. Henson acknowledged the message and cleared out the contact before shutting down and zeroing the radio. He memorized the pertinent information and then carefully burned his notes.

He wished that there was time for at least a brief nap or a quick bite, but there was not a second to waste. He grabbed a few items from an old pine armoire and then went back outside to his motorcycle. Filling the Motoguzzi's tank from an old jerry can, he reflected that it was probably going to be another very long night.

11:30 p.m.

Nestor Glades knelt on the port side of the MC-130 Combat Talon's broad tailgate. He glanced down to verify the altitude—ten thousand feet—from the altimeter mounted on the top of his reserve parachute. Leaning slightly outside of the aircraft, he looked forward and observed the flickering lights of Cap-Haïtien in the distance.

Under cover of darkness, a three-man advance team of the Rapid Response Force was making a HALO—High Altitude, Low Opening—free fall infiltration jump into Haiti. Their primary mission was to locate the two missing crew members. Glades was confident that would merely be a matter of tracking them until they were found holed up in an evasion hide site.

The remainder of the RRF—fifteen men—was staging at Homestead Air Force Base, near Miami. Their infiltration plan was still being finalized, but their main task was to pinpoint and recover the spacecraft. Of course, since the advance team was literally jumping right on top of it, locating it should be a simple endeavor. On the other hand, quietly sneaking it out of there might be a trick. Ideally, the entire mission would be conducted as a clandestine affair, with the Haitians—and the remainder of the world—totally unaware of the American presence on their soil. A totally clandestine operation was certainly preferable, but if necessary, the RRF could execute a much more overt—and appropriately violent—rescue operation if the circumstances dictated.

After being involved in the RRF's training over the past few months, Glades had been pressed into service as a jumpmaster for the free fall jump. Although the team made ten or more practice jumps every month, none of their members had successfully graduated from the Army's strenuous HALO jumpmaster school, so they relied on Glades to perform the task.

As jumps went, this should be a fairly easy operation, not unlike a stringently controlled jump at the Fort Bragg schoolhouse. Since Haiti lacked a sophisticated air defense radar network, there was little risk incurred in making the high altitude incursion into their airspace. But there was no room for complacency, since this was a genuine operational mission—the RRF's first—and they were jumping into a sovereign foreign nation without State Department sanction or any other legitimate authority.

Glades would be delighted if he could just perform his jumpmaster duties, close the tailgate, and return to Homestead for a cold beer, but he didn't have that luxury. Just three hours earlier, General Fels, an officer he greatly respected, personally asked Glades to accompany the team as a technical advisor. Although his duties were vague at best, his singular task was to ensure that the team didn't do anything stupid, at least to the extent that he was asked to prevent any stupid actions that might result in the team's compromise and/or annihilation. Glades hoped for a short and uneventful mission, where he could remain quietly in the background while the team took the lead and did their job.

The aircraft's loadmaster held up two fingers, almost like a peace sign, indicating that they were two minutes away from the drop zone. Glades nodded, adjusted his black leather "bunny hat" helmet, and signaled for the three jumpers to stand up and come to the rear of the plane. He led them through the all-important final gear check, making sure that their pack closing pins were properly positioned and that the dim red lights on the backs of their parachute packs and altimeters were working properly. He pivoted to let one of the men check his pins and pack light.

After positioning the three men at the sheer edge of the tailgate, he knelt down to spot the jump. Using a portion of the tailgate's hinge as a bombsight, he verified the red and green marking lights on the drop zone. Buffeted by the slipstream, he leaned slightly into the aircraft and signaled steering corrections to the loadmaster, who relayed them to the pilots.

Seconds later, satisfied that they were on the right heading, he turned his head to lock eyes with the three men standing abreast on the tailgate. He slapped his gloved hand downwards on the cold aluminum and brought it back up with an upraised thumb. Making sure that they returned the gesture, he briskly pointed into the dark void. In an instant, without hesitation, the trio exited as one, falling away into the darkness. Glades smiled at the loadmaster and then dove after them. Arching his back, he fell through the turbulent slipstream and allowed himself a brief moment to enjoy the sensation of free fall.

When the moment passed, he counted the pack lights and made sure that the three men were correctly tracking towards the green and red marking lights in the pitch darkness below. Less than a minute later, as they opened their canopies at the designated altitude, he hesitated an extra couple of seconds to make sure that he would be the lowest man under canopy, so that he could lead the less experienced stack of jumpers into the drop zone.

Just four minutes after he had left the plane, Glades glided his MC-3 Para-Commander canopy into the small landing area, gently tugging at his steering toggles to make minor corrections. Thankfully, all four men touched down in the burned oval, almost on top of the marking lights, instead of landing in the less forgiving undamaged sugarcane. After he rolled up his canopy and recovered the rest of his drop gear, Glades breathed a sigh of relief; the little team had made it safely to the ground, so maybe his job was done.

16

ZANJ NWA

Dubuission Homestead, Haiti
1:23 a.m., Saturday, March 14, 1970

Major Lewis, who had been a captain when Henson initially went through the assessment training at Aux One-Oh, was now in command of the Rapid Response Flight. After the jump, once Glades accounted for the men, Lewis took control of the mission. The humid air was still, and a brilliant mantle of stars decorated the sky. Except for the periodic buzz of mosquitoes and an occasional rustling of leaves, the only sound was a dog plaintively yelping in the distance.

In the darkness, Lewis convened the group to review their plans for the coming hours. They lay flat on their bellies so that their heads and upper bodies were concealed under a poncho; the ground cloth's grommeted edges were drawn down tightly to prevent any leakage of light from the red-filtered flashlight that Lewis used to illuminate the map and aerial photos. In minutes, the poncho was clammy with the condensation from their exhaled breaths.

Henson looked at the sweaty, camouflage-streaked faces dimly illuminated in the red glow. It was almost like an Aux One-Oh reunion; the four new arrivals—Lewis, Glades, Ulf Finn, and Steve Baker—had been present on the day when he had failed the ejection pod search. While he had since reconciled himself that he had already been earmarked to leave the Air Force and transition on to Apex, he wasn't convinced that Lewis hadn't gone out of his way to make his departure a miserable experience. In fact, he suspected that Lewis had resented his presence at Aux One-Oh from the outset and had endeavored to trick him into inextricably losing his cool.

"So you've looked at the vehicle?" whispered Lewis.

"Yes, sir," answered Finn, the team's safing and recovery expert. "But I don't have a good plan to extract it yet. We'll probably end up coming in here with a helicopter. Those damned locals sure didn't help matters by busting the thing up."

Not exactly an accurate statement, thought Henson. With the vehicle broken up as it was, they now had the option of ferrying the pieces out overland, at least to a location where it could be transferred to a truck and brought to the port. Using a helicopter was a risky proposition, one that would surely attract unwelcome attention.

"But everything else is intact?" asked Lewis. "Especially the high value items?"

Trying to quietly swat an annoying mosquito, Finn nodded assent. "I've checked the interior, sir. Everything is where it's supposed to be. I've disabled all the pyro and removed the film and tapes. With your permission, sir, I'll set charges as a contingency, so we can blow it in place, if need be, on your order."

"Do it," said Lewis. "Anything else?"

"Yes, sir. The crew must have departed in a big hurry, probably to escape the fire. Both survival kits are still stowed in the vehicle, so they just have the clothes on their backs. No kits, no over/under gun, no radios. I strongly suspect that they'll eventually circle back here. I wouldn't be surprised if we see them before the sun peeks up."

"Noted," replied Lewis softly. "Henson, you said you discovered tracks on the far side of the ridge, going downhill, right?"

"Yes, sir. And the Dubuission brothers said they followed the crew down the ridge, roughly to the west, for about thirty minutes," answered Henson.

"Would you care to explain why you didn't just go ahead and follow their trail?" asked Lewis.

"Because I felt the priority was to confirm the location of the crash site as quickly as possible, sir," replied Henson. "Especially to pass the word that the crew wasn't here at the site." He didn't understand why he was being compelled to justify his actions; surely there would be time for an inquisition later, once the crew was recovered. Besides, his exact instructions were to conduct a hasty search, and once he physically located and reported the crash site, he was explicitly directed to stop the search and make preparations to receive the advance team.

"Also, just so I have clarity on the circumstances," said Lewis quietly. "You were running the CRZ and providing terminal guidance when the crew made their final approach. Weren't they following your guidance when they overshot the landing site and crashed here?"

"You're right. I was running the CRZ, sir. By *myself*. But there were a lot of extenuating circumstances that led to them landing here. It couldn't be helped."

"I'm sure," commented Lewis. "Okay, men, here's my plan. Everyone stays put until daylight. Hopefully, Finn is right, and the crew will circle back after dawn breaks. If they don't, Finn and I will move down the west side of the ridge to chase their tracks, hopefully to wherever they're currently holed up. Baker, I think it's most likely that they will return, so I want you to remain here in case they need medical attention. Sergeant Glades, I also want you to stay here, in case any new information comes in from Homestead."

"How about me, Major?" asked Henson.

"You'll hang out here as well, Henson. I have everything under control. I don't anticipate needing your assistance, but I don't want you drifting around anywhere, either."

"Sir, the commander of the *Fad'H* northern garrison is expecting me for breakfast at seven. I've established good rapport with him, and I think it's best that I be there."

"Well, far be it from *me* to infringe on your social calendar, Henson." Lewis sniffed. "The truth is, you've already caused enough damage, so consider yourself free to flit off on to your breakfast date. Come back if you have an update. Otherwise, stay out of our way."

"Will do," replied Henson, trying to conceal his hostility.

"If there are no questions," concluded Lewis, stifling a yawn, "one hundred percent stand-to at zero five hundred. Today will be a busy day, so take turns getting some rest. That's all, gentlemen. Henson, remain."

The men other quietly slid out from under the damp poncho and padded off into the darkness, leaving Henson and Lewis behind.

Lewis switched off the flashlight. They were so close that Henson could feel the Major's moist breath in his face. "I would strongly caution you to watch your insubordinate tone," warned Lewis. "If you're still harboring some sort of latent grudge from Aux One-Oh, Henson, then you need to get past it. Immediately."

Henson took a deep breath, exhaled, and softly replied, "Major Lewis, I'll show you the respect you're due, but you also need to remember that I'm not subject to your command. I'm a *civilian* contractor. At this point, we have the same mission: to deliver these two men safely home. To that end, I'll do whatever you ask."

Household of Colonel Mendoza, Cap-Haïtien, Haiti
6:55 a.m., Saturday, March 14, 1970

Henson tapped at the ornately carved mahogany front door of Roberto's home. The maid, a dour-faced and portly Haitian woman, answered and then led him to the parlor. Roberto was there, attired in crisply starched fatigues, buckling his Sam Browne belt.

"I must apologize," said Roberto, gesturing at a chair as he leaned over to lace his boots. "'But we cannot sit for breakfast this morning. My cook is making me a sandwich to take with me; perhaps he could fix one for you as well. Fried egg and bacon? Matthew, you must sample some of this bacon while it's still fresh. It's absolutely delicious."

"*Mesi.* That sounds very good," replied Henson, sitting down on a leather-upholstered ottoman. "But what's wrong, Roberto? Is there trouble somewhere?"

Roberto shook his head. "There's an awkward situation brewing in Dondon. I'm placing my soldiers on standby at the garrison, in case we have to respond quickly."

"What's happening?"

"Matthew, I should not tell you this, but my executive officer has a nephew who attends voodoo rituals at the Peristyle de Beausujour, in the northern outskirts of Dondon, right alongside Highway Three. He said the peristyle's *houngan* is holding two *blans* there. That's apparently why we didn't receive reports of the *loups-garous* from Dondon, because they've captured them, or *blans* they suspect of being *loups-garous*, and they don't want the word to slip out."

Henson's head spun with the dire implications. "That's terrible for the *blans,*" he blurted, shaking his head. "What could possibly be worse?"

"What could be worse? *Plenty*," asserted Roberto. "Their ordeal has only begun. *Loups-garous* are greatly feared, because they can wreak so much havoc. It raises the *houngan's* stature immensely if he is able to subdue just one, and now this *houngan* has *two* of them in custody. If he is able to permanently eliminate them as a threat to the community, then his power will be indisputable. He will be the most dominant *houngan* in Dondon, without question."

"Eliminate their threat? He plans to *kill* them?"

"He *could*, but I seriously doubt that he *would*. So long as they are held in human form, the *loups-garous* are a valuable commodity. My exec's nephew overheard that the *houngan* intends to present them as a gift to a very powerful *mambo* who lives in Mirebalais."

"Why would he do that?" asked Henson.

"Mostly so he could gain political favor and wider influence," explained Roberto. "You see, if voodoo were an earthquake, Mirebalais would be the epicenter. The Saut-d'Eau waterfalls are near there; they are a sacred site for voodoo practitioners and Catholics alike."

Roberto continued. "This particular *mambo* rules the roost in Mirebalais, and she's also a personal spiritual advisor to Papa Doc Duvalier. She's a notorious *loup-garou* herself; when Papa Doc summons her, she transforms herself into a bird and flies to Port-au-Prince."

As Roberto turned to comb his hair in a mirror, Henson stole a glance at his Timex watch. "When do you think this will happen? And how will it happen?"

"When? Within the next few days, but possibly as early as tonight," avowed Roberto, looking over his shoulder. "When he's ready, the *houngan* will hold a special ceremony to summon the *loas*—the spirits—to assist him in bringing the *loups-garous* under his ultimate control. Afterwards, the townspeople will be invited to witness the *loups-garous* being placed in a hut, which will be guarded by members of the peristyle."

Roberto pivoted towards Henson. "In the early morning hours, when everyone is sound asleep, there'll be some sleight of hand, and the two *loups-garous* will be spirited away to Mirebalais. In the morning, a chicken and a goat will be found in their place."

"The townspeople will rejoice, since they have been delivered from the tyranny of the *loups-garous*. The *houngan's* stature will be greatly enhanced, because he has forced the *loups-garous* to permanently assume the form of harmless animals. So long as they are manifested as a chicken and a goat, the townspeople can keep a watchful eye on them. Of course, if they've really been a nuisance, it's just as likely the animals will be slaughtered and eaten."

"Can't you do something about this?" demanded Henson.

Roberto shrugged and replied, "I know better than to become mired in the affairs of *houngans* and *mambos*, particularly in this situation. If that *mambo* is denied her *loups-garous*, and Papa Doc hears of it, then it would only be a matter of time before the *Tonton Macoutes* tap on my door. And Matthew, while I believe that most voodoo is just harmless hocus pocus fakery, I genuinely fear the *Tonton Macoutes*. *All* Haitians fear the *Tonton Macoutes*."

"But . . ."

Roberto held up his hand. "I share your concerns for these *blans*, whoever they are, but this is a terribly complicated situation. There have been no reports of missing *blan* tourists or workers, so these two have probably snuck across the border illegally and are up to no good. So whatever reckoning awaits them, however horrific, they've brought it upon themselves."

Henson struggled to contrive a way to seek the assistance of the *Fad'H* officer while revealing as little as possible. "Roberto, I have a confession," he confided. "When I transmitted my daily report last night, my company instructed me to be on the lookout for two Americans. I think that they may be pilots. There's a reward if I find them and assist them in getting out of Haiti. I'm sorry for not saying anything previously, but it's a lot of money, and I thought . . ."

"Hah!" blurted Roberto, slapping his thigh. "American *pilots*! I suspected as much. Let me share my theory, Matthew. I have contacts in the Cuban DGI intelligence service, and they tell me that U-2s and other American spy planes routinely fly over Cuba. Anyway, these two likely had an equipment failure, ejected, and came down in Haiti. That would explain the apparition that people saw; it was probably one or both of them descending under parachute."

"You may be right," replied Henson. "Roberto, since we know where these men are, why don't we take advantage of this situation? We could split the reward money."

"And how much would that be?" asked Roberto, polishing his sunglasses with a handkerchief.

Mentally counting the money remaining in his bag, Henson swallowed and replied, "The reward is ten thousand dollars, in US currency. Roberto, I would be willing to give you the full amount if you would see fit to help these men. All you need do is send your men to fetch them."

Roberto laughed. "Funny. Matthew, I don't send soldiers to *fetch*. That's a dog's task."

"I'm sorry. I chose my words poorly. I surely didn't mean to insult you, friend."

"No matter. Did I not just tell you that the *Tonton Macoutes* would likely become involved if anyone interfered with this transaction? There's *no* hiding from them. What would you expect me to do? Take that pittance, scramble across the border, and seek asylum in Santo Domingo?"

Roberto continued. "Eventually, I would pay for my indiscretions, regardless of where I landed. Besides, Matthew, regardless of how much money the CIA . . . I mean, regardless of how much money *you* were to offer me, even if it was enough to buy refuge in Panama or Brazil, I could never leave this place. Despite all of its faults, *this* is my home. Haiti is where I belong."

"Then what do we do?" demanded Henson.

"We? *We* can do nothing, friend. You can do as you wish, but I must do as I am told. If Headquarters instructs me to secure these men, then I will. Now, if I can make a recommendation to you—climb on your motorcycle, race back to Morne Bossa, switch on your short-wave, and report what has happened."

"I will do that," vowed Henson.

"Your State Department should become involved as swiftly as possible, make the necessary apologies, and do what they can to hasten their release. I cannot fathom why that has not happened already. And, Matthew, one more thing."

"Yes?"

For once, Roberto fleetingly showed emotion—frustration bordering on anger—and said, "Matthew, this is a *tremendously* awkward situation for me. You need to make it abundantly clear to your handlers that I know how they operate. My soldiers and I are compelled by oath to repel foreign invaders. If *anyone* arrives on these shores to take these men by force, then we *will* fight them. Of course, my little garrison can offer scarce resistance against a battalion of US Marines, but I assure you that we will draw blood and plenty of it. Do you understand?"

"I understand," said Henson solemnly.

"I hope that you do. Matthew, I know that you are obligated to assist these two unfortunates because they are your countrymen, but I caution you: tread lightly."

Outside, a jeep's horn beeped twice. The maid entered the parlor and handed Roberto two lunch pails; he gave one to Henson and said, "Here is the breakfast I promised. Now go, friend. Do what you think is right, not just today but always. *Bon chans. Adeiu.*"

"*Adieu.*"

Dubuission Homestead, Haiti
9:15 a.m., Saturday, March 14, 1970

After sending a dispatch by shortwave, Henson climbed on his Motoguzzi, sprinted south on Highway Three, negotiated the tricky road from Grande-Rivière-du-Nord, parked the bike, and then ran up the trail. Arriving at the crash site, he quickly discovered that the news had beaten him there. Glades and Baker sat in the shade of a tree overlooking the Gemini-I wreckage, mulling over the details of the freshly received message as they examined the map.

Baker tore the cellophane wrapper off a bar of compressed oatmeal from a compact "LRRP" ration. He munched on the cereal bar as he used a shaving brush to apply "LSA" lubricant to his silenced MAC-10 submachine gun. Glades continued to study the message, as if there might be some details or nuance he had overlooked on the first reading. The motionless morning was oppressively hot and humid, like a steaming damp blanket draped over the earth.

"Man, this commuting is killing me," groused Henson, flopping quietly on the ground next to the three men. "One more fast lap and I'll need a new set of kidneys."

"Nice to see you, Matt," said Baker, twisting the cap closed on the green LSA tube. "We just received your latest dope. It came in on the wire from Homestead about five minutes ago. It sure doesn't sound like welcome news. Let me give you a quick update. We heard from Lewis about thirty minutes ago. They said they followed a distinct track and blood trail down the hill to Highway Three, but after they crossed the road, the track petered out."

Baker continued. "Lewis was working his way north towards the river to see if they can pick up the track again. Right now, they're out of radio range. They're supposed to make radio contact again in an hour, but once we tell them the news, it's still going to take them at least an hour to make it back here, and that's if they break brush and haul ass. This ain't easy terrain."

Henson looked at his watch. "So we cool our heels for the next two hours?"

"At least," noted Baker. "But the good news is that Homestead told us our reinforcements are coming in. The airport at Cap-Haïtien is closed from dusk to dawn. Around two o'clock tomorrow morning, two C-130s will land there with rest of the guys and four jeeps. Homestead said that the current plan is for them to come straight here, link up with us, grab the crew, and then we all head north on Highway Three. There's a US Navy ship about five miles offshore right now. After we have the crew, we'll work our way to a pick-up site on the coast northeast of La Petite Anse. When we radio the ship, they'll send boats to ferry everyone out."

"That's no good," declared Henson. "The *Fad'H* commander—Colonel Roberto—said that the crew may be moved out of Dondon as early as tonight. They'll be way to the south of here, in a more secure location, so we might not have a chance to grab them after tonight."

Glades shook his head. "The message from Homestead said nothing of the sort."

"Then something obviously got lost in translation," said Henson. "Look, I may have something to fold into the mix. On the way here, I thought about this situation, and I think I may have another way to grab the crew. Care to hear it?"

Baker answered, "We're not gainfully employed right now, so testify, brother."

Henson quietly cleared his throat. He took off his Panama hat, placed it on the ground, and then set his canvas shoulder bag beside it.

"In order to explain this scheme, I have to lend you some background first. Prior to this crash, I hadn't worked up in these mountains before, so I've never met any of these people. For some reason, when I first meet them, they just seem to be scared to death of me. It bothered me for a while, and then I realized what was going on." Building from those details, Henson explained his plan.

When Henson finished, only Glades responded. Nodding in affirmation, the Ranger sergeant quietly said, "Gutsy plan, but I *like* it. It's audacious, but I'm a big fan of audacity. But there's one gaping hole. We haven't physically confirmed this location in Dondon, nor have we verified that the crew is actually being held there."

Glades compared the aerial photo to the precisely drawn details of the topographic map. He pulled a pair of binoculars from his rucksack, set a bearing on his lensatic compass, took a sip of water from his canteen, and stood up. "I'll wander down the hill and take a look," he said.

"I'll go with you," said Henson.

"No need," asserted the Ranger, shaking his head as he adjusted his web gear and checked his weapon. "I move faster alone. I'll be back in two hours. If I'm not, or you're gone, then I'll see you back at Eglin in a week or so." He turned and faded quietly into the woods.

Baker quietly whistled and said, "I've been in some dicey situations and have hung around some bad dudes, but *that* rascal just scares the crap out of me. I'm not sure he's human."

"Well, he's right at home here," replied Henson. "It's like a land custom-made for him."

The men stood and walked over to the partially disassembled Gemini-I spacecraft. As they examined it, Henson checked his watch and noted, "Hey, we've got a little time on our hands . . ."

Studying the landing gear jutting out from the overturned spacecraft, Baker ran a finger across the metal bristles protruding from the bottom of the skid. "Yeah," he said. "There's plenty of time for both of us to dictate our last will and testament."

"Actually, I was thinking of something else. You mind looking at something? I would really appreciate your medical opinion."

Baker grinned and replied, "Medical opinion? Oh, this doesn't bode well. Matt, please tell me you haven't acquired some social disease down here. I'm not set up to treat the clap."

"Nothing like that. Two brothers own this place. Jean and Henri Dubuission." Henson pointed towards the north. "They live in a hut up there, past the sugarcane. I told them to stay put there until I said otherwise. Anyway, I would like you to take a look at one of them."

"You really think that's a good idea? You don't think that they might figure out something is going on if I went up there?"

Henson laughed quietly and commented, "Well, considering that a spacecraft landed in their backyard and set their crops on fire, and then four guys dropped out of the sky on top of it the next night, I suspect that they already know something's up."

"Point taken. So what's on your mind?"

"Something's been bothering me," replied Henson. "Henri supposedly has leprosy, and while he looks awfully hideous, I'm really not sure."

"*Leprosy?* Oh man, *nasty* business, Matt," Baker said, frowning. "But just by sheer coincidence, you happen to be talking to an authority of sorts. When I was going through Special Forces medic training with the Army, they sent us to Louisiana, to the National Leprosarium in Carville, for a few weeks of on-the-job training."

"Special Forces? The Green Berets sent you? Why on earth would they do that?"

"Several reasons. First, it was billed as a civic action mission, sort of like a big Eagle Scout project. Second, Special Forces works in a lot of places where leprosy is still endemic, like Vietnam and Africa, so it was an opportunity for the guys to see it there before they saw it in the field. I think it was to toughen us up some, and maybe teach us a little compassion as well."

"So since you know so much about it, what causes leprosy?"

Slowly walking around the spacecraft as he studied its details, Baker replied, "Leprosy? It's bacterial. Fortunately, about ninety-five percent of the world's population is naturally immune to the bacteria."

"Can it be cured?" asked Henson. "Is there *anything* that can be done for Henri?"

Lightly running his hand along the pocked surface of the Gemini's blunt heat shield, Baker answered, "Cure? No, but some new treatments look promising. Back at Carville, the docs had a lot of success with a sulfone drug called Promin. But before you lift anyone's hopes, let me warn you—if your guy really has leprosy, there's not much that can be accomplished in this environment. I seriously doubt that Promin or any of those other drugs are available down here."

"Well, I would still appreciate it if you took a look."

"Okay, but why the rush?"

"I figure that if we don't do it now, we won't have another chance. As it is, Lewis will blow a gasket if he finds out. I figure it's better to beg forgiveness later than to ask for permission now."

"Say no more," declared Baker, swinging his M5 medical bag onto his broad shoulders. The large canvas bag was jammed with medical supplies and drugs. He picked up his submachine gun. "Lead on, my black brother. Take the radio with you, just in case."

A few minutes later, Henson and Baker walked into the hut where the brothers waited. Upon seeing Henri, Baker burst out laughing. Jean cursed in anger as Henri sobbed. With tears streaming down his cheeks, the tormented outcast turned his face in shame.

"*Great* bedside manner, Dr. Schweitzer," smirked Henson, removing his hat and wiping sweat from his brow. "Can you make him feel *any* worse about himself?"

"Hey, I'm not laughing at him. I'm laughing because I'm relieved," declared Baker as he composed himself. "This isn't leprosy, Henson. Not *even* close. Man, I can't believe that this guy has been stuck on this damned hill for this. What a damned shame."

"If it's not leprosy, then what the hell is it?"

"Well, dumbass, if you had ever read your Bible, you might recall that lepers were covered with sores and boils. This is neither. It's just a big damned lump." Baker sat cross-legged on the dirt floor and examined the massive growth protruding above Henri's right eye.

"Okay, it's definitely a lump. Is it cancer? A brain tumor?"

"Henson, you're just not cut out for the medical field, are you? You just don't have a very good grasp of the obvious. If our friend had a brain tumor this big poking out of his skull for the past fifteen years, do you really think he would still be alive right now?"

"Good point."

"This is a tumor of sorts, but it's benign. It's called a lipoma. Lipomas are common in the Caribbean, especially among men of African descent. It's simply a large semi-solid fat deposit under the skin." Baker placed his fingers on the mass and moved it around under the skin. "But I must admit, as lipomas go, this one's a real humdinger. Biggest I've seen, by far."

"So can anything be done with it?" asked Henson. "Can a surgeon remove it?"

"Surgeon? I suppose, but since there's not one on this hill, I guess *I'll* just have to do it. Actually, it's a fairly minor procedure. I could do it with a can opener, but since I left my church key back at the house, I'll just have to use my field surgical kit."

"How long?"

Zipping open his M5 bag, Baker laughed. "About forty-five minutes for the forehead, if there are no complications. A little over an hour if we do a twofer. I'll do the back one first. If it's not a lipoma, I don't want to mess with his face and cause any additional harm. We'll be back down the hill long before Lewis returns. Just one thing, though. Just to make sure, would you mind asking Henri if he wants them taken off?"

Henson crouched down and spoke quietly in Henri's ear. Henri smiled broadly and nodded. Jean's mood softened considerably, and he smiled as well.

Laying out surgical instruments, Baker said, "I take that as an affirmative reply. Look, I can do this with just a local anesthetic. I have lidocaine for that, but it would be nice if we could use a general to calm him down a little more. Can you ask his brother if they're hiding a bottle of rum somewhere? Maybe for special occasions? I think this would certainly qualify."

Henson spoke to Jean, and the Haitian left the hut. "How much does he need to drink?"

"Not much. Two or three fingers should do the trick."

Jean returned with a bottle and a metal cup. Baker opened it and took a whiff. "Oh man, this is some noxious stuff. I'm sure it packs a potent wallop. Go ahead and administer him a snoot. Afterwards, scrub up good with that Phisohex. You're going to assist. First surgery?"

Henson nodded as he poured rum in the cup.

"Then you owe me a case of beer when we get back."

Handing the liquor to Henri, Henson asked, "What if Lewis gets back early and catches us up here doing this?"

Baker chuckled. "That's highly unlikely. Besides, have you told Lewis about the leprosy?"

Henson nodded. "I did."

"Then Lewis isn't setting foot up here. Surely, the notion of catching lep would petrify him. Trust me, Matt, mum's the word and no one will be the wiser. Except our buddy Henri here, since we're granting him a shot at a normal existence for the rest of his days on earth."

Henson escorted Jean outside as Baker prepared his instruments. Minutes later, with the slightly intoxicated Henri laying face down on the straw mat, Baker shaved the scalp around the lipoma before liberally swabbing the area with brown Betadine "monkey blood" disinfectant solution. He injected lidocaine to numb the site, then drew a slightly oval-shaped mark with a ballpoint pen. With practiced exactitude, following his oval reference mark, he deftly used a scalpel to make an incision over the mass. As Henson squeamishly held the surgical wound open with a retractor and forceps, Baker carefully dissected and excised a glistening, orange-yellow gelatinous mass about the size of a billiard ball.

After tying off minor bleeders and suturing the incision closed, Baker swathed the wound in gauze and then shifted his attention to Henri's deformed face. "One down," he commented. "We have five minutes before it's time to make a radio check with Lewis and Finn. After we raise them and give them the latest news, we'll roll Henri over and swing for the fences."

Peristyle de Beausujour, Dondon, Haiti
10:38 a.m.

Although Carson was certain that the Haitian police or some other authorities would eventually come, nothing significant had transpired since they had been brought here yesterday morning. Striving to make effective use of his time, he memorized the details of the building. Hoping that rescue was imminent, he gradually shifted Ourecky and himself so that they were clearly visible from all potential access points that assaulters might use.

There was a window nearby; when his guard looked lax, he furtively stood up and looked out. He could easily hop through the glassless window, make it to the safety of some woods about a hundred yards away, and be gone from here before anyone was the wiser. Acutely aware that it would entail leaving Ourecky behind, he quickly discarded the escape option.

In his brief moments of lucidity, Ourecky was lethargic, but most often he was all but unresponsive. Carson was still concerned that Ourecky had incurred a serious head injury in the crash. Should he force Ourecky to remain awake? Should he let him sleep? In any event, it didn't seem to matter; at this point, the engineer was more inclined towards unconsciousness, and there was little Carson could do to dissuade him from dozing off.

Since yesterday, the stream of curious visitors had died down to an intermittent trickle. Besides the two elderly women who attended to them, and the two guards, there was a fearsome-looking black man who orchestrated the proceedings. About six feet tall, with closely shorn gray hair and a wisp of a beard, the authoritative Haitian looked to be in his early sixties. Stripped to the waist and wearing red trousers sheared off just below the knees, he looked strong and supple, with taut muscles and sinews laced tightly over a powerful frame. About six other regulars drifted in and out of the building.

Supervised by the sinister overseer, they appeared to be decorating the large room for some sort of festivity. They chanted, swept the dirt floor, placed fresh candles in sconces, and adorned the rafters with garland and fetishes.

The insufferable heat had sapped Carson's appetite as well as his strength. He no longer ate, but had managed to accumulate a small hoard of bananas and mangoes, just in case they later made a run for it. But although escape was still an option, the most likely scenario was that someone would eventually come for them.

Convincing himself that it was just a matter of time until rescuers arrived, he decided to board his imaginary submarine. As he stared at a maniacal grin of a skull painted on the adjacent wall, Carson progressively slammed shut and dogged the watertight doors, eventually isolating himself in the snug control room. And there he waited.

Dubuission Homestead, Haiti
10:54 a.m.

The second phase of the surgery proceeded smoothly. After making a large incision, Baker removed another mass, even larger than the first, from Henri's forehead. "Let me close this," he said, preparing sutures. "I'm going to take my time, to make sure that he heals up with minimal scarring. When I'm done, I'll call for you to bring in his brother."

Henson went outside and waited with Jean. With his arms wrapped around his knees, Jean squatted by the hut's low entrance, nervously rocking back and forth on his heels. A few minutes later, they heard Baker's voice from within the hut. "All done. Bring him in, Matt."

Jean entered the hut and knelt beside the straw mat where his twin dozed. Baker lifted the gauze so he could admire his handiwork. Seeing his brother's transformed face, Jean wept.

"Sounds like he's happy with the results," commented Baker, peeling off his latex gloves. He opened a leather-bound case filled with glass vials. He opened a vial and shook several pills into a small paper envelope.

"This is tetracycline," explained Baker. "Henri needs one every six hours for the next week. And here's a few aspirin tablets, since it's a sure bet he'll wake up with a killer headache. I'll leave some soap, a tube of Betadine, gauze and tape. Tell him to keep both incisions clean and lubed up with monkey blood. Tell him to make sure that Henri sleeps on his side for the next couple of nights, and to watch over him, just in case he pukes in his sleep."

As the two men stood up to leave, Jean grabbed Henson's shoulders and spoke excitedly. Henson called after Baker, "Do you have a mirror? He wants Henri to see his new face."

"Mirror? Mirror?" Initially, Baker frowned, and then he grinned as he patted his chest pocket. He fished out an emergency signaling mirror, knelt down beside the straw mat, and held the mirror so that his patient could appraise the results of the surgery.

Barely conscious, Henri gazed briefly at his reflection and smiled weakly. "*Mesi anpil, zanj,*" he muttered. "*Thank you very much, angel.*" With a serene expression on his face, he promptly fell asleep.

"I think our work here is done," commented Baker, tucking the mirror back into his pocket. "Time to truck on down the hill. I should be a doctor, dude. I would *never* miss a tee time."

By the time they made their way through the sugarcane back to the spacecraft, Glades had returned from his solo jaunt into the outskirts of Dondon. "Care to explain what you two boys have been up to?" he asked curtly. "Lewis is on his way. He'll be here shortly."

"We know," blurted Henson. Knowing that Glades had them dead to rights, he explained about the Dubuission brothers and their treatment of Henri's "leprosy." Expecting Glades to be furiously irate, he was surprised when the hard-bitten Ranger just smiled and chuckled.

"What's so funny, Nestor?" asked Baker. "We thought you were going to rip our heads off."

"Well," answered Glades in his West Virginia drawl. "I just never figured you two for the benevolent kind. I've got no bones to pick with you. It was the right thing to do."

"Thanks," said Henson. "That means a lot, coming from you."

"You know, my mother used to tell me that kindness has a memory," said Glades. "And an act of kindness passes from person to person, generation to generation." He scratched his head and added, "Who knows what you two may have set in motion?"

"So did you make it to the peristyle?" asked Henson anxiously. "What did you see?"

"I did." As he described it, Glades's reconnaissance corroborated Henson's report; he pinpointed the voodoo peristyle where the crew was being held and briefly watched Carson through a window. "And it definitely looked like they were gearing up for some sort of big shindig," he concluded. "That matches what your Haitian colonel had to say."

Over the course of the next few minutes, on a patch of bare ground, using ration boxes, string, and other scrounged materials, Glades constructed an intricate model that replicated the peristyle and surrounding area. He carefully shaped dirt with his hands to accurately depict the lay of the terrain, and used twigs to represent vegetation.

Drenched with sweat, Lewis and Finn arrived shortly afterwards. Glades acquainted Lewis with the layout of the peristyle compound. Cross-referencing the map and aerial photographs to his model and sketches he made during his surveillance, he pointed out the salient terrain in the vicinity of the building, including approach routes and crucial observation points.

Asking few questions, Lewis analyzed the materials and then spent approximately fifteen minutes gathering his thoughts. Finally, he said, "Based on what Sergeant Glades has reported, here's my tentative plan. Glades, I want you and Finn to move down the hill to establish a surveillance position on the voodoo temple."

Lewis continued, "According to Homestead's plan, the main body and jeeps will come in after midnight." He pointed at a spot on the map. "Baker and I will link up with them here, on Highway Three between Menard and Dondon. We'll give them the last minute details from the surveillance team, and then I'll lead the main element on a raid to rescue the crew. We'll exploit the element of surprise, shock, and overwhelming firepower to neutralize any resistance. I guess we'll give these Haitians an opportunity to see just how effectively their voodoo magic stands up to copper-jacketed lead. Current time hack is 11:45. I'll write up the rest of the plan and transmit it to Homestead. Any questions?"

"What if Henson is right, sir?" asked Finn. "What if these two pilots are moved to Mirebalais tonight, before the main body arrives for the raid?"

"I don't think that's likely to happen," stated Lewis. "Homestead certainly has the most current intelligence. Besides, I don't see any other way to do this."

"Uh, Major, Henson has concocted another plan," said Baker. "It's kind of radical, but you might consider it as an alternative."

"No," said Lewis flatly. "We're going to do this my way."

Glades sat on the ground nearby, using a P-38 can opener to open a small tin of peanut butter. He leaned towards Lewis and spoke quietly. "You said you were going to rely on the element of surprise, shock, and overwhelming firepower to pull this off. Considering the circumstances, do you really think that's a good concept?"

Livid, barely able to keep his voice low, Lewis snarled, "Don't *dare* stoop to lecture me, Sergeant. This is an *operational* mission, not another canned training exercise, so I don't *need* your technical advice. Besides, since you always seem to hold up Ranger School patrols as the gold standard, I intend to execute this operation *exactly* how we were trained in Ranger School."

"Not quite," replied Glades calmly. "I know exactly what they teach at Ranger School. No one there ever taught you to fire on innocent civilians."

"But they're *not* innocent civilians," countered Lewis. "They've abducted two American pilots and they're holding them hostage."

"So says you, Major. But let me remind you that those two men aren't even supposed to be here, and unless something has significantly changed in the past few hours, the US government isn't acknowledging their existence. Moreover, that applies to us as well, so no one has any legal standing in this matter, except the Haitians who live here."

Glades continued. "So before you indiscriminately open fire on women and children, you might want to reflect on a place called My Lai. The Americal Division mowed down a mess of unarmed civilians there, and *those* officers are going to stand trial. And My Lai happened in a war zone where the rules are a lot less vague. Trust me, Major, you don't want to kill innocent civilians. *Ever.* It's really frowned upon, so I strongly recommend that you revise your plan before this operation escalates into a very ugly international incident."

"Fine, Sergeant Glades," snapped Lewis. "If you have another way to do this, then why don't you enlighten me?"

"Me? Honestly, I don't have anything useful to contribute at this point," replied Glades. "But your man Henson has a plan. You might want to listen to it."

"I've heard it, Major," interjected Finn. "I think it will work. And if it does, we'll walk out of here without firing a shot; plus Henson has a scheme to sneak the vehicle out as well."

"Okay," mumbled Lewis. "Henson, let's hear it."

As he listened to Henson outline his plan, Glades contemplated Lewis and the nature of officers like him. Men will enthusiastically follow a good and selfless leader, even to their certain deaths, but they despised and feared the ambitious, ego-driven careerists who seemed to be the bane of the modern military.

Lewis was a classic case; his every action appeared to be driven by career advancement. He seemed much less concerned about rescuing the pilots than he was obsessed about what kind of medal he would be awarded for leading the mission. Back at Homestead, he'd actually fretted whether the foray into Haiti would qualify as a combat mission, one that could warrant awards for valor, or whether it would be classified as merely a peacetime action.

The prospect that Henson might have a better plan—a workable plan—was likely too much for Lewis to fathom. He just couldn't seem to grasp that a good idea might exist outside the realm of his own thoughts; he was too egotistical to listen to his subordinates. If that wasn't dangerous enough, he seemed locked into the notion of "one size fits all" tactics, even in situations like this, where there was no good solution to be had.

Glades thought of the conflict raging in Vietnam, and the thousands of good men killed because they were unfortunate enough to be placed under the command of unimaginative careerists like Lewis. In a similar vein, he thought of the scores of officers and NCOs killed by soldiers who were just too frustrated, angry and afraid to even concern themselves with the potential consequences of their homicidal actions. He wondered how many officers went to sleep believing that they were the latest incarnation of Patton, only to be jolted awake by the rustling of tent canvas. For many of the fragged, probably their last conscious thought was a question: What could I have *possibly* done to compel one of *my* own soldiers to kill me?

Chewing on a dry twig, Glades was absolutely sure that Lewis would summarily reject Henson's plan in favor of his own, and that was a scenario that he truly dreaded. Regardless of whether they were successful in wresting the two pilots from their captors, they still had to make it to the coast, and a confrontation with the Haitian military was virtually inevitable once the rescue team made their presence known. Glades had been around long enough to know that it was plenty easy to have your ass soundly kicked by a fourth-rate army like the *Fad'H*, particularly if you were outmanned and outgunned.

Additionally, there was absolutely no way that he could allow Lewis to wantonly kill innocent civilians. In the world of Nestor Glades, there were certain rules that were inviolable. Of his sacrosanct tenets, two were foremost: Glades would never intentionally harm a civilian, and he would never harm anyone—prisoners included—he had taken under his protection.

Glades knew how this game would end if Lewis refused to alter his playbook. He didn't want to subvert the officer's authority, but in his pocket he carried a handwritten letter from General Fels. The letter was like a fail-safe mechanism to address the worst case scenario; once executed, it would explicitly relieve Lewis of command and place Glades in control of the mission. Tapping his chest pocket with dread, he was sure that he was only moments away from opening the envelope and presenting the letter to Lewis.

So what happened next shocked the otherwise unflappable Nestor Glades. Lewis snapped his fingers quietly and gestured for him. "I've listened to Henson's plan," said Lewis. "And although I don't think it's perfect, it sure seems to be the best way to skin this cat."

"There are still a few loose ends," said Henson. "This afternoon, I need to go into town to scare up a vehicle and gather up some odds and ends. I'll also need at least one more guy with me for the take-down." He turned to Glades and asked, "Would you mind?"

"I would be honored," answered Glades. "But it's still your show. Major, are you sure you're comfortable with all this?"

"I am," replied Lewis. "We'll finish up the plan, brief the men, and execute. With any luck, we'll eat breakfast at Homestead tomorrow morning."

17

HOUNGAN

Aerospace Support Project
3:35 p.m., Saturday, March 14, 1970

Distraught, Tew anxiously paced the floor while Wolcott intently studied the latest report from the rescue staging site at Homestead Air Force Base. According to Fels, the situation was dire, but his men on the ground in Haiti had formulated a solid plan, and he urged that everyone show patience and restraint until they had an opportunity to execute.

Tew dug in his desk drawer for some antacid pills. Finding the bottle empty, he cursed under his breath and slammed the drawer closed. "I *cannot* believe that with all our sophisticated technology, we have two men stranded in this backwards country, held captive by a sorcerer convinced that they're some kind of werewolves, and that he's intent on trading them off to curry political favor," he bemoaned. "And here we are, the most *powerful* nation on the face of the planet, yet we only have three troops on the ground to resolve this situation."

Looking up, Wolcott slipped off his reading glasses. "You forgot Henson and that Army guy, Glades. That totes up to *five*, Mark. Besides, the other fifteen will go in tonight."

"You're right. What a *difference* that will make," snorted Tew sarcastically. "Who needs an airborne division or a carrier battle group? Five men? We have a veritable *army* at our disposal. Look out, world! What a fiasco. I suppose the only issue that remains is whether we leave the rescue to Fels and his men, or whether we call Hugh Kittredge."

Reflecting on that contingency, Wolcott groaned quietly. General Kittredge had agreed to let them handle the situation, so long as there was a reasonable expectation of a favorable outcome. But if it became apparent that they were in over their heads, they were expected to phone Kittredge without delay.

In turn, Kittredge would call the president, and the State Department would swiftly throw its weight behind securing the release of their two men in Haiti. But it was an action that would not be without consequence. No matter how much money was spent or how quietly this distasteful matter was handled, Wolcott could not envision an outcome in which the world would not learn about their secret operations in orbit.

"Well?" asked Tew.

Wolcott slowly cracked his knuckles, which emitted a disconcerting sound like pea gravel being crunched underneath the weight of heavy hobnailed boots. "Say the word, Mark, and I'll make the call," he declared, lightly placing his hand on the phone receiver. "But we both know that when I make the call, Blue Gemini is over. *All* over."

"So what would you have me do?" demanded Tew. "Moreover, Virgil, what would *you* do?"

"Me? What would *I* do? We're right against the goal line and it looks mighty ugly, but Isaac Fels is confident that his boys can make the winning play. I say we trust them to do their jobs."

Tew slumped into his chair and nodded his head. "You're right, Virg. We should trust them."

Dubuission Homestead, Haiti
5:15 p.m., Saturday, March 14, 1970

After the plan was finalized, Henson rode into Grande-Rivière-du-Nord to do some last minute shopping. Three hours and roughly two hundred dollars later, he had procured two second-hand dark suits and various oddments, along with the short-term rental of a Volkswagen van.

At Lewis's request, Glades had sculpted another terrain model to complement the first, except the new mock-up depicted an expanded slice of the geography in the area. Glades was happy to oblige. He was especially glad that Lewis had seen fit to listen to Henson, and that the two had managed to blend together the best parts of both of their plans.

Looking over his notes, Lewis stood up, quietly cleared his throat and said, "Okay, gents. We're going to spiel through yet another round of one-two, buckle your shoe. He leaned over the first terrain model and pointed at the peristyle. "This is the target. There's only one entrance, on the north side of the building, and one window, on the east side. Sergeant Glades stated that the pilots are situated in the southeast interior corner of the building."

Shifting to the second terrain model, Lewis gestured at the patrol base location. "We're here. Once we finish this chalk talk, Henson, Glades and Baker will move to the hide site, where the VW bus is stashed. Finn and I will head down the hill to establish a surveillance site."

Lewis glanced at Henson. "Once we're settled in position and have a good feel for the comings and goings, we'll call you with an update. After that, Henson, the game is all yours. You initiate when you're ready, and you call the plays. You're up."

Henson stood and used a stick to point at the vehicle staging site. "We'll linger here until everyone's in position and ready for us. Once we're set, we'll roll north along the river, then turn southwest on Highway Three and straight on to the objective. Baker will be driving and will stick with the vehicle for the duration."

Donning his Panama hat and mirrored sunglasses, Henson continued. "Once we arrive at the peristyle, Sergeant Glades and I will bust in like we own the place. If my plan works, we'll grab the pilots and hustle straight out. We jump back in the van and then split to the north."

Lewis nodded his head and said, "Hopefully, everything will go smoothly and Henson will make the grab. If so, once his element departs to the north, Finn and I will quietly stooge on the site for at least another twenty minutes. At this point, our role is to strictly delay any pursuit. If anyone jumps in a vehicle and it looks like they're going to pursue, we'll discourage them by disabling their vehicles. Finn, unless I direct you otherwise, you'll shoot tires only."

"Tires only, sir. I got it," groused Finn, caressing the burlap-wrapped stock of his M21 sniper rifle. "Tires only. Whoosh, whoosh . . . bump, bump . . . make car stop. No shoot driver. Shoot driver *bad*. I finally have an opportunity to shoot this puppy for real, and I'm restricted to popping holes in rubber."

"Just think of it as prophylactic marksmanship," said Lewis, smiling. "Again, if everything goes according to plan, Finn and I wait patiently for twenty minutes and then fade back up the hill. We'll take your motorcycle to the pick-up site. Okay, Henson, you're up again."

Henson said, "Once we have the two pilots, we'll drive straight to the pick-up point on the coast. We'll conceal the VW in this bamboo grove," he said, gesturing at a point about a mile short of the coastline. "I'll come back for it after the pick-up. If the pilots are relatively stable, we'll stay put and wait for you all to show up. If Baker makes the call that they need immediate medical attention, then we radio for an early pick-up boat. In any event, we anticipate you'll be about two hours behind us."

"Okay," Lewis said. "Returning back to the peristyle. If you go in and encounter significant resistance . . ."

"Then we yank the guns out of our bags and crawfish out," said Henson. "Once we're outside, we'll break hard to the left, towards the VW and the road, and get out of your way so you can cover us by pinning everyone down. From that point, if we can overcome the resistance with your help, we'll grab the pilots, jump in the van and skedaddle north. If not, we'll wait for the main body to arrive and play it by ear."

Lewis covered the remaining details of the scheme and then asked, "Any questions?" With no response, he shook hands with Henson. "Good plan, Matt. Let's make this happen."

6:50 p.m.

As Henson checked the Volkswagen's engine and brakes, Glades and Baker smeared their faces with a thick mixture of loam camouflage paint and soot from sugarcane ash. They worked together to ensure that their Caucasian coloring was completely obscured, even using their fingers to work the makeshift greasepaint into their nostrils and deep into their ears.

"So, Matt, what do you think?" asked Baker, grinning fiercely. "Dark enough to suit you?"

"Well, if I didn't know any better, I would have taken you for a couple of blackface minstrels," commented Henson. "All we need is a banjo and a washboard, and the three of us could stage an old-timey vaudeville act."

He checked the safety on the M10 submachine gun loaned by Finn. It wasn't the feared Uzi machine pistol favored by the fearsome *Tonton Macoutes*, but it was sufficiently close enough in appearance that most casual observers probably couldn't tell the difference, particularly in bad light. Besides, he was fairly sure that few Haitians had witnessed an actual Uzi and were still alive to describe it. He tucked the stubby M10 into his shoulder bag. With the gun's added weight, the bag hung ominously at his side.

Glades double-checked his own M10 before pulling a dark suit jacket over his white guyabera short. Donning a Panama straw hat like Henson's, he said, "Like you said earlier, when we hit the door, both of us need to move calmly and deliberately, like we own the whole damned universe. The game's over if we act nervous or antsy. Got it?"

"Got it," answered Henson.

Peristyle de Beasujour, Dondon, Haiti
7:48 p.m., Saturday, March 14, 1970

Struggling to comprehend his surroundings, Carson questioned whether he was asleep or awake; with every hour that passed, his world seemed to spiral further downward into a progressively surreal nightmare. People had been arriving for hours; just when he thought the cramped space could not possibly accommodate any more bodies, a few more packed in.

Some sort of voodoo rite was in full swing. A small ensemble of musicians gathered to play on a bizarre collection of instruments. Four chanting men pounded on makeshift drums, while several others blew into pipes and hollowed bamboo joints, creating weirdly ethereal music that Carson found frightening and yet somehow captivating.

Barely clothed men and women swayed and gyrated to the hypnotic music, gravitating towards the decorated pole that was the hub of the event. Moving in a rhythm-induced trance as the drums pounded at a feverish tempo, some twirled and shimmied, while others, apparently imitating savage animals, stooped over and grunted.

The frenzied celebration was lit by candles. In the sputtering light, the ghastly wall murals seemed to come to life. The hideously smiling specters pulsated, as if anxious to emerge from the orange-painted concrete to join the festivities.

Small clay pots smoldered with a mixture of charcoal and vile incense. The air was so thick with odors that Carson could scarcely breathe. He stayed close to the floor, where the air was at least slightly cooler, protecting his burned hand and leaning to his side to shield Ourecky.

Most of the revelers kept their distance, but some—particularly a few young men with rum heavy on their breaths—drew close enough to prod him with sticks. He saw that some of the men casually flaunted machetes, and dreaded what the night might hold in store.

With crazed eyes, perched on an elevated wooden platform, the bare-chested sorcerer with frayed red pants held sway over it all.

Moving with an almost hyperkinetic intensity, he frequently looked at his captives, wearing a sinister grin that sent a wave of fear down Carson's spine. He trusted that they would be rescued, but now he wondered if their rescuers would arrive in time to deliver them from this evil.

9:02 p.m.

Seated outside the Volkswagen bus, they waited. In a hushed voice, Lewis transmitted over the radio: "Assault, this is Recon Six. We are in position? Are you set? Over."

"Recon Six, Assault is set and waiting," answered Henson.

"Roger. Be advised that there are now approximately fifty adult subjects inside the structure and six outside," said Lewis. "No vehicles on site. No visible weapons except machetes. Over."

"Recon Six, this is Assault," replied Henson. "We copy fifty subjects inside and six outside. No weapons visible. Over."

"Assault, roger on all. The field is yours. Execute at your discretion. Over."

Henson keyed the mike and said calmly, "Assault. *Roger*. Stand by."

Like commuters in a suburban car pool, the three men climbed aboard the waiting van. Baker took the driver's seat. Henson sat in the right seat and Glades crouched in the back.

"Ready?" asked Henson, looking towards Baker.

"Ready as I'll ever be." Baker turned the key and the Volkswagen's motor rumbled to life.

"How about you, Henson?" asked Glades, adjusting his woven straw satchel. "Ready?"

"To be honest, I'm starting to have some misgivings."

"Henson, you have a good plan. Cease with the negative thoughts. Shut up, put your head down, and execute. This is the moment of truth, gentlemen. Ready now?"

"Ready," answered Henson. "Roll it, Baker."

Baker jammed the vehicle into gear and pulled down on a short dirt track and onto the gravel surface of the northbound road that paralleled the river. Henson keyed the radio and transmitted, "Assault is moving."

"Assault, this is Recon Six," answered Lewis in a whisper. "I copy you are moving."

Henson clicked the transmit switch twice. So as not to attract undue attention, the Volkswagen bus putted along at a virtual snail's pace, not exceeding twenty-five miles an hour on the gravel road. Baker occasionally had to beep his horn and weave to avoid strolling pedestrians, as well as stray goats and chickens wandering in the dark.

"I *cannot* believe that I'm riding in the back of a damned Volkswagen bus to conduct a raid," said Glades, holding the back of Henson's seat to maintain his balance. "Man, I could probably *walk* faster than this damned thing."

9:24 p.m.

"This is Menard. You're about a minute out," stated Glades, consulting the map as they drove.

Henson nodded and said over the radio, "Recon Six, this is Assault. We're one minute away. Got an update? Over."

Lewis's voice came back over the radio. "Assault, Recon Six. No change. Still estimate fifty subjects inside structure. Be aware that there are approximately fifteen children playing outside, about twenty feet in front of the entrance. Pilots are still located in same position inside structure. Over."

"Good copy," replied Henson. "See you shortly. Out."

A few moments later, Glades announced, "That's it, two hundred meters ahead on the left."

Driving as if he was casually pulling into the parking lot at a neighborhood grocery store, Baker downshifted and pulled into the grassy area in front of the peristyle. Coasting around in a wide buttonhook maneuver, he braked to a smooth stop so the blunt front end of the Volkswagen bus was pointed out towards Highway Three, ready for an immediate departure. Pulling out the parking brake, he left the engine idling. "Good luck," he said.

Henson drew a deep breath, swung his door open, and stepped down from the micro-bus. Slipping on his mirrored sunglasses, he swaggered towards the peristyle with malicious purpose. Although he was sure that Glades was the best man to play the part, he still wasn't entirely confident that the subterfuge would work.

Stealing a quick glance back over his shoulder, a chill passed through his body. Glades looked almost precisely like the first *Tonton Macoute* he had seen at the airport when he initially arrived in Haiti. Henson knew that the Army Ranger was essentially just a hillbilly from West Virginia, but now even he could not be convinced that he was not a stoic pillar of pure black evil.

In front of the peristyle, multi-colored pennants popped and fluttered in the breeze. As Lewis had reported, about a dozen children were playing outside. Seeing Henson and Glades approach, the children screamed like crazed banshees and scattered like leaves in the wind.

Striving to maintain focus, Henson was surprised at the music and chants emanating from the peristyle. The rhythms tugged sharply at him, drawing him in, like it was sucking him into a vortex. It was unnerving, like the eerie music somehow was triggering a connection to his distant past, rippling through space and time and his ancestors' souls to remind him of his African heritage. Pushing those thoughts aside, he arrived at the door and stepped inside.

As he crossed the threshold, a woman's shrieking disrupted the proceedings. *"Tonton Macoutes! Tonton Macoutes!"* Glades entered on Henson's heels and then took a position just inside the door, ensuring that Finn would still have a clear field of view from the hillside.

"Atansyon! Silans!" bellowed Henson. *"Attention! Silence!"* Crying and pleading, the revelers backed away from Henson, scurrying into the far corner until they were one tightly compressed mass of sweating bodies. A tinkling crescendo rose as abandoned machetes and rum bottles hit the floor. He brusquely shoved stragglers aside as he strode towards the center pole.

"We have been informed that you are illegally holding property of the State," announced Henson in Haitian Creole, shifting his shoulder bag slightly as he patted it lightly with the outstretched fingers of his left hand. "We are here to claim it."

He advanced to the corner and stood over Carson and Ourecky, glaring at them like they were a pair of disobedient hounds. He chuckled and bent forward at the waist. Reeling his hand back, he slapped Carson hard across the face; the stinging sound reverberated through the room. He placed his palm on Carson's forehead and leaned closer, as if he was inspecting a consignment of merchandise for defects or blemishes. He pulled open Carson's lips and examined his teeth, and then turned the pilot's head to look into his ears. Softly, he whispered, "Air Force rescue team. We're here to take you home. Follow my lead."

He repeated the process with Ourecky, who was only marginally conscious at best. Suddenly, Henson saw a momentary flash of recognition in his eyes.

"*You?*" mumbled Ourecky feebly.

"Yeah. It's *me*," replied Henson quietly, subtly nodding. One at a time, he yanked the men to their feet. As Carson assisted Ourecky, Henson pushed them roughly towards Glades and the door. The thronged Haitians were motionless, groveling in abject shock.

Shoving the pilots before him, Henson had almost made it to the door when he heard an enraged voice. It was the *houngan*, bellowing a single word like a scathing curse, "*Blan!*"

"*Blan!*" grunted the *houngan* again, nimbly jumping down from his vantage point. Suddenly emboldened, he spoke in rapid-fire Haitian Creole, exhorting his followers to resist the strangers in their midst. Still frozen in fear, most of the crowd did nothing, but five large Haitian males apprehensively stepped forward to obstruct Henson's escape route.

"*Blan!*"

Slowly pivoting about, Henson expected the *houngan's* ire to be focused on Glades. After all, it was the weakest part of the ruse, attempting to pass off a white man as not only black, but as a dreaded *Tonton*

Macoute. So as Henson finally turned to face the *houngan*, what he didn't expect was the black wizard's gnarled finger to be pointed at *him*.

"*Blan*," sneered the *houngan* scornfully, wagging his finger. Slowly becoming reanimated, muttering among themselves, the *houngan's* followers were obviously divided in their perception of the unfolding drama. Most appeared convinced that their spiritual leader had taken abrupt leave of his senses, but a sizeable faction of loyal disciples obediently rallied to challenge the interlopers in their midst.

Almost instinctively, Henson's hand flashed towards his shoulder bag. Out of the corner of his eye, he discerned the glint of metal in the flickering light of the candles; at least a few of the peristyle members had reclaimed their discarded machetes. He and Glades had their submachine guns, but they were really only meant as a show of force to make good an escape. Now Henson saw that it was going to be an ugly scene, with no guarantee that they could get the pilots through the door unharmed, and equally unlikely that he or Glades would be unscathed as well. He realized that there was only one chance of surviving this showdown. It was time to improvise, so he ceased *acting* like a *Tonton Macoute*, and instead *became* one.

Slipping off his mirrored sunglasses, shifting his feet into a boxer's stance, he aggressively leaned towards the sorcerer until their broad noses touched. Their sweat mingled and dripped to the floor. A tense eternity passed as the two men stood motionless, like ebony statues chiseled in deep relief, glaring at each other. Watching tiny muscles twitch in his adversary's wizened forehead, Henson knew exactly what the *houngan* was seeing as he peered into his eyes: Dark pupils like opaque glass, entirely devoid of human expression.

"*Should I take your soul from you now?*" hissed Henson menacingly, laying the palm of his hand against the *houngan's* bare chest and pressing deep into his dark flesh. "*Or should I come back to collect it later?*"

Trembling, the terrified *houngan* recoiled back into the crowd, as if he had been yanked by invisible marionette strings. Henson threw back his head and laughed, partly out of theatrical necessity and partly out of sheer relief. Confident that the *houngan* was no longer a viable threat, he turned, and slowly walked towards the door. The panicked Haitians surged to escape from his path; Moses and his legendary staff could not have parted the Red Sea any faster.

Glades took control of Carson and Ourecky; playing his role to the hilt, he slapped the backs of their heads as he propelled them outside into the night. Henson followed, and they calmly trotted back towards the idling micro-bus.

They were twelve feet from the van when Ourecky suddenly faltered. Clutching his abdomen, he groaned and collapsed to his knees. Glades and Carson reacted immediately, hoisting him by his armpits and swinging him into the cargo space of the Volkswagen. Glades jumped in and slammed the door shut as Henson took his place in the front seat.

"Good in back?" asked Henson, pulling the M10 from the bag.

"We're up," replied Glades calmly.

"Go, Baker," said Henson softly. "Take it easy. Don't get in a rush."

Baker shoved the Volkswagen into gear and they lurched onto Highway Three, headed towards the north. Henson called over the radio, "Recon Six, this is Assault. Headcount five, moving north. Be advised one subject appears to be seriously injured. Will advise. Over."

"Assault, Recon Six," replied Lewis in a hushed voice. "Good work. Copy you are moving with headcount five. Copy injured subject. Be advised that no locals have emerged from building. They still appear to be bewildered. Over."

"Roger. Stand by." As Baker drove north on the gravel highway, Henson turned around and asked, "You're Carson, right?"

Carson nodded and pointed down. "Right. And this is Ourecky."

"We know." Henson tossed a rubberized waterproof bag, about the size of a laundry bag, to Carson. "Here. There's a uniform and boots in there for you. There's some other stuff as well. There's a ground-to-air survival radio, a survival kit, a barter kit with gold coins, and an American flag so you can identify yourself to aircraft. Go ahead and climb into the uniform when you can. We have another bag for your buddy. Same stuff."

"Thanks," said Carson, untying the bag. "Are you guys *really* the rescue team?"

"Yeah."

"I guess I was expecting a helicopter."

"Sorry to disappoint," answered Henson. "Our helo's in the shop. Listen, we're taking you to a pick-up site. It's on the coast, about half an hour north of here. There's a US Navy ship waiting offshore. If we're stopped and have to abandon this vehicle on the way to the pick-up site, stick with us. We'll take care of carrying Ourecky. Are you okay? No physical problems?"

"I'm okay," replied Carson. "And thanks for breaking us out of there, except for the part where you slapped the hell out of me."

"You're welcome," said Henson. "So, how's he looking back there? Is he coming around?"

"I think he's slipping into shock," said Glades, struggling to keep his balance as he examined Ourecky. "I would swear he's been shot, but there aren't any holes in him."

"We had a rough landing," explained Carson. "His head hit the control panel."

"Head injury. Not good," stated Baker as he shifted gears. "Not good at all."

"Okay," said Henson, looking back into the cargo space. "As soon as we can, we'll stop and put Baker back there."

With a long straightaway ahead of them, with no vehicles or pedestrians in sight, Henson said, "Okay. Stop here and switch."

Downshifting, Baker braked to a smooth stop. He and Glades traded places, just as they had previously rehearsed. Baker unzipped his M5 medical bag and immediately went to work. Guarding his abdomen, his ashen face contorted in pain, Ourecky moaned. Glades jumped into the driver's seat and quickly got the vehicle rolling north. Henson oriented Carson to the map, showing him where they were and where they were going.

Baker slashed open Ourecky's long underwear shirt with his survival knife and then quickly examined his abdomen and torso. Ourecky winced as the medic lightly kneaded his stomach. "Oh, this ain't good," noted Baker. "It looks like he's bleeding internally."

"How can you tell?" asked Henson, swiveling his head to look in the back on the van.

"See that faint bruising around his belly button?" answered Baker, illuminating Ourecky's abdomen with his flashlight. "That's a sign he's been bleeding into the peritoneum, his gut cavity. His gut is also rigid and tender, that's why he's protecting it like he is. He's probably been leaking into it for a while, but likely had a good clot formed that kept the bleeding in check while he was stationary. But once he started moving, it obviously got jarred loose."

Baker used a stethoscope to listen to Ourecky's chest. "This ain't good, either," he observed. "His breathing is shallow and his pulse is rapid and thready. I'm guessing that all that blood pooling in his gut is starting to impinge on his diaphragm, so it's going to make it that much harder for him to breathe."

"What can you do?" demanded Henson.

"Not much," replied Baker, trying to keep his balance in the slick bed of the van. "I can tap his gut to relieve some of the pressure. I can stabilize him, at least temporarily, but he really needs to be seen by a real surgeon as quickly as possible."

"How long can you keep him alive?" asked Carson anxiously.

"I don't know," replied Baker. "There's no way I can tell the extent of his injuries. He could last two or three hours, or he could punch out in thirty minutes."

"Surely there's a doc on board the ship. Do you think he can make it there?" asked Glades, slowing down as he beeped the horn at several children walking in the middle of the road.

"I seriously doubt it," replied Baker. "And they said it would take at least an hour for the boats to make it ashore, depending on the tides, and possibly longer to transit back to the ship."

"Can't you do anything else?" pleaded Carson, gripping Ourecky's hand. "Can't you do a transfusion? Maybe give him some fluids of some sort?"

Baker placed his hand flat on Ourecky's abdomen. "It wouldn't help much. I've got serum albumen and other fluids, but any additional volume that we pump into him would just spill into his gut and press harder against his diaphragm. His only shot is to get to a surgeon *now*."

Using a small penlight, Henson analyzed the map. "Sergeant Glades, there's a dirt road about a half-mile ahead on the left," he said. "It's the back route into the airport. Take it." He picked up the radio handset and spoke into it. "Recon Six, this is Assault. Over."

"This is Recon Six," replied Lewis in a very low voice. "Go ahead."

"Six, our medic is telling me that one of these guys won't make it to the ship. He's in very fragile condition. I think I can send him out by air, if I can divert to the airport. Okay by you?"

"Do it," replied Lewis. "We're going to displace in about ten minutes and head up the hill. Be advised that no one came after you. Right now, they seem to be just milling around in front of the building. They don't look too happy with themselves. Over."

"Got it," replied Henson. "Be careful getting to the pick-up site."

"What the *hell* is that stench?" asked Baker. He grabbed Carson's hand, sniffed the salve, and examined it. "Man, that's ripe! Did you stick this paw in a frying pan or something?"

"Burned it right after landing," answered Carson. "Don't worry about me. Take care of *him*."

Cap-Haïtien Airport, Haiti
10:10 p.m.

Seeing Taylor's Maule parked alongside his building, Henson breathed a quick sigh of relief. In the back, clutching his bloated stomach, Ourecky made gurgling sounds and writhed in pain.

"He's in a very fragile state," advised Baker. "We need to evacuate him now."

Even before Glades braked completely to a stop, Henson jumped out and sprinted towards the door. Pounding on the warped plywood, he shouted, "Taylor! Let's go! Up and at 'em."

At least two minutes passed as Henson continued to bang at the door. Finally, yet again naked, a barely conscious Lydie answered the door. "*Kisa, blan?*" she demanded, obviously agitated to be so disturbed from a sound slumber.

"*Mesye Taylor, souple,*" replied Henson.

She turned her head and mumbled, "*Blan.*"

Pulling on a pair of blue jeans, Taylor came to the entrance. "What, Henson? What in the hell are you doing out here this time of night?"

"I need to rent your plane. Right *now.*"

"Didn't we just have this same conversation?" asked Taylor. "Can't this wait until morning?"

"*No,*" said Henson emphatically. "There's a badly injured man in the van. He needs to be flown to the closest American medical facility as quickly as possible."

"Is he shot?" asked Taylor.

"No, but apparently he has severe internal bleeding and is going into shock. Can you do it? I need an answer right *now.*"

"Is he military?"

"Could be."

Taylor nodded knowingly. "If he's military, your closest bet is Guantánamo Bay Naval Base in Cuba. That's about two hundred miles west of here, about an hour and a half. Second option is Puerto Rico, but that's about four hundred miles east. Either way, it's going to cost you. As soon as I touch down on US soil, it's fairly likely that my plane will be confiscated. Assuming that I even come back here, I'll need money to uh . . . *acquire* a new plane."

"How much?"

"Twenty thousand," replied Taylor flatly. "US greenbacks only. No Haitian *gourde.*"

Henson was in no mood for negotiating, nor was there time to strike the best bargain or fret about whether he would be reimbursed. "Ten thousand. That's *all* I have. Take it or leave it."

"Done," answered Taylor. He rubbed his eyes and whistled for his two guard dogs. "Stick your guys in the plane. I'll be ready in ten minutes."

Henson turned and gestured to load the plane. Glades pulled the Volkswagen around so that Carson and Baker could move Ourecky quickly.

"Wait," said Taylor. "How many guys are flying out with me?"

"Three," replied Henson.

"Nix that. I'll take *two*: the injured guy and a medic. The less weight I haul, the faster I fly. I'll yank out the seats on the right, so he can lie flat with his feet sticking back into the cargo area."

"I'll take out the seats," offered Carson. "Where are your tools?"

"In my workshop, on the left side of the shed. And while I'm getting my stuff together, there's also a hand-driven pump out there. Top off my tanks from those drums over there."

"I'll handle that," said Glades.

As the two men worked to ready the plane, Henson pulled the rubberized pouch from his back pocket and counted out fifties and hundreds.

Moments later, Taylor reappeared with a faded blue canvas gym bag. He took the money that Henson offered. Without counting it, he stuffed several bills into the bag and handed the majority of the cash back to Henson. "You still have my card?" he asked.

Henson nodded.

"Then write me at my Miami address. I'll send you instructions on where to send the rest of the money." Taylor jammed some aerial charts into the bag and added, "Matt, one more thing. *If* I do come back, I won't be coming back here. I had already planned to move my shop to Port-au-Prince, so this is good timing. This place is paid up for another two months. If you want to use it, it's yours."

"Thanks. How about Lydie? Is she going with you to Port-au-Prince?"

"Lydie? No. She's paid up through the month, so she can cook and clean for you if you want. After that, she'll probably want to go back home to Limbe."

Henson nodded. He looked at Lydie, enchantingly beautiful in the glimmering light of the partial moon. Mercifully, she had thought to cover her nakedness with her makeshift T-shirt dress. He heard the sound of a car horn beeping far away, and looked towards the west. He saw a string of headlights in the otherwise dark streets of Cap-Haïtien and knew that it had to be a military convoy; someone had probably tipped off Roberto about the rescue.

"It's the *Fad'H,*" observed Taylor, pulling on an old baseball cap and a leather jacket. "It looks like Colonel Roberto intends to pay us a visit." He nonchalantly started into his pre-flight checklist. "You boys need to hit the road as soon as my tanks are filled."

Glades spun the pump's rotary handle as fast as he could as Baker and Henson loaded Ourecky onto the floor of the plane. Henson spoke a few words to Lydie and then took over for Glades on the pump. The young Haitian woman nodded and ran back into building.

"I'll finish this," said Henson, grinding away at the pump. "Take Carson and head for the pick-up site."

"You won't make it there by the time Lewis gets there," said Glades.

"I'm staying. After things settle down, I'll head back to my place to call the States on the radio, and then I'll hoof it up to the bamboo grove to grab my bike."

"Suit yourself," said Glades. He whistled at Carson and pointed at the van. He extended his black-painted hand out to take Henson's. "Hey, it was a pleasure working with you, Matt."

"Likewise. I hope to see you again someday. Hopefully not too soon, though."

After saying an anxious goodbye to his unconscious friend, Carson walked up and spoke to Henson. "Thanks."

"My pleasure. Now, you two need to leave."

As the Volkswagen putted off to the south, Lydie brought a faded quilt out of the shed. She covered Ourecky before snugly tucking the blanket in around him. She hugged Taylor, obviously knowing it was for the last time. With tears in her eyes, she retreated back into the building.

Watching the convoy's lights grow ever nearer, Henson strained to transfer the fuel as fast as he could. Taylor, on the other hand, continued through his checklist at an almost leisurely pace. Aggravated, wanting to hasten the process, Henson asked, "You can't do that any faster?"

"I'm getting ready to fly two hundred miles over open water in a single engine airplane," replied Taylor. "So excuse me for not rushing, but I'm not about to do a half-assed pre-flight. Besides, I'll be finished before you're done cranking that knob, and I know I can beat Roberto's guys at this point, even if they're driving at full tilt, so just relax, Henson. I've done this before."

Taylor took off in the nick of time. As Henson disposed of the tools and passenger seats, he glimpsed the darkened Maule soar off in the distance towards the south. At the same time, three *Fad'H* trucks and a jeep arrived on the airport grounds from the north end.

Standing before the building, he watched as the convoy screeched to a stop. With junior officers and sergeants barking orders, a phalanx of soldiers swiftly dismounted and encircled the building. Henson's exhausted brain spun as he tried to concoct a plausible cover story, but nothing gelled.

"*Bonswa*, Matthew," said Roberto, stepping down from his jeep and swaggering towards the building. "A bit late to be out and about, isn't it? Shouldn't you be in Morne Bossa?"

Having no ready excuse for his actions, Henson simply replied, "*Bonswa*, Roberto." Certainly it was an awkward situation, but at least there was nothing incriminating for them to discover. The only loose end was Lydie, and if she talked . . . well, on the bright side, Carson and Ourecky were on their way home, and with any luck Lewis and his crew would still make it to the pick-up site without interference. *If that leaves me to face the music*, he thought, *then so be it*.

He heard the door swing shut behind him; glancing back, he saw Lydie step out into the moonlight. Smiling serenely, she sidled forward to stand next to Henson. She placed her hand in the small of his back and contentedly leaned her head against his muscled shoulder.

"*E bonswa*, Lydie," added Roberto, grinning as he tipped his cap.

For once, Henson was speechless, completely stumped by Lydie's gesture.

"So when the cat's away, the mice will play, eh?" observed Roberto.

"What?" sputtered Henson. "*Lydie?* Oh no, it's not like . . ."

"Whatever you say, friend," replied Roberto, winking. "And here I was, absolutely convinced that you were involved in some sort of clandestine operation, only to discover that you *were* involved in some sort of clandestine operation. So where is Mister Taylor this evening?"

"On an errand, I suppose," answered Henson truthfully. "I think he's on his way to Port-au-Prince, making a few stops along the way."

"Port-au-Prince? Oh. I had heard rumors that he was courting some *blan* woman there and would eventually move. Just as well. Truthfully, I've never been very fond of him, so . . ."

"What brings you out this evening, Roberto?" interjected Henson, watching the soldiers milling about next to the building.

"Just a routine drill," answered Roberto, examining a stopwatch. "I wanted to make sure that my troops are ready to respond immediately to the airport, just in case the Americans attempt to airland forces here. Eighteen minutes and nine seconds from a dead start. Quite good, no?"

"Quite good," affirmed Henson.

"Well, I'll leave you two be," said Roberto. "Perhaps dinner this week? Tuesday evening?"

Henson nodded.

"Maybe someday, when it's no longer too awkward, you might bring a friend with you."

"Perhaps."

"*Adieu*, Matthew," said Roberto. "*Adieu*, Lydie."

"*Adieu*," replied Lydie sweetly, nuzzling Henson and taking his hand.

Roberto turned and spoke to one of his lieutenants, and in moments the soldiers embarked in the trucks and the convoy was gone as quickly as it came.

Henson stood with Lydie, watching the string of red lights slowly dwindle to the north. Then he kissed her on the cheek and said, "*Mesi anpil*, Lydie."

"*Pa dekwa*," she replied, grinning. "*You're welcome.*"

"*Adieu*," he said, slinging his canvas bag over his shoulder. He turned and walked away to the south, towards Morne Bossa and his shortwave radio.

"*Adieu*," she called after him.

Wright-Patterson Medical Center, Ohio
4:30 p.m., Sunday, March 22, 1970

Breathing heavily, with her heart pounding in her chest, Bea anxiously walked down the hospital corridor clutching her infant child. Just an hour ago, Mark Tew had called to tell her that Scott and Drew had crashed at a remote test site two days earlier, and Scott had been seriously hurt.

Her emotions were a mix of fear and anger. She was terrified about the potential extent of Scott's injuries; when she demanded answers, Tew had been vague and evasive, only saying that Scott had undergone three surgeries before he was considered stable enough to be transported back to Ohio. She was furious that more information wasn't forthcoming, and she was angry with her husband for obviously taking too many risks.

She walked past several open wards, finally coming to an individual room normally reserved for senior officers. Stepping inside, she first saw Carson. His left hand was heavily bandaged, but otherwise he looked none the worse for wear.

Seeing her husband, her apprehension was dispelled by sheer relief. He looked like hell, but he was alive. Bare from the waist up, he wore blue pajama bottoms. There was a large bruise and scar above his right eye. A jagged incision, about a foot long, ran diagonally across his swollen belly. Sewn closed with thick black suture threads, the scar was puffy at the seams and obviously fresh. A transparent drainage tube, about the thickness of a finger, poked out of his right side. Three bottles of intravenous fluids hung on a rack next to his bed. Scarcely able to breathe, she stood at the foot of his bed, staring at him, watching his chest slowly rise and fall.

"Bea?" said Carson, moving to stand next to her. "Are you going to be all right?"

"Is he awake? Do you think he can hear me?" she replied, hugging Carson with one arm as she cradled the baby with the other.

"He drifts in and out," answered Carson. "I'm so sorry about Scott, Bea. It was my fault. I misjudged the landing, the left side landing gear collapsed, and we caught fire. I'm so sorry."

"You can be sorry later. Is your hand going to be okay?"

"Oh, sure. Just a minor burn. The docs said it will heal up in no time."

She sat in a chair next to the bed. Several medical charts hung from hooks on the wall, but her eye was drawn to a clipboard that held a neatly typed hospital transfer form; the document, which bore a heading of "Naval Hospital - Guantanamo Bay, Cuba" indicated that a "Major John Smith, USAF" had been cleared for priority medical evacuation to the hospital at Wright-Patterson Air Force Base.

Cuba? she thought. *What kind of "flight tests" could they have been doing in Cuba?* She remembered that Cuba had been prominent in the news just a few years ago, when America and Russia had almost been drawn into a nuclear war over missiles being shipped there.

"What's this?" she asked, pointing at the clipboard. "Who's John Smith?"

"That?" replied Carson furtively. "Uh . . . that was probably left there from the last patient."

"So you're telling me that you and Scott didn't crash in Cuba?"

"Bea, I *swear* to you that we didn't crash in Cuba. Honest."

Bea decided to drop the inquiry. Carson seemed sincere enough, but it really didn't matter since she wasn't going to hear a straight answer anyway.

Ourecky moaned softly and slowly opened his eyes.

Bea leaned over him, kissed his check and said, "I'm here, Scott, but this isn't the way it's supposed to be. You were supposed to come see *me* in the hospital."

Woozy, Ourecky tried to sit up. "Bea," he said softly. Carson bent over and turned the crank to raise the head of the bed slightly.

"Scott, meet your son," said Bea, holding the baby towards him. Reaching towards Ourecky with a tiny hand, the infant was awake and inquisitive.

"He's perfect," muttered Ourecky, barely conscious. "He's beautiful. Bea, I . . ." His eyes closed as he sank back into a drug-induced stupor.

"Like I said, Bea, he fades in and out," said Carson, standing upright. "It's the pain medications. He'll be back around before too long. Just be patient." He leaned towards the baby, grinning and cooing. "Scott's right. He is perfect. Does he have a name yet?"

Bea smiled. "Funny you should ask. Just before you two left on this last trip, Scott insisted that I name him after his best friend."

"Who's that? Someone from Nebraska?"

Bea laughed softly. "Not quite. Drew, allow me to introduce Andrew Carson Ourecky."

"He's named after *me*? You named *him* after me?" asked Carson incredulously.

"That's what Scott wanted," she replied. "I hope you're not too offended, Drew, but we've decided to call him Andy, at least until he gets a little older."

Staring at the baby, Carson asked, "And he's okay? No . . . problems?"

"He's perfectly healthy," answered Bea. "Ten fingers and ten toes, two little ears and a little button nose. I don't know what all the fuss was about. So, do you want to hold him?"

"I've never held a baby," admitted Carson. "I'm scared to death I might drop him."

"Well, he has Czech blood in him, so he's pretty damned tough." Bea kissed the baby on his forehead and held him out to Carson. "Just try not to drop him directly on his head."

Carson awkwardly cuddled the infant close to his chest. Captivated by his namesake, he looked at the infant and smiled.

Dubuission Homestead, Haiti
9:15 a.m., Friday, April 10, 1970

Taking care to remain well clear of Dondon, Henson remained in Haiti for another month to tie up loose ends. To his great relief, he had been informed that he would be reimbursed for all of his expenditures in the quietly executed rescue mission.

He closed out his Morne Bossa site and relocated to Taylor's old place. Since he was a man of simple needs, well accustomed to fending for himself, he gave Lydie a generous severance before returning her to her family in Limbe, about twenty miles southwest of Cap-Haïtien. Besides, although he found Lydie attractive and immensely desirable, Henson resolutely adhered to Abner Grau's admonition to never mix business and pleasure.

In what was certainly one of the slowest logistical exercises in modern history, he orchestrated the piecemeal movement of the dismantled spacecraft from the Dubuission brothers' hilltop. To accomplish this task, Henson hired the services of a flatbed truck, eight porters, and four watchmen.

The flatbed delivered building supplies—wood, nails, and corrugated roofing tin—to a rented staging site near Grande-Rivière-du-Nord. Every day for two weeks, pushing an oversized cart mounted on a truck axle and tires, the porters negotiated the rough road from town to the Dubuission property.

In the morning, they carried in construction supplies and in the evenings, they hauled out big crates of ore samples. The flatbed delivered the samples to the airport, where Henson locked them in the big shed thoughtfully provided by Taylor. A small contingent of armed watchmen—all former Fad'H soldiers vetted by Roberto—maintained a constant vigil on the building.

It took the porters an entire day to drag out the largest piece, wrapped snugly in a canvas tarp, which weighed roughly half a ton. Once all of the ore samples had been brought to the building at the airport, a chartered DC-3 arrived from Miami to transport them to the headquarters of Apex Exploration in Dayton, Ohio.

Well-paid by Haitian standards, the porters and watchmen were delighted to be of service; moreover, they were more than content to keep their mouths shut about the mineral deposits discovered on the Dubuission farm. Well, *most* of them kept quiet, anyway.

A man of his word, Henson provided the Dubuission twins with not just enough corrugated metal to put a roof on one house, but he had arranged for the delivery of sufficient roofing tin to outfit another house as well.

Jean Dubuission had quickly found success in the romantic department, particularly after the word circulated that a wealthy *blan* mining venture was *extremely* interested in the brothers' property. Assuming that Jean would soon be wed, Henson had provided enough materials for a second dwelling, since it just wouldn't be proper for Henri to live under the same roof as a married couple.

With his official tasks accomplished, Henson had packed up his gear and was due to fly to Miami the next morning. But still, there was yet one more thing to be done. Returning to the Dubuission farm for perhaps his final visit, accompanied by the brothers, he walked out into the burned field where the Gemini-I had come to rest. In the natural cycle of the earth healing itself, green shoots were already sprouting up in the midst of charred cane stubble.

In the bright morning sunlight, he noticed something odd. Taking off his hat, he stooped down to see scrape marks in the ground where the black dirt had recently been turned by a shovel. "What's this?" he demanded, pivoting around to face the brothers.

"Sorry," replied Jean sheepishly. "About a week after you left with the last piece, two European *blans* came up here to dig up some soil. They gave us some money. A few days later, they came back and offered us a *lot* of money for our land, enough to make us the richest men in the Department du Nord. So, anyway, we are thinking about buying some property closer to town and perhaps building our homes there. We've been on this hill a long time, you know."

"Are you angry with us?" asked Henri shyly.

"It's *your* land. Do with it as you wish," replied Henson, smiling as he replaced his hat. In truth, he already knew about the transaction. Roberto had confided that a Belgian mining company had secretly taken samples there and had found minute traces of valuable metals—including nickel and beryllium—in the soil. He had relentlessly hounded Henson for the past week, insisting that he quit fumbling around and secure the mineral rights to the Dubuission property before the greedy Belgians snatched them away. "Just don't ever tell anyone about what happened here. Fair enough?"

"But of course!" replied Jean.

Henson walked with them to their humble hut, the place where Baker had performed his impromptu miracle. Remembering that day, he looked at Henri's face. Although still shiny and pink, the scar on the Haitian farmer's forehead was healing well. Only a week ago, following instructions that Baker had left for him, Henson had removed the stitches.

"Henri, have you ever seen the ocean?" asked Henson. "It's very beautiful, and it's only a few miles north of here."

"I've been on this hill *all* of my life," said Henri, sighing as he rubbed his scar and shook his head. "I was a leper. Matthew, you know that. You and that *blan zanj* cured me."

"Then I would like the honor of taking you on your first excursion into the world," declared Henson. "We'll take a ride to Cap-Haïtien and perhaps drive around some towards the west as well. Jean, would you be opposed?"

"No. I would be honored, Matthew, after all that you have done for us."

"So, Henri, have you ever been on a motorcycle?" asked Henson, straddling his Motoguzzi.

"No, I haven't," answered Henri, climbing on behind him.

"Well, it's sort of like life," replied Henson, kick-starting the bike. "Just jump on and hold tight."

18

KOCHEVNIK

Burya Test Facility
Kapustin Yar Cosmodrome, Astrakhan Oblast, USSR
7:17 a.m., Wednesday, May 6, 1970

Yohzin arrived at the meeting place—a defunct launching facility—that General Abdirov had specified. Over the phone, his friend had said that he would not have time for their usual picnic lunch, but he wanted to meet all the same.

Yohzin motioned for Magnus to remain in the car and instructed his driver to pull away a short distance and wait for his signal. The secluded site resembled a ghost town in a Western cowboy movie. A pair of wary vultures gnawed at a fox's decaying carcass while more of the scavengers circled overhead, impatiently waiting their turn. An old telemetry antenna swayed against slackening guy wires. A rusting rocket gantry creaked in the wind. Parked at the base of the structure, a derelict armored car—with scorched paint and melted tires—was evidence of an explosion and subsequent fire during a refueling accident three years ago.

As he waited for Abdirov, Yohzin loosened his stiff collar. The sky was clear and the sun shone brightly. The grass and clover were ruffled by a gentle breeze, which also wafted a faint scent of kerosene. Curious, he leaned his head back and sniffed; the distinctive odor was an almost certain clue that a nearby rocket was being fueled in preparation for an imminent departure, but he was not aware of any impending launches.

Abdirov arrived, slowly climbed out of his sedan, and also directed his driver to park elsewhere. The two men greeted each other. Seemingly in a hurry, Abdirov looked at his wristwatch and said, "Let's take a hike, so I can show you our most momentous development."

The two men slowly walked about four hundred meters to the west, crossing a line of low hills that concealed the remote area—the former *Burya* test facility—that was Abdirov's domain. As they walked, Yohzin noticed that the scent of kerosene grew progressively stronger. He also heard men shouting and the clanking of tractors and other machinery.

The two men clumsily ascended to the top of an earthen berm that surrounded a disused fuel depot for test rockets. Handing him a pair of powerful naval binoculars, Abdirov pointed towards a launching pad roughly a kilometer away.

Yohzin was startled by the spectacle before him. Swallowing deeply, he was absolutely confident that he should not be witnessing this sight. Abdirov obviously assumed a great risk by taking him here. The pad held a venerable R-7 "Old Reliable" rocket, topped by an intermediate stage and an aerodynamic shroud encasing its payload. At the base of the launch gantry, men and machines worked feverishly to ready the rocket for a departure that was obviously imminent.

Adjusting the focus of the powerful binoculars, Yohzin spotted some surprising details. A stubby launch escape rocket, an almost certain clue that the spacecraft would be manned, protruded from the top of the sleek shroud. "Is that a *Soyuz*, Rustam? Is it manned?"

"It is," replied Abdirov proudly.

Yohzin swallowed. The rumors were apparently true; clearly, Abdirov was actually set to launch men from his secret facility at Kapustin Yar. "For what purpose?" he asked.

Abdirov answered, "We're sending up three men tomorrow to evaluate some new equipment and procedures, to prepare for establishing a permanent station in orbit. The station will be called *Krepost*. It will be launched on a UR-500 booster and is designed to remain in orbit indefinitely. We will launch the crews and supply ferries from this pad."

"So it's a reconnaissance mission?" asked Yohzin, handing back the binoculars. "Similar to the *Almaz* that Chelomei is building?"

"*Nyet*," replied Abdirov, frowning. "*Krepost* will *not* be a reconnaissance mission. Listen, it took us a decade, but we are finally making good on Khrushchev's big promise."

"Khrushchev's promise?" asked Yohzin. "*What* promise?"

"After Titov orbited in his *Vostok*, Khrushchev told the West that we would eventually put nuclear weapons in space, so we could drop them on our enemies whenever we chose. *That* was his promise."

"I don't think that was a promise, Rustam; it was more like a threat."

Abdirov chuckled. "Trust me, with Khrushchev, a threat *was* a promise. Anyway, Gregor, the *Krepost* will be armed with a powerful nuclear warhead. The General Staff had wanted it to be at least a hundred megatons, but the weapon currently in development will likely yield between thirty and fifty megatons. Once it is in orbit, we will rotate crews to keep it company. And that's not all: *that* station is just the first of many more. Soon, we'll have an arsenal in the sky, so it won't matter whether the Americans can build better bombers, more rockets, or submarines that we can't detect. We'll be able to rain hellfire on them at our discretion, regardless of whether they land the first blow and destroy all of our weapons on Earth."

Yohzin restrained an urge to gasp. "You're serious, Rustam?" he asked.

"Deadly serious."

"Why are you showing me this?" asked Yohzin, handing back the binoculars.

"Because the General Staff has finally granted me approval to have you transferred to my faculty," declared Abdirov. "Think of it . . . we will finally be working together again, *and* we will be sending men into space!"

It was a truly momentous occasion, thought Yohzin. And then he recalled that he had many other commitments, not the least of which was to deliver fresh information to the Americans, and that commitment entailed travelling to Moscow every month. "But what of my additional duties with the GRU?" he asked. "Will I have adequate time to fulfill those responsibilities?"

"That's all over," declared Abdirov. "As part of the arrangement that I made with the General Staff, I stipulated that you be relieved of your extra duties so you could focus entirely on preparing the *Krepost* for flight."

Yohzin frowned.

"Are you disappointed?" asked Abdirov. "I would have thought you would be delighted."

"Well, I am, but because of my background, the GRU counts on me to . . ."

"Don't worry, little brother," said Abdirov, displaying his contorted half-smile. "There's no need to deceive me. I know *precisely* why you are always so anxious to travel to Moscow every month."

Yohzin swallowed. "You do?" he asked quietly.

"Sure. I know that you are enthralled with the ballet and theater. Look, I will still be going there for briefings at least once a quarter, so I'll just make sure that you accompany me, and we can also make plenty of room for Luba and your boys. You'll be working plenty hard for me, so I'm not going to deprive you of the arts that you so love."

"*Spasiba.*"

"There's something else you should know," added Abdirov. "The network of *Krepost* stations will eventually be linked to *Perimetr*, once that system is made operational. Even if all of our strategic leadership is decapitated by the Americans, the warheads will still find their targets."

Although he wasn't fully cleared to know all of the details, Yohzin was conscious of an ongoing debate concerning *Perimetr*, a proposed automated firing apparatus for strategic weapons. Those who knew of it also referred to *Perimetr* by its macabre nickname: The Dead Hand.

In Yohzin's opinion, the entire Dead Hand concept was insane, but adding a space-borne component was even more deranged. Shaking his head in sheer amazement, Yohzin asked, "But, Rustam, tell me, are you truly comfortable with placing nuclear weapons in space?"

"Of course," vowed Abdirov. "Why wouldn't I be?"

Yohzin looked at his friend's disfigured face. Shuddering, he suddenly grasped that there could be no man more perfect for this mission. Abdirov had been to the far shore of the River Styx, and although he had returned—barely—he was indelibly scarred by his trial. So if he had suffered the ravages of fire, he obviously wasn't reluctant to sentence the enemies of the Soviet Union to a similar fate.

Burya Test Facility
Kapustin Yar Cosmodrome, Astrakhan Oblast, USSR
6:15 a.m., Friday, May 8, 1970

Soviet Air Force Major Pavel Dmitriyevich Vasilyev climbed down from the GAZ-66 van that had delivered the three-man crew to the launch pad. He was designated as the First Flight Engineer, the second-in-command of the flight that was scheduled to launch this morning.

As he gazed upwards at the R-7 rocket waiting on the pad, Major Petr Mikhailovich Travkin, the mission's Second Flight Engineer, stepped down after him. They were close friends with almost identical military pedigrees; after initially flying fighters, the two had graduated together from the Soviet test pilot school at the Central Scientific Research Institute in Chkalovsky in 1967 and were selected for the *Krepost* program a year later. Both were thirty-two years old. Unlike most of their fellow cosmonauts-in-training, both were married. Their spouses—Irina and Ulyana—and children had remained at Star City, home of the Yuri Gagarin Cosmonaut Training Center near Moscow, during their husbands' temporary assignment at the Kapustin Yar Cosmodrome. Their families had grown close; at Star City, they met often to dine or socialize together.

Their *Soyuz* flight—code-named *"Kochevnik"*—was a ten-day mission to evaluate the weapons deployment control systems that would eventually be installed in the final version of the nuclear-armed *Krepost*. A secondary goal was to ensure the men's adaptability to living in space, as well as their capacity to function effectively as an integral and harmonious crew. Since the planned *Krepost* missions would be of thirty to sixty days duration, living in intimate quarters, the question of their compatibility was crucial. It was something that could not be left to chance; the only alternative was to subject the three-man crew to an acid test above the atmosphere. This journey was a perfect opportunity for them to not only hone their skills, but also to forge a close rapport as a team.

As he considered the R-7, topped by the aerodynamically shrouded spacecraft that would be their home for nearly two weeks, Vasilyev was tremendously excited about the prospects of flying into space, but he also felt great trepidation. His reluctance had little to do with the treacherous dangers they would face, but rather grew out of his concerns with the mission commander, Lieutenant Colonel Vladimir Felixovich Gogol.

To imply that Gogol was an odd duck would be a gross understatement. He was short-tempered, belligerent, and abrasive. Short, squat, and heavily muscled—built like a stump with arms—the Ukrainian officer possessed the physique of an Olympic wrestler or power lifter. His grizzled face was like a gnarled potato left too long in the ground, replete with moles and warts. As best as Vasilyev could determine, there was absolutely no logical explanation for Gogol's selection as a military officer, much less a cosmonaut, except perhaps some mysterious connection to someone with significant political influence.

Taken at a glance, Gogol's qualities were few, and his shortcomings many, but there were some facets not so easily discernible. Although he could probably pass himself off as a rural bumpkin, entirely content to perch on the high seat of a Belarus tractor, tilling soil to plant a new crop of beets at a collective farm, Gogol's rough exterior concealed a shrewd intellect. He was immensely tenacious and by far the toughest man that Vasilyev had ever met.

Despite his myriad quirks, Gogol also happened to be the most experienced cosmonaut in the *Krepost* program, if not one of the most accomplished cosmonauts in the Soviet Union. While the other eleven pilots had never even ventured above the stratosphere, Gogol was already a veteran of three ten-day missions in space. Secretly flown in parallel with the Soviet's six Vostok flights, the long duration flights were purely military missions to practice reconnaissance procedures and to test specialized equipment designed to detect ballistic missile launches. When several less stouthearted cosmonauts declined the task, Gogol enthusiastically volunteered to fly in a Vostok stripped of its heavy ejection seat, but packed with an extensive battery of optical equipment, sensors, and related gear. Vasilyev smiled to himself; perhaps Gogol's selection to pilot the dangerous solo missions could be explained by another trait: besides being durable and intelligent, he was also immensely *expendable*.

Like an ugly and awkward mule that could somehow paradoxically match pace with thoroughbreds, Gogol was not trotted out into public view, so as not to deflect the spotlight from those more deserving. Aside from the secret nature of his missions, he was not nearly as photogenic or polished as the glorious cosmonauts—like Gagarin, Titov and Leonov—whose handsome faces graced billboards throughout the hero-worshipping Motherland.

But his flights were not without incident. In 1964, he had literally vanished after his Vostok's retro rockets misfired at the conclusion of his third mission. Initially presumed dead, he later reappeared, as if by magic, after spending the past forty months in China.

After his retros fired several seconds late, Gogol had endured a violent ballistic reentry in the tumbling reentry module of his Vostok. After crashing to Earth in a remote region dominated by snowcapped mountains and vast desolate plains, he struggled to make sense of his strange surroundings. Initially, just as he was trained, he waited patiently for the arrival of rescue forces, but none came, and he heard nothing but static on the radio. After days passed, he decided to take matters into his own hands.

A training manual could be written from his post-crash actions. Referring to the mission clock on the Vostok, he estimated his longitude and then determined his approximate latitude by building a makeshift sextant from sticks and string. He realized that he had likely come to Earth in northern China, a Communist country not exactly hostile but not particularly friendly, either.

He spent two days hiding his Vostok reentry module by physically rolling the spherical capsule—which weighed in at roughly two thousand kilograms—over two kilometers and into a deep ravine. He destroyed all sensitive equipment aboard before concealing the spacecraft with rocks and vegetation. He burned his pressure suit and any paraphernalia related to spaceflight and took care to only retain those items from his survival kit that would be associated with a conventional pilot. For clothing, he fashioned a makeshift burnoose from white parachute fabric, which he wore over his cotton long underwear. Then, he took off on foot across the expansive desert.

After wandering for days, he was found and adopted by a family of Mongolian nomads and spent the next two years sharing their felt *ger*, yogurt, and mutton. He helped them watch over their sheep and assisted them with tending to their camels and other livestock. He had hoped they would eventually wander far enough north that he could depart from them and slip across the border, but had no such luck.

Eventually, as he recounted to his interviewers during his lengthy debriefing, he realized that he would have to trek north on his own but knew that he would only be successful if he travelled in the summer months and with adequate preparation. As he learned rudiments of the language, he interacted with the various tribes that his adopted family encountered during their travels. He bartered scraps of parachute cloth to accumulate a larder of compact, nutrient-rich foods similar to jerky and pemmican. He made bladders of sheep's gut to transport water, sewed together a simple rucksack of sheepskin, and fabricated a pair of makeshift snowshoes just in case he encountered treacherous weather as he crossed the high mountain passes. As he traded and interacted with other tribes, he gradually learned more about the terrain and conditions to the north and west.

In particular, he focused on one old nomad's fearful description of a "road of iron ribbons" that stretched as far as the eye could see, from where the sun rose up to where it went to sleep at night. According to the nomad, the strange road was regularly traveled by fearsome "belching dragons." Gogol was confident that the wizened old man had witnessed a railroad north of the Soviet border, and if he could make it to those tracks, then they would guide him to human habitation and eventually home.

But unfortunately for Gogol, word passed amongst the nomadic tribes, describing the Caucasian stranger who had mysteriously appeared in the expansive wastelands. As spring arrived in 1967, just days before he had intended to begin his final trek home, those rumors attracted the attention of a passing Chinese Army patrol. Curious, they sought him out, and he was eventually flown to Beijing.

After weeks of intensive interviews, he convinced his Chinese interrogators that he was merely a fledgling Soviet pilot who had gotten lost on a training mission. He claimed that he had been unaware that he was deep over Chinese airspace when he ejected from his Yakovlev interceptor after it sputtered out of fuel. Reinforced with the weathered artifacts from his survival kit, Gogol's story was so consistent and convincing that the Chinese eventually concluded that he was the stupidest man to ever sit in a cockpit. Assuming that he was of no potential value, they repatriated him to Soviet Union in early 1968. Apparently anxious to divest themselves of the errant pilot, the Chinese did not make any demands in exchange for his safe return, but did send a bill for his lodging and meals, which the Soviets promptly paid. Coincidentally, Gogol arrived in Moscow on the very day—March 27—that Yuri Gagarin died in the crash of a MIG-15 near the town of Kirzhach.

Quietly feted as a Hero of the Soviet Union, Gogol was assigned to the *Krepost* effort at its inception. After working with the *Krepost* cosmonauts-in-training at Star City, he hand-picked Vasilyev and Travkin for his crew. From the very outset, he made it abundantly clear that he was the boss and would not tolerate insolence or insubordination in any form. Anxious to fly in space, and abundantly proud to be selected for the very first crew, Vasilyev and Travkin knuckled under and made every effort to please their gruff and eccentric taskmaster.

As difficult as it was to work with Gogol, Vasilyev was assured that his resilience and engrained survival skills would be invaluable if they landed in a remote wilderness area. Given the mission of the *Krepost*, that was a highly likely probability. When and if they deployed their warhead, it would almost certainly precipitate an all-out nuclear exchange between the East and West, if it wasn't already underway. Consequently, by the time they were ready to return to Earth, there was no guarantee that there would be much of the Soviet Union left, so they might be compelled to land almost anywhere in the world. Soviet cosmonauts already received considerably more extensive survival training than their American counterparts, an investment that had already paid off repeatedly in instances where cosmonauts had to wait hours and sometimes days for rescue forces to arrive in remote areas, and *Krepost* cosmonauts-in-training received roughly twice as much survival training as "regular" cosmonauts.

Having witnessed his prowess on survival training excursions, Vasilyev was confident in Gogol's ability to survive in all extremes, from pole to pole and all points in between. Gogol's physical toughness was mythical, the stuff of legends. Several months ago, after some soldiers died of exposure during maneuvers, the Soviet General Staff became concerned that their modern soldiers were becoming soft and losing the field skills that had served the Soviet Army so effectively during the Great Patriotic War. The General Staff instituted training to reinforce basic survival skills, with the centerpiece being a simple exercise in which every soldier—of every rank, from the bottom to the top—was expected to spend a night sleeping in the frigid cold of Soviet winter, with only a standard issue woolen greatcoat and a canvas poncho/groundcloth—*plash-palatka*—to keep them warm.

To demonstrate that they were equally as hardy as common soldiers, and to perhaps remind them that they were not prima donnas but still military men subject to the orders that must be obeyed by all,

Gogol and the *Krepost* cosmonauts-in-training were directed to undergo the nocturnal drill. The order was met by considerable grumbling and grousing. After all, they were carefully selected men who were headed to an entirely new frontier, and they wouldn't be wearing the anachronistic greatcoats into orbit. All but Gogol griped about the exercise. In the morning, the shivering cosmonauts-in-training emerged from their flimsy cocoons, exhausted from their fitful and fruitless attempts to sleep. As they foraged for dry wood to build a campfire, they found Gogol sleeping soundly on the ground, covered with a layer of fresh snow, snoring like a hibernating bear, with his neatly folded greatcoat and *plash-palatka* lying beside him.

Still, there was something significantly off about Gogol. Surely, the coarse cosmonaut was the absolute epitome of manliness, but Vasilyev had never once seen him in the company of a woman. It certainly wasn't that Gogol lacked for opportunity. Probably not by accident, the *Burya* facility's staff had a disproportionately large complement of attractive female workers, and the *Krepost* cosmonauts had their pick of the litter. It was tacitly but abundantly clear that any nurse or female technician was theirs for the taking, regardless of whether the object of their affections shared in the attraction. After all, cosmonauts-in-training were manly men engaged in patriotic pursuits, and they richly deserved any ready opportunity to vent off pent-up pressure.

While Vasilyev and Travkin remained faithful to their wives, most of their brethren did enthusiastically partake in the offerings—frequently with wild abandon—using and tossing away women like soiled tissues. Claiming that they were obligated to live life to the fullest since they would likely perish in the coming months, the neophyte cosmonauts were perpetually on the prowl, like lions stalking lambs from a captive herd.

So how could Gogol's aloof behavior be explained? It was sometimes the subject of quiet conversation, and the explanations offered were as numerous as the colors of the rainbow. Some claimed that despite his coarse exterior, Gogol was a steadfastly conscientious pilot, a shining example of Socialist virtues, focused entirely on his critical mission. Further, some reasoned that Gogol's self-discipline was why he naturally selected Vasilyev and Travkin for his team, since the three men effectively lived like monks at Kapustin Yar, nightly retiring to their solitary rooms to study mission documents and technical manuals.

On the other end of the spectrum, some asserted that Gogol was just genuinely anti-social. Still another explanation, based on rumors—wholly unsubstantiated—was that he had married into the nomadic family during his sojourn in the desert, and still yearned for his Mongolian bride. Vasilyev chuckled at the thought; he doubted that the Neanderthal could pine for anyone or anything except perhaps an unending supply of chow, cigarettes, and vodka.

But Vasilyev was still puzzled by Gogol's behavior. Once, late one night, momentarily under the influence of too much brandy, he had confided to Travkin his suspicion that Gogol might not be attracted to women because he was more comfortable—perhaps *too* comfortable—in the company of men.

Travkin immediately—and correctly—seized on Vasilyev's implication that Gogol might be attracted to men rather than women. Shocked, he scoffed that Gogol could not be a homosexual because he was not even remotely effeminate. Besides, he averred, a queer could not possibly become a cosmonaut, since any such tendencies would have been caught during the endless batteries of psychological testing that they were compelled to endure. To end the conversation, Travkin cautioned Vasilyev to *never* voice his theories again. While homosexuality was forbidden in the Soviet Union, it wasn't exactly unheard of, but it was folly—and a criminal offense—to accuse a senior officer of such sordid behavior without substantive evidence. In accordance with Travkin's advice, Vasilyev dropped the subject and they had not spoken of it since.

Still gazing up at the drab-painted R-7, Vasilyev heard a massive belch behind him. He pivoted to watch Gogol clamber down from the van. The stubby butt of a *Zolotoye Runo*—"Golden Fleece"— cigarette dangled from Gogol's thick lips; he was a notorious chain smoker. After enduring a noxious and incessant pall of smoke during their simulated flights, Vasilyev was elated that his commander would be compelled to abandon his annoying habit, at least for the duration of their mission.

Gogol paused, yawning as he scratched his crotch. In a cosmonaut tradition initiated by Yuri Gagarin just before his inaugural flight, he strolled to the rear of the drab-painted vehicle, unzipped the fly of his flight coveralls, and urinated on the right rear tire. Vasilyev and Travkin obediently followed suit, adding their yellow streams to their commander's.

Gogol zipped his fly, discarded his nearly spent cigarette and immediately lit another. The launch pad supervisor ordered him to extinguish it; he responded by blowing a cloud of smoke in the exasperated officer's face. Anxious to get aboard the *Soyuz*, diligently struggling to appear nonchalant, Vasilyev and Travkin made small talk and fidgeted as they waited for their boss to finish his smoke.

Crushing the butt with his boot, Gogol said, "Let's go, my little kittens." He grinned; after losing most of his natural teeth in a training crash early in his flying career, his smile was now a gleaming array of stainless steel caps.

Soyuz "Kochevnik," On Orbit
11:50 a.m. GMT (Greenwich Mean Time), Friday, May 8, 1970
GET (Ground Elapsed Time): 7 Hours 5 minutes / Revolution # 5

After reaching orbit and spending a few hours in space, Vasilyev felt like he could relax slightly. Their *Soyuz* spacecraft appeared to be functioning nominally, and they had been tentatively cleared to continue with their entire mission profile. The code name of their spacecraft and mission was "*Kochevnik*"—"*Nomad*"—derived from Gogol's operational nickname, rooted in his wandering interlude in Mongolia.

Vasilyev had heard plenty about the *Soyuz* even when he was training back at Star City, and much of what he had learned gave him ample cause for concern. Now considered fully operational, the *Soyuz's* history had been marked with significant growing pains, including a catastrophic episode on the very first manned flight of the spacecraft. Just three months after the terrible Apollo 1 fire that took the lives of three American astronauts—Grissom, White and Chaffee—cosmonaut Colonel Vladimir Komarov died after his spacecraft's parachute failed to open. Tragically, even before Komarov's flight, the first generation model—7K-OK—of the *Soyuz* was known to have been rife with technical problems, but Soviet party leaders apparently insisted that the mission go on as planned.

Designed by the esteemed Korolev bureau, the *Soyuz* spacecraft consisted of three interlocking components, stacked one atop the other. The first section was a cylindrical Service Module, which itself consisted of two parts. The first was a pressurized compartment that housed life support equipment, electric power supply equipment, various electronic instruments, and communications gear. The second portion, which was not pressurized, was located at the base of the Service Module and contained the propulsion systems—including the associated propellant storage tanks—necessary for maneuvering in orbit and returning to Earth. Two wing-like arrays of electricity-producing solar panels, one on each side, protruded from the Service Module.

Mounted on top of the Service Module was a bullet-shaped Descent Module, which the cosmonauts occupied during the launch and reentry phases of the flight. Covered with a heat-resistant exterior to protect the capsule and its contents from the brutal heat of reentry, the vehicle was also equipped with two parachute systems—primary and back-up—as well as a unique solid fuel rocket system that fired just prior to landing, so as to cushion the Descent Module's impact with the ground. Besides conveying the cosmonauts safely to Earth, the Descent Module could function as a temporary shelter, on both land and sea, in the event that rescue forces were delayed.

Definitely not designed for comfort, the interior of the Descent Module was woefully cramped. During ascent and descent, the three men were packed in as tightly as canned mackerels, seated almost on top of one another in custom-made form-fitted couches. Space was at such a premium that there was not adequate room for the three men to wear pressurized space suits; they wore flight coveralls instead.

Although it was a sophisticated spacecraft, the Descent Module's snug cabin resembled a disorganized storage locker; large boxes and bags of equipment, mostly survival gear to protect them in various environments, were strapped to the walls. Just getting aboard the Descent Module was a complicated and carefully choreographed ballet. Once embarked, moving around was virtually impossible.

Atop the Descent Module was the Orbital Module. The module was shaped like a stretched sphere and contained their galley, toilet, life support equipment, communications and various other instrument consoles.

On a normal *Soyuz* flight, the Orbital Module would contain an array of scientific experiments, but the *Kochevnik* mission was not for science, but to evaluate the systems and procedures for deploying an "Egg," the cosmonaut's nickname for the massive nuclear warhead that would eventually be fixed to the aft end of the *Krepost* station. To this end, an elaborate mock-up of the *Krepost's* weapons deployment system was mounted at the stern end of the Orbital Module.

During launch, the *Soyuz's* Descent and Orbital Modules were shrouded by a fiberglass aerodynamic payload fairing, which was topped by a launch escape tower, both of which were discarded on their way to orbit. In the event of a mishap during the ascent phase, the launch escape system—a solid fuel abort rocket—would fire, yanking the Orbital and Descent Modules from a malfunctioning R-7 booster. After reaching an altitude of approximately fifteen hundred meters, the launch escape system would fall away, and a special parachute—designed to open rapidly in such circumstances, would bring the crew softly—at least relatively so—to Earth.

Since arriving on orbit, the trio had left the Descent Module and set up shop in the Orbital Module. Compared to the Descent Module, the Orbital Module was spacious; it certainly didn't seem so during their simulated flights on the ground, but now that they were in orbit, given the benefit of weightlessness, it seemed almost enormous.

They rarely returned to the Descent Module, except to sleep. Although it was cramped quarters for three men, it was relatively comfortable for one and afforded some privacy. After the inner hatch was closed and the lights dimmed, it provided a welcome sanctuary from the noise and radios in the Orbital Module.

Their schedule was set so that one cosmonaut rested while the other two were alert and responsive at any time. There was considerable overlap built into the plan, as well as adequate time for non-mission activities—physical exercise, eating and personal hygiene—included in every day. At this juncture, Travkin was off watch. Leaving Gogol and Vasilyev to man the Orbital Module, Travkin was down below in the Descent Module. Although still charged with residual adrenaline from the launch, he was trying to snatch a catnap.

To this point, as they completed their initial chores after reaching orbit, the three men had been immensely busy. There had been no time for sightseeing, so Vasilyev was somewhat surprised to find Gogol at one of the observation ports, staring down at the Earth through a large pair of binoculars.

Looking over Gogol's shoulder, Vasilyev realized that they were making a daylight pass over Mongolia, and wondered if the grizzled cosmonaut might be actually be yearning for someone below. "Looking for somebody?" he asked.

"Just looking," growled Gogol, turning away from the window and heading for the food pantry. Vasilyev was constantly amazed at his commander's casual efficiency in this strange environment. While he and Travkin were still getting accustomed to weightlessness and would likely be clumsily struggling for days to come, Gogol was absolutely in his element. Physically powerful and possessed with unusual dexterity, he was perfectly adapted to weightlessness. Like a cross between an octopus and orangutan, he instinctively gained tremendous leverage by using a free limb to latch onto any nearby anchor. While the neophyte cosmonauts would usually launch themselves into an uncontrolled tumble whenever they tried to use muscle power to accomplish a difficult task, Gogol could apply his considerable strength to virtually any chore. As awkward as he might appear back on Earth, he was at home here.

After Gogol finished his snack, Vasilyev asked, "Do you want to update the mission log? There are several entries that you need to initial."

"*Nyet.* There's no need to rush with stupid paperwork. Anyway, I've been craving a smoke." Gogol took a pack of cigarettes from a chest pocket of his blue coveralls, extracted one, and stuck it in his mouth.

Suspecting that his commander had taken leave of his senses, Vasilyev blurted, "You can't smoke up here!"

"Sure, I can." He used a small screwdriver to loosen a switch on a control panel, and then connected some sort of homemade device to the switch's exposed wires. Vasilyev could see that the gadget consisted of a resistor, a rheostat, and a small tight coil of Nichrome wire. It was very much like the cigarette lighter that was a common feature in Western automobiles.

"But it's too dangerous!" declared Vasilyev. He was convinced that Gogol was playing some sort of practical joke on him, but he still cringed with dire thoughts of being immolated on his first flight in space.

"Hah! I smoked all the way through my second and third flights," boasted Gogol. He leaned his face towards the device and twisted the rheostat dial. The Nichrome coil glowed a bright orange. Gogol calmly lit the cigarette, inhaled deeply, and closed his eyes as he savored the smoke.

Fearful that the cabin would erupt in an inferno, Vasilyev gasped. He was astonished by Gogol's cavalier attitude about the volatile hazards. He grabbed a fire extinguisher from a bracket mounted next to the communications panel.

Happily puffing away, Gogol grinned at Vasilyev's apprehensive discomfort. "You're not in peril, kitten," he said. "Watch." To allay Vasilyev's fears, he used a simple demonstration to show why it wasn't dangerous. He let go of the lit cigarette, allowing it to float in the middle of the cabin. Within moments, the slowly burning tobacco consumed all of the oxygen in its vicinity, and the cigarette snuffed itself out. He relit the cigarette and revealed that as long as he drew on it occasionally, just slightly enough to surround the burning portion with a fresh supply of oxygen, he could keep it burning indefinitely.

Laughing, Gogol extracted a book of paper matches from his pocket. He struck a match and let go of it; the match burned for a few seconds, in an undulating orb of blue flame, and then likewise died of asphyxiation. He lit a second match; within seconds, it sputtered out as well.

From that point on, throughout the mission, much to Vasilyev's consternation, the commander slowly puffed on a cigarette almost whenever he was conscious. Thankfully, since Gogol slept a lot, even during their watch periods, he wasn't awake that often. He obviously believed that he'd paid his dues on his previous flights and was now entitled to a life of leisure, leaving the chores to the pair of rookies. On a positive note, he seemed to be in a much better mood in orbit; in contrast to his dour demeanor on Earth, he was almost jovial.

Soyuz *"Kochevnik,"* On Orbit
7:10 p.m. GMT (Greenwich Mean Time), Friday, May 8, 1970
GET (Ground Elapsed Time): 14 Hours 55 minutes / Revolution # 9

"Emergency action!" blared Gogol from the Descent Module.

"What the hell?" asked Vasilyev. He immediately felt for the key on the chain around his wrist and saw that Travkin, obviously in a panic, was doing the same. *Emergency action?* It made no sense; they were currently out of radio contact, so Control could not possibly have transmitted an emergency action message.

"Five seconds to retro rockets!" declared Gogol, floating up through the inner hatch from the reentry module. "Prepare yourselves for reentry, kittens!"

Gogol bit off a chunk from a small loaf of dark bread, glanced at his wristwatch, and counted down, "Three . . . two . . . one . . . zero." Exactly on the mark, he let rip an enormous fart. "Retros functioning normally!" Travkin and Vasilyev gagged as the small cabin's atmosphere was filled with an overwhelming stench.

As he listened to Gogol's guffaws, Vasilyev contemplated the consequences. Given the marginally efficient air scrubber of the *Soyuz*, already struggling to keep pace with a backlog of stale cigarette smoke, they would likely suffer the reeking aftermath of the commander's flatulence for another hour or longer.

*Soyuz "Kochevnik," * On Orbit
10:12 p.m. GMT (Greenwich Mean Time), Friday, May 8, 1970
GET (Ground Elapsed Time): 17 Hours 57 minutes / Revolution # 12

Gogol rummaged through the pantry and found himself a suitable snack, which happened to be the entire contents of Travkin's lunch meal. After stuffing his face, he jammed his body into a quiet corner and then immediately fell asleep. Within seconds, he was snoring loudly.

"I'm so hungry," complained Travkin quietly. "That bastard eats all of his chow and most of mine also."

"I'll share my lunch with you, Petr Mikhailovich," offered Vasilyev. "I won't let you waste away. Ulyana would be furious with me if you died of starvation."

Minutes later, an alarm blared, reminding Vasilyev that there was one crucial chore than Gogol could not duck.

"*Emergency action message!*" declared Travkin, checking the communications panel.

As loud and obnoxious as it was, the wailing klaxon failed to wake Gogol. Vasilyev shook him to jolt him from his stupor. It was the first emergency action drill that they had received, and woe to Gogol if they failed to act in a timely manner. "Emergency action message," he announced, leaning forward so that his mouth was directly over the commander's left ear.

"*Damn it!*" said Gogol, rushing towards the communications console. "*Emergency action?* Why the hell didn't you wake me?"

Gogol arrived at the console just in time to hear Control retransmitting the initial message: "*Kochevnik! Kochevnik!* Action . . . Action . . . Action . . . Enemy attack in progress . . . Deploy device on first available contingency target . . . Deploy device on first available contingency target. Authorization code: Six-Five-Zero-Three-Three . . . Authorization code: Six-Five-Zero-Three-Three."

As he had been conditioned to do in countless drills at Kapustin Yar, Vasilyev immediately memorized the all-important deployment authorization code. Even as Gogol composed himself, Travkin had already jotted down the message. Mere seconds after transmitting, Control demanded verification that they had correctly received the message.

Travkin held out his notes. Gogol plugged the jack of his communications headset into the console, nodded at Travkin, and keyed the microphone. "Control, this is *Kochevnik* commander. I verify receipt of emergency action message to deploy device on first available contingency target. My authorization code is Six-Five-Zero-Three-Three."

"Correct," replied Control succinctly. "Execute."

Vasilyev smiled. *And now the fun begins.* The deployment drill was a timed event, in which several steps, requiring the cosmonauts to work in unison, had to be executed correctly in a set timeframe. The entire process was electronically captured by an onboard recording machine, so that the exercise could be reviewed later on the ground. It was widely assumed that Gogol's trio was destined to be the first crew to occupy the *Krepost*, but that might not come to pass if they could not properly drop the Egg. A failure would be an indelible blot on their record.

Gogol immediately moved to the mock-up weapons deployment station installed in the stern end of the orbital module. It did not contain all the hardware of the actual station that would be installed on the *Krepost*, but it did contain a fully functioning version of the targeting computer, which was based on Gemini spacecraft computer technology stolen from the Americans. During an actual deployment, Gogol would man the deployment station, and the two other men would move to two lock stations dispersed throughout the station. In order to deploy the Egg, the last step required that each man had to enter an activation code—memorized before the flight, and unique to him—and turn a special key at their individual station. The three activation keys had to be rotated no more than one second apart.

The timing and spacing between the stations—at least on the actual *Krepost*—was specifically intended to prevent the possibility of two men overpowering the third. Even if they had the third man's key and activation code, it would be virtually impossible for one man to turn two arming keys within the span of a second.

"Key check," ordered Gogol, holding out the arming key that he constantly carried on a chain around his left wrist. Travkin and Vasilyev held out their left hands to solemnly display matching keys. Vasilyev nodded and said, "Key check good. Open the code locker."

Travkin swiftly spun the dial on the code safe's combination lock, opened the safe, and extracted the targeting book. He traced his finger down the target list, double-checked the mission clock, and announced, "Comrade Commander, the first available contingency target is Dallas, Texas: Latitude 34 Degrees 47 minutes North, Longitude 96 Degrees 47 Minutes West."

Gogol examined the target list. "I verify that the optimal target is Dallas, Texas: Latitude 34 Degrees 47 minutes North, Longitude 96 Degrees 47 Minutes West. I will enter the information."

As the commander, it was Gogol's sole responsibility to enter the target into computer, even though all three men had repeatedly practiced this aspect of the drill back at Kapustin Yar. The computer would automatically guide the Egg through the process of leaving orbit all the way to its intended destination. Once the coordinates were locked in, the targeting computer would execute a series of calculations and then assume total control of the station, at least until the Egg had physically separated in an independent reentry capsule.

After an actual Egg deployment, the three-men crew would remain aboard the *Krepost* as long as they could safely do so, then depart on the *Soyuz*. The timing was crucial, since the deployment sequence also initiated an automatic timer for a series of explosive charges that would be dispersed throughout the *Krepost*. To ensure that the vacated station would not remain intact, potentially subject to inspection and exploitation by the Americans, the self-destruct charges would detonate one hour after the Egg had detached, causing the *Krepost* to implode upon itself. Consequently, as soon as they completed the last step of the process, at least one of the men would immediately rush to the dormant *Soyuz* to initiate the complicated power-up process while the other two scrambled to collect the materials—mission logs, documents, personal belongings—that would accompany them home.

The Egg deployment process could only be interrupted by entry of a recall code—sent up from Control—into the targeting computer. *Would there even be adequate time to receive a recall code?* Maybe. Although this was notionally an immediate deployment, the phrase "immediate deployment" was actually a misnomer, since the deployment could be exceedingly prompt—even less than a minute, in some extreme circumstances—or could take an hour or longer.

Gogol hurriedly entered the coordinates and turned the red key that would lock the coordinates into the computer. This was really supposed to be a two-man procedure, with Vasilyev verifying the coordinates before the key was turned. In any event, a red light blinked on, indicating that the computer had not accepted the coordinates as entered.

"*Govno!* Shoddy damned gizmo!" blurted Gogol. "Must be some sort of short circuit!" He banged on the console with a flashlight, re-tried the process, but the computer still did not accept the coordinates.

"Would you like me to try?" asked Vasilyev.

"*Do it!*" ordered Gogol frantically. It was a significant departure from protocol, but Gogol obviously realized that he was running so close to the acceptable time margin that they would fail the exercise if he didn't yield the computer operation to Vasilyev.

Gogol held out the target list as Vasilyev dutifully entered the coordinates into the computer. Once he was confident that they were entered correctly, he looked back and asked, "Would you verify the target coordinates, Comrade Commander?"

"*Da.* I verify the target coordinates," stated Gogol, looking at the numerical readout over Vasilyev's right shoulder as he compared the displayed digits to the target list. "Lock the fix."

"I am locking the fix," replied Vasilyev, swiveling the red key that locked the coordinates into the computer. This time, a green light flashed rather than a red one. "Platform aligning. Stand by for deployment data estimate."

"Standing by."

Two minutes later, a second green light pulsed on the computer display. He checked it, and announced, "Comrade Commander, platform is aligned. Eleven minutes to braking rockets."

"I verify eleven minutes, Pavel Dmitriyevich," stated Gogol. "Go to your lock station and stand by for deployment."

Eleven minutes? thought Vasilyev. As he kicked lightly off the stern bulkhead to float back to his assigned station, he whistled shrilly through his teeth. In a real deployment, they would barely have sufficient time to retreat to the *Soyuz*. It would be an ugly situation at best; they would likely have to physically separate from the *Krepost* before even completing the power-up sequence. The procedure, known as a cold start, would afford them no recourse if the *Soyuz*'s systems did not correctly regain power.

"Insert arming keys. Rotate arming keys on my mark," declared Gogol.

"First Flight Engineer's arming key inserted. Standing by for your mark," replied Vasilyev.

"Second Flight Engineer's arming key is inserted. Standing by for your mark," called out Travkin.

"Five, four, three, two, one, mark." The three men turned their keys as if one, and a third green light flashed on the console, indicating that the exercise was successful and executed within the time limits.

"And *that*, kittens," declared Gogol, floating free and displaying his shiny metal grin as he lit yet another cigarette. "Is how it's *done*."

Soyuz "Kochevnik," **On Orbit**
10:14 p.m. GMT (Greenwich Mean Time), Saturday, May 16, 1970
GET (Ground Elapsed Time): 8 Days 17 Hours 59 minutes
Revolution # 140

Up until the eighth day, their flight had been a resounding success; the three men had existed harmoniously as a crew, completing all tasks and exercises in a timely manner, once Vasilyev and Travkin accepted the reality that Gogol had effectively come up here for an extended camping trip. Things were going well, and they had all but validated their compatibility as a tightknit unit. And then something happened that would forever alter the equilibrium between the three men.

Vasilyev had another hour to go before he went off watch. His bloodshot eyes watered from the clinging pall of cigarette smoke that permeated the spacecraft. He was anxious to go into the darkened Descent Module and close them, after he had a bite to eat and a chance to clean up.

As of an hour ago, the weapons deployment exercises were officially concluded, so now they were in the process of preparing for their return home. As part of the close-out tasks, he was disassembling the aluminum panels of the deployment console mock-up, so he could recover the critical computer as well as the recording machine. He would carefully pack these two items in a foam-padded box and then stow them in the Descent Module for the return trip.

Turning a wrench, he heard Gogol's voice from the far end of the Orbital Module and turned his head to see Travkin next to the inner hatch. Gogol floated beside him; his left hand grasped Travkin's shoulder.

"I'm changing the watch schedule," announced Gogol. "You'll stand the next watch. Consider this a stress test, so I can see how well you function without adequate rest."

Vasilyev groaned quietly. He had been awake for over nineteen hours; since he had been doing his chores as well as Gogol's, he was on the ragged edge of tired. And besides, Travkin had just come off watch less than two hours ago. *What gives here?*

"Like I said, you'll stand the next watch," explained Gogol. "Petr Mikhailovich and I will go sleep down below, where it's quiet. You'll stay up here and make sure that damned computer is secured, so we can carry it home with us."

Even though his thoughts were fogged by near-exhaustion, everything suddenly made perfect sense. Beside Gogol, Travkin grimaced, silently pleading for Vasilyev to do something, *anything*. He didn't appear to be the least bit enthusiastic about snuggling with Gogol in the cramped Descent Module.

Gogol yawned broadly, grinned, and said, "And I'm *very* tired, kitten. I need my rest, so don't bother us. Just bang on the hatch when your sand runs out."

"You're changing the watch?" snapped Vasilyev. "On what authority? The watch schedule was set by Control before launch. We're obligated to adhere to it as written."

"I am the *commander* up here, and I will change the watch schedule as I see fit!" ranted Gogol. "It's my prerogative, and you shouldn't dare challenge me."

Obviously recognizing what was in store, Travkin's eyes begged for Vasilyev to intervene.

Recognizing that little could be gained by direct confrontation, Vasilyev took a deep breath and composed his thoughts. "Fine," he said calmly, gambling on an indirect tack to circumvent Gogol's scheme. "You're *absolutely* correct, Comrade Commander. You are well within your authority."

"*Hah!*" answered Gogol. "I'm glad that you came to your senses long enough to recognize that."

Closing his eyes, Travkin looked like he was anxious to just disappear.

Vasilyev consulted the communications schedule, studied his wristwatch, and then officiously stated, "The next communications pass opens in forty-two minutes. I'll call Control to let them know we're altering the watch schedule, on your order, and then amend the logs accordingly. You can initial the entries after you wake up."

His face turning beet red, Gogol glared as he gritted his steel teeth. After over a minute of uncomfortable silence, he looked over his shoulder and spoke to Travkin, "Go help him with that damned computer, kitten. I want it properly stowed before I wake up."

Turning back to Vasilyev, Gogol added in a menacing tone, "I caution you to watch your step, Pavel, if you ever want to fly again. And be *very* cautious with your tongue when we get home. This ship is *my* oyster, and what happens up here stays up here."

Cursing, Gogol retreated into the Descent Module and slammed the hatch behind him.

Vasilyev breathed a sigh of relief. If nothing else, the short-lived clash erased any of his lingering doubts concerning Gogol's proclivities. Judging by the relieved expression on Travkin's face, the Second Flight Engineer now subscribed to his theory as well.

Now that some critical questions had been answered by Gogol's behavior, the matter shifted from the theoretical to the practical. Even as they heard the roar of snoring from the Descent Module, the pair of flight engineers kept their voices to a quiet whisper, conspiring about how they should react if a similar situation arose again.

Vasilyev was confident that the two of them could subdue Gogol if a scuffle ensued, but he also knew that it would be an entirely different matter if either one of them had to contend with the commander alone. Like infantrymen covering each other in a pitched battle, they vowed to stick together for the rest of the flight, always remaining in the same module, even if they were off watch. They cataloged the various objects—wrenches, fire extinguishers, helmets—that could be brought into play as weapons, if need be.

And Gogol was correct; neither of them would ever fly again if they elected to disclose the incident after they returned to Kapustin Yar. After all, it would be *his* word—the testimony of a Hero of the Soviet Union and a highly experienced cosmonaut—against theirs. After all was said and done, they would be fortunate if they were even able to fly a target drone at a gunnery range, much less a spacecraft.

Even though the matter was not spoken of again, and Gogol obviously restrained himself from making any more untoward advances, the remainder of the flight was an uncomfortable stalemate. But since they were slated to fly with Gogol again, on *much* longer missions, Vasilyev knew that he could not delude himself into believing that it might not happen again. Even though they were effectively yoked to Gogol, it was also readily apparent that they would have to be constantly prepared to protect themselves from him.

Filyovsky Park, Western Administrative Okrug, Moscow, USSR
9:35 a.m., Sunday, May 31, 1970

"So, General, what is so urgent that we have to meet in person?" asked Smith. "This extremely risky."

"My circumstances are about to change," explained Yohzin. "I've been relieved of my secondary duties with the GRU, so I won't have the same freedom of movement. I'm certain that I will still be dispatched here to Moscow, although not as frequently. Once a quarter, at most."

"But your primary duties at Kapustin Yar will not change?" asked Smith apprehensively. He looked to be on the verge of panic.

"Not at all. My superiors have just determined that I should concentrate my attention on my primary task, the testing of medium-range ballistic missile prototypes, so the RSVN leadership has decided that they will no longer loan my services to the GRU. As you might imagine, this caused quite some consternation with the GRU, but I'm confident that they will get over it."

Even though he was absolutely sure that the Americans would clamor for information about the *Krepost* if they knew he had access to the program, Yohzin had already decided that he would not divulge anything concerning Abdirov's nuclear-armed space station. In fact, he doubted that the Americans were even aware of the *Krepost*. So if they didn't know about it, he certainly wasn't going lead them to the door and direct their eye to the keyhole.

There was another reason that he was reluctant to reveal the *Krepost*. In a strange twist of conscience, although Yohzin now felt few qualms about selling out his country, he just could not bring himself to betray his old friend, Abdirov. Consequently, he had conjured up a scheme to provide the Americans with a steady diet of timely information about the medium-range ballistic missile prototypes testing program he had recently relinquished. He convinced the program's new director to let him read the technical reports, for the sake of continuity, just in case he was eventually compelled to appear at any form of audit or inquiry. It was an easy sell; despite the program's excellent record under Yohzin's leadership, the new director was almost overjoyed that he might be willing to shoulder some of the blame if problems arose in the future. So, even as he concealed Abdirov's *Krepost* from them, Yohzin should have a steady supply of other fresh grist for the Americans' mill.

"You will still be able to furnish us with the same quality of information as before?" asked Smith.

"Of course."

The American breathed a sigh of relief. "General Yohzin, I don't think I have to tell you that my bosses have been delighted with what you are able to deliver," he admitted.

Delighted? thought Yohzin. Of course he knew that they were delighted. How could the Americans not be overjoyed? They had obviously developed an insatiable appetite for the information he regularly delivered on a gilded platter. Surely, Smith here would probably build the rest of his career from this coup. But ironically, the intelligence that he was feeding them, as substantial as it was, didn't hold a candle to the secrets that he wouldn't share.

"While we appreciate you bringing this matter to our attention in such a timely manner, it's just a minor setback," said Smith. "If you're not able to come here on a routine basis, then we'll need to devise some other mechanism to communicate."

"Minor setback?" asked Yohzin. "Did you not hear me say that I will be spending virtually all of my time at Kapustin Yar? And don't bother to suggest that we communicate by some form of clandestine wireless. Everyone knows that the GRU has a substantial array of RDF equipment there, and they constantly monitor radio frequencies."

Everything Yohzin said was true, but he had another underlying concern. In his opinion, the only acceptable means of passing the information was when he visited Moscow. That was significant in other regard; although verbally negotiated but not formally written, his contractual agreement with the Americans was that he received a substantial payment—which was deposited in an account that would be available to him when he and his family went into exile in the United

States—for every transaction. To keep the accounting as simple as possible, for the purposes of the agreement, he was paid every time that he made a drop in Moscow, which had previously happened at least once a month. Now, he would be very fortunate if he made it to the capital city on a quarterly basis. Granted, he was already well on his way towards amassing a handsome sum that would finance an American Ivy League education for both of his sons through the doctorate level. He wasn't greedy, but he did want to ensure that he was adequately compensated for his efforts.

"No, Comrade General. Sincerely, this is merely a technical problem. We have people at Kapustin Yar who can get your information out."

Yohzin chuckled. The American was either very innocent or very arrogant, or perhaps just extremely uninformed. "*Bosh,*" he said. "You cannot be serious. The security at Kapustin Yar is absolutely airtight. You have *no* one there. It's not even remotely possible."

"We *do* have people there," replied Smith. "Granted, they aren't of James Bond's caliber, but they are certainly sufficient for this task. We'll devise a new dead drop for you to pass your information to them, and then they will convey it to us."

Clicking his tongue like he was admonishing a child, Yohzin shook his head. Servicing a dead drop at Kapustin Yar was not even worth discussing. As much as the Americans might have been accustomed to the steady flow of valuable information, a dead drop at the cosmodrome would entail far too much risk. For Yohzin, personally, the rewards could never be commensurate with the hazards.

"A new dead drop would be no simple task," said Yohzin." The GRU regularly conducts aggressive surveillance on everyone at Kapustin Yar, regardless of position or rank. They are sure to discover a new dead drop in short order. The GRU at the cosmodrome are mostly goons, but they are highly motivated goons who are very competent at their work. They are vigilant to a fault; it will take considerably more than a little subterfuge and sleight-of-hand to distract them."

"Agreed," noted Smith. "So, do you have any thoughts on the matter?"

"I'll think about it," answered Yohzin, watching Magnus yawn broadly and twitch his ears. He knew that the communications issue was a greater problem than the Americans could possibly anticipate. He saw a potential opportunity to exploit the situation and seized upon it. "Let me ask you, Smith, if I come up with a reliable means to pass information, then would you be willing to pay a bonus into my accounts?"

"Certainly, General. Without question."

19

MANDATORY RETURN

Mission Control Facility
Aerospace Support Project, Wright-Patterson Air Force Base, Ohio
2:25 p.m., Friday, June 12, 1970

Hobbling into the Mission Control on wooden crutches, Ourecky was met with a wave of boisterous applause. It was his first visit there since Mission Four. In the interim three months, he had endured several major surgeries and intervening periods of recuperation.

Carson trailed shortly behind. After shaking scores of hands and exchanging greetings with the controllers, the two studied the mission status screen projected at the center front of the large room. Crew Three was currently upstairs, executing Mission Five. The large graphic display showed that Jackson and Sigler were roughly half an orbit—approximately forty-five minutes—away from intercepting a Soviet military communications relay satellite.

"How was Walter Reed, pard?" asked Wolcott, strolling down from his glassed-in office at the rear of the facility.

"Pretty much the same as always, Virgil," replied Ourecky. "Hard beds and bad food. But they are serving a new flavor of Jell-O at least."

Ourecky had recently returned from a two-week stay at the premier Army hospital, where he had been treated for peritonitis, an inflammation of the lining of the abdomen, as well as surgery to further repair his liver, which had been partially ruptured in the crash. Because his liver was not yet fully functioning, his face was puffy and jaundiced.

"New flavor of Jell-O? Let me guess," replied Wolcott, lighting a cigarette. "Lemon? It looks like you've been eatin' it by the barrelful."

Gingerly sitting down behind a vacant console, Ourecky laughed and said, "No, Virgil, I come by this coloration honestly. The docs say it should clear up in a week or so. They're fairly certain that they've fixed all my plumbing."

"So you're headed home for a while?"

"I am. Bea has me penciled in for a lot of babysitting duty. Of course, I'm not sure how effective I'll be, because I'm pretty sure that the baby can move faster than me."

Carson propped Ourecky's crutches against the console. Drifting away from the conversation, he sat down next to Heydrich's workstation. "How are they looking, Gunter?"

"Not good," stated Heydrich, adjusting his glasses and tapping charred tobacco from a well-worn briarwood pipe. "Parch just doesn't have your finesse. He doesn't manage his burns as well. He consistently ends up with residual IVI's to null out, so he's depleted a lot of fuel just correcting errors. I'm keeping my fingers crossed, but I'm not holding out very much hope for them."

"Man, I feel downright terrible for those guys," said Carson, shaking his head and frowning. "Is there *any* chance they can complete the intercept?"

"Possibly," replied Heydrich, working a slide rule as he scribbled down numbers. "But highly unlikely. At this point, they've burned so much fuel that they're hovering right on the mandatory return threshold. I can't lay all the blame at Jackson's door, though. Sigler has flubbed his calcs at least four times so far, so that just compounded the problem."

Squeezing a tennis ball, Carson studied Heydrich's figures concerning the mandatory return threshold. According to the mission rules, no matter how close they came to completing the intercept, there had to be sufficient fuel set aside to maneuver back into position for reentry.

Once their maneuvering fuel dipped even slightly below the threshold, they were obligated to terminate intercept operations and shift their focus to safely returning home. Of course, all of this was contingent on the accuracy and honesty of the numbers; if the crew was overstating their PQI's—Propellant Quantity Indicated—they might have already exceeded the margin.

"Wow. They're mighty close to flying on fumes," observed Carson. "Are you absolutely sure that they're correctly reporting their PQI's? If they're gunning hot to make the rendezvous, they might be fudging the numbers."

"*Ja.* Perhaps," replied Heydrich, fingering his glasses up on his nose. "But if that's the case, Jackson will be headed to the moon on his next flight, because Virgil will kick his ass at least halfway there, and lunar gravity will pick up the rest of the tab."

"Well, if we give them the benefit of the doubt and accept these numbers as the whole truth and nothing but the truth," said Carson, flexing his sore fingers. "Is there *any* chance that Virgil and Tew will bend the mission rules for them, even slightly?"

Heydrich shook his head and chuckled. "After all the close calls you've been through, Drew, especially that last one? Not a chance in hell. Personally, I think that the rules have been bent as much as they ever will be, so I only see by-the-book flying from here on out."

Carson heard a tapping sound behind him and looked up to see Wolcott gesturing for him to come up to the back office. He did so, joining Wolcott as he gazed out over the nerve center through the big plate glass window. As they watched through the glass, Ourecky gingerly pulled up his shirt to display his collection of scars to the flight controllers. After they had all marveled at his new wounds, he pulled out his wallet and yanked out a string of baby pictures.

"He's proud of that baby," observed Carson. "He has to be the most obnoxious new father *ever*."

"Well, shucks, he sure went through hell to make that baby's acquaintance," said Wolcott, turning away from the window. "With all he's been though, you might reckon he's indestructible."

"Far from it, Virgil. He's just as fragile as the rest of us, and he sure knows it now."

"Amen to that, hoss, but I don't think he comprehends just how close he came to kickin' the can," observed Wolcott, shaking his head as he took a long drag from his cigarette. "The surgeon at Guantánamo said that if he had landed on their operating table just five minutes later, he would have bled out completely. As it was, he was on the table for twelve hours, all touch and go. *Everyone* at Gitmo donated blood for him, at least those folks who were compatible. He could probably take Iwo Jima all by his lonesome, with all the danged Marine blood that's coursin' through his veins."

"If I can ask, Virg, what happened that night? After we dropped off Scott at the airport, he flew to Cuba and I caught a boat to Miami. I've never heard any of the other details."

"You ain't heard the particulars?" asked Wolcott. "Ultimately, Ourecky owes his life to three people. That dude Henson apparently went to a *lot* of effort to call Homestead to let them know Ourecky was headed to Cuba. Homestead notified Gitmo, and Gitmo had an ambulance standing by at the airfield and *all* of their surgeons waiting at the hospital. The PJ—Baker—kept your buddy alive on the flight from Haiti, and that was no mean trick in itself."

Wolcott continued. "And last but not least, pard, he's got that bootleg pilot to thank. That guy—Taylor—knew they were burnin' a mighty short fuse, and he punched straight through Cuban airspace to arrive at Gitmo as quickly as possible. The Cubans scrambled two MIGs to intercept, but it didn't faze him. If he had flown the standard approach route in that crate of his, it would have been another twenty minutes to Gitmo. Ourecky wouldn't have survived."

"Do you know what happened to Taylor?" asked Carson. "Did they seize his plane at Gitmo? Henson implied that he was involved in some fairly shady dealings."

"They confiscated his plane initially," answered Wolcott, nodding his head. "Once Base Ops ran his license and paperwork, he spent the night in the brig. As I hear it, the next morning the base commander received a phone call from someone highly placed up on the food chain. By the time Taylor was dropped off at the flight line, his plane had been spot-cleaned of every drop of blood, his maintenance paperwork was completely up to date, his engine was tuned, and he had a full load of clean fuel in his tanks. He filed a flight plan for Port-au-Prince, and went on his merry way, no questions asked. He even had two F-4's escort him out to international airspace."

"Good."

"So how's Ourecky knitting up, pard? Have you been keeping a close eye on him?"

"He's doing well, all things considered. He's suffering a lot of pain in his gut, especially after all the surgeries and the last round of infection, so he's still popping pain pills like Pez candy. Bea and I constantly pester him about it, though. I'm sure he'll be weaned off them in a few weeks."

Wolcott chuckled. "Oh, that's a hoot, hoss. It's pretty danged sad when a guy is henpecked by his missus *and* his left-seater. Who can he turn to then?" He took a deep draw from his cigarette and asked, "So when will he be ready to go upstairs again?"

"Honestly? I don't know, Virg. Physically, he should be back up to speed in three to four months, but he has some tall psychological hurdles to clear before he's ready to go back up, if he'll ever be ready to go back up. And truthfully, Virg, he's not the only one."

"You'll burn through it, pard. So will he. Both of you will eventually recover the gumption to climb back up on the horse that threw you. I know; I went through some pretty dark interludes back during the War. I lost over half my crew on one mission. After we limped back to England, I spent over an hour in the shower, scrubbin' off bits and pieces of my co-pilot. It was kind of like that song from *South Pacific*, except that I *literally* did wash that man out of my hair."

Grimacing, Wolcott continued. "That was definitely my blackest moment, but even after that, I went back up. There just comes a time when that sense of duty takes over, and you do what has to be done. That's the way it was for me, anyway."

"I hope that's the way it is for Ourecky and me."

Wolcott nodded, leaned back in his chair, and said, "Here's a hypothetical question, pard. Let's assume that Ourecky takes a wee bit longer to heal up than you do. The next shot is in September. Jackson and Sigler are still on tap to catch that elevator. If you're healed up by then, how would you feel about flyin' with Sigler in the right seat? Could you two geehaw?"

Carson thought about it a moment and answered, "Honestly, Virg, I don't know how to explain it, but I just don't think I could make it happen with Sigler. Something clicks between Ourecky and me, kind of a yin and yang balance, and it works. I know I'll eventually go up again, but I would prefer to go up with Ourecky, provided he's cleared to go as well."

"That's what I thought," Wolcott said. "Well, in any event, we ain't sending you back up until you're completely mended. *Both* of you."

Carson looked at a clock on the wall. "They're in their contact window. We should be receiving the word anytime now, whether they made the intercept or whether . . ."

". . . they ain't," said Wolcott, stubbing out his cigarette in a glass ashtray. "I spied you talking to Gunter, pard. I assume that he lent you the hot scoop about their PQI's."

"So you don't expect them to make the rendezvous?" asked Carson.

"Nope," replied Wolcott, shaking his head as he toyed with a chipped pearl button on his starched denim shirt. "If the truth be known, I didn't much expect them to make the intercept when the hold-down bolts blew and they left the pad at the PDF."

"If you weren't confident in them, Virg, why did you send them up?"

"First and foremost, pard, despite what happened to you and your buddy during your Caribbean vacation, we still have a schedule to meet. The trains have to keep leavin' the station at the appointed times. Now, I pleaded with Mark to request a hold on the launch schedule, but he wouldn't hear of it. He was bound and determined to send up Jackson and Sigler, even though we both knew that they weren't ready."

"Why?"

"Why, pard? Simple. He's insistent on hoistin' some of the load off of you and Ourecky. For him, I think that a big part of the process was proving that Jackson and Sigler could deliver the mail. So we went round and round about it, and I finally folded my cards and let it happen."

"I still don't understand why, Virgil."

"Carson, did I ever tell you that I was raised on a cattle ranch in Oklahoma?" asked Wolcott, tilting his Stetson back on his bald head.

"Not directly, but I was certainly aware of it."

"Well, pardner, what with all the movies and TV shows about cowboys, I think people tend to form a lot of romantic notions about ranchin'. They probably think it's all like one big endless episode of *Bonanza*. In truth, ranchin' is hard, backbreakin' work. Moreover, it's a damned business like any other business, but there are some oddities to it. For example, long before you drive the cattle down to the stockyard, you can predict almost to the penny what you're going to make off the herd and whether it's going to be a good year or a losing year."

Gazing through the glass, Wolcott continued. "Every time you lose a head, that's another notch cut out of your profit margin. My daddy used to fret all night when bad thunderstorms blew in. He would tally the lightning strikes, and when the storm passed, he could tell you how many carcasses we would find out on the pasture after the sun rose. And damned if he wasn't smack dead on most of the time."

Wolcott took a sip from his coffee and added, "So, long before we saddled up for the last drive, even though there might be some minis-cule fluctuations in beef prices, my daddy plainly knew what payout would be waitin' at the teller's window at the stockyard. My point is that you still have to keep plugging away, even when you already know the outcome, and even more so when it's an outcome that ain't necessarily in your favor."

"I understand," Carson said. "But you're still planning to send them up on Six?"

"I am, pardner," replied Wolcott, nodding. "Not that it's my decision to make. Mark Tew is just dadblamed adamant that if those rascals don't get it right this time, then they'll get it the next. So I suppose I'll just keep on placating him, until we run out of rockets or until you two are ready to fly again."

"Sure seems like a waste of hardware."

"Yup. Yea and verily," answered Wolcott. "But let's ride another trail for the moment. When do you expect the flight surgeons to clear Ourecky to fly?"

"To orbit?"

"No, pard. When do you expect that he'll be sufficiently recuperated to go back up with you in a T-38?"

"Probably mid-September, sir. October at the latest."

"Good. When you feel confident that he's ready to ride cross-country with you, I have a job for you two," said Wolcott. "It's more of a public relations venture than an operational mission, but we need some positive public relations in our favor, so I'd like to send you two out to knock on some doors on behalf of Blue Gemini. I'll read you into the specifics when the time comes."

"Sounds good, Virg."

A collective groan rose from the floor below. Although they couldn't hear his words, they turned to the window to watch as a somber Heydrich read aloud from a Teletype print-out.

Moments later, Heydrich knocked on the door and stuck his head in. "Virg, they botched the intercept. They finished the closing maneuver too high, so they're sitting less than a quarter-mile away above the target. And they've already exceeded the mandatory return threshold."

As an anguished look crossed his weathered face, Wolcott nodded.

Carson imagined the frustration that the guys upstairs must be feeling. The Soviet satellite was probably hovering right in front of their nose, taunting them. Eventually, it would progressively drift away since it was moving slightly faster in a lower orbit.

A quarter-mile probably seemed inconsequential, and he was sure they were tugged by the temptation to continue maneuvering, but closing that gap would expend virtually all the OAMS thruster propellant they had left. Not only would their safety margins for reentry be erased, but there also wouldn't be sufficient fuel to adequately execute a close-in inspection or deploy their Disruptor. *What a tremendous waste,* he thought.

"What do I tell them?" asked Heydrich.

"Bub, there ain't *nothing* to tell them," drawled Wolcott. "Shucks, Gunter, you know the rules. Abandon all intercept operations and maneuver for reentry. That simple. They know that, so they should already be executing without us having to cite chapter and verse."

Heydrich nodded. "I just wanted to hear it from you, Virgil."

"And you did. One more thing, pard: hustle someone upstairs to inform Mark Tew. Make sure he knows that we're shuttin' down this rodeo."

20

ON THE ROAD WITH DIONYSUS

Aerospace Support Project
8:36 a.m., Friday, September 11, 1970

In the throes of a terrible cold, Wolcott was overcome by yet another hacking spell. He considered lighting another cigarette, hoping that would calm his throat, but unwrapped a Vicks medicated throat lozenge instead. If there was ever a week *not* to be under the weather, this was definitely it. Mission Six was due to launch in five days, and there was yet much to be done to prepare for it.

Despite their last fiasco in June, Jackson and Sigler were bound again for orbit, this time to target a suspected OBS platform launched two months prior from the Soviet launch complex in southern Kazakhstan. If the pending mission wasn't enough to keep Wolcott on pins and needles, virtually everyone assigned to the Blue Gemini was stricken with the same debilitating cold. In particular, Gunter's mission control crew had been decimated by the passing illness; less than a third of his staff was available to work, and some of his key players would probably have to double-shift until their reinforcements rose from their sickbeds.

Listening to Merle Haggard's gravelly voice singing "Okie from Muscogee," Wolcott smiled to himself. Tew did not tolerate music—*especially* country music—in the office. But at present, Tew was laid up at home, felled by the cold that he had probably passed to Wolcott, so Merle crooned at full volume from the cassette player on his desk. Savoring the last strains of the song, he listened for a loud click that signaled the end of the tape and then turned the cassette over to listen to one of his personal favorites—"Mama Tried"—of Merle's repertoire.

Although the pre-launch details of Mission Six were pressing, Wolcott took a few moments to review some intelligence data. Wiping his sore nose with a tissue, he flipped open a folder and perused a stack of photographs.

Although Wolcott had no way of knowing, an agent had placed his life at grave risk to hurriedly snap the images at the *TsKBEM* "Central Design Bureau of Experimental Machine Building" aerospace facility in Podlipok, Kaliningrad. They showed a full-scale mock-up of the three-man *"Salyut"* space station currently under development.

Although the program was guarded in the usual secretive folds that enveloped all Soviet aerospace progress, the *Salyut's* mission wasn't unduly different than that of the planned Skylab space station that NASA intended to launch after the Apollo moon missions drew to a close.

Sniffling, Wolcott scrutinized a report that revealed troubling details about the otherwise innocuous *Salyut*. According to the file, another design bureau—*OKB-52*, under the leadership of famed Soviet aerospace engineer Vladimir Nikolayevich Chelomei—was laboring on yet another manned space station.

The second effort—tentatively named *"Almaz"*—was shrouded in even greater secrecy than *Salyut*. Initially, US intelligence officials suspected that *Almaz* and *Salyut* were competing projects by the rival bureaus, particularly since the *OKB-52* organization had been ordered to transfer a significant amount of *Almaz* technology to the *TsKBEM* bureau to expedite development of the *Salyut*. But the intelligence report alluded to a more sinister purpose for *Almaz;* it wasn't merely a competing design that had been left by the wayside in favor of *Salyut,* but a totally different vehicle intended solely for military purposes.

Masters of obfuscation, the Soviets apparently were using *Salyut* as a cover for *Almaz,* since *Almaz* would be launched in roughly the same timeframe as *Salyut.* One detail of the report caused Wolcott to gasp in dismay; the *Almaz* was to be outfitted with a 23mm Nudelman automatic cannon, apparently to repel US space vehicles sent up to intercept it. *The Soviets arming a manned space station?* It was just too much of a coincidence to be a coincidence.

Merle was silenced in mid-song as the cassette tape came to an abrupt end. Re-reading the section on the automatic cannon, Wolcott listened to the annoying tick-tick-tick of the wall clock on the opposite wall. *Were the Soviets on to them? Was there a leak within the Project?*

Cap Nellis Air Force Base, Las Vegas, Nevada
9:25 p.m., Monday, October 5, 1970

As he listened to the drone of turboprop engines, Eric Yost was a *happy* man, bound for his personal version of heaven, courtesy of the US Air Force. He was returning to the States after a fifteen-month stint at a DEW—Distant Early Warning—Line radar site in Greenland. As of tomorrow, his Air Force service would be complete, so a lifetime of military retirement checks and free medical care awaited him.

Ironically, Jimmy Hara was responsible for his resurrection. Although Hara had made it abundantly clear that he could turn Yost over to appropriate authorities to be prosecuted for treason, he was willing to cut Yost a break if he promised to never again speak of Morozov, Hangar Three, or anything else that happened at Wright-Patterson. If he did ever see fit to mention any of those things, declared Hara, he would be sitting in a cell at Leavenworth before the sun went down and would remain there for the remainder of his natural life.

As he dispatched Yost to Greenland, Hara admonished him to keep his nose clean—no drinking and no gambling—for the duration of his purgatory tour, and when he returned to the States, he could retire as if nothing had ever happened. The only other provision that Hara specified was that Yost would never again set foot anywhere in the vicinity of Wright-Patterson or Dayton; Hara advised him that it would be wise to steer entirely clear of the entire state of Ohio, unless Yost felt compelled to settle his gambling debts with his loan shark creditors.

Because he was returning from a remote site and had no family anxiously awaiting his homecoming, the Air Force offered to fly Yost to any installation in the United States—other than Wright-Patterson—for his retirement out-processing. Since he had no inclination to see snow or be cold ever again, a warm and welcoming climate was his primary consideration.

For a while, he had considered retiring in Hawaii but realized that living on an island would be a little too claustrophobic for his tastes. Finally, he settled on Nellis Air Force Base; located in the desert immediately proximate to Las Vegas and its abundance of glittery casinos, it was the ideal place for Yost to begin his new life. He smiled as he looked out the circular window of the C-130 transport that was delivering him and two heavily crated F-4 jet engines from New Jersey to Nellis. Glimmering neon lights beckoned in the distance.

In his first six months of his Greenland ordeal, he had adhered to Hara's instructions and diligently salted away every hard-earned nickel of every single paycheck. Like an Arctic hermit, he stayed on the radar site compound, never venturing out to the nearby Danish community a few miles away. During that period, his existence was simple, almost monk-like. He minded his business, pulled his guard shifts, ate all of his meals at the chow hall, and slept. Despite a wealth of enticing opportunities, he even abstained from drinking and gambling.

Then, seven months into his stay, he finally surrendered to the urge and threw his ante into a barracks poker game. The rest was history; it was as if he *couldn't* lose. In time, he had his run of the garrison; he could do anything he wanted, except leave.

He lived like a king, dwelling in a Quonset hut normally occupied by ten men, along with a native woman who cleaned, cooked and otherwise catered to his every need. When he finally left Greenland, *everyone* at the camp—officers, enlisted, civilian contractors, Danes, natives, you name it—still owed him money.

He patted the duffle bag jammed under his webbing seat. To imply that he was flush with cash would be a tremendous understatement. Except for his shaving kit and a couple of intricately carved ivory figurines to commemorate his Arctic sojourn, the duffle was *absolutely* stuffed with money. He bore nothing else but the uniform on his back and aspirations of hitting it big—*really* big—in Vegas.

Giddy with anticipation, he felt a phantom itch and rubbed the gap once occupied by his left index finger. He had lost the digit to frostbite after falling asleep at his guard post on his third month at the radar site.

After amputating the blackened finger, the doctor recommended that he be evacuated to the States, but the camp commander was under strict orders to keep him up North, except in truly life-or-death circumstances. Looking back on the episode, he smiled at the commander's stubbornness; Yost was now a wealthy man as a result of his protracted stay, and a sizeable chunk of his net worth came straight out of the commander's coffers.

Holding out his hand to examine his remaining digits, he mused on whether it was worth the sacrifice, and then assured himself that it was. After all, it was just a finger and he had nine more. But as he snugged his lap belt for the final approach to Nellis, Yost could have no inkling what turmoil his missing finger would eventually bring.

Over Wilber, Nebraska
10:30 a.m., Wednesday, November 4, 1970

"Hey, Scott, wake up!" announced Carson over the intercom loop. "Look over there to your right, about two o'clock. That's your hometown, isn't it? Isn't that Wilber?"

Rudely snatched from a sound stupor, Ourecky glanced down at the snow-covered landscape as Carson pushed the T-38 into a descending right bank. Suddenly feeling queasy, he picked out the squat gray facade of Saint Wenceslaus as a landmark and confirmed their location. "Yeah, Drew, that's Wilber. Home sweet home."

"Hey, if you vector me in, we'll bust a low pass on your parents' farm. Wouldn't that be a hoot? You can call them later to tell them it was you."

"Not a good idea," replied Ourecky, wiping drool from his chin. He took a swig of pink Pepto-Bismol and then stashed the empty bottle. He swung his oxygen mask into place and locked the bayonet clip to secure the rubber facepiece over his mouth and nose. Hoping to clear his head, he sucked in the cool flow for a few seconds before adding, "It would just scare the crap out of the cows. Papa would spend the rest of the afternoon collecting them out of the cornfields and herding them back into the pasture. Let's just focus on getting to where we need to go, okay?"

"Okay," replied Carson, resuming his heading as he put the T-38 into a gradual climb. "Still feeling under the weather?"

"Yeah. It's been an excruciatingly *long* damned week." Ourecky looked at a photo scotch-taped to his instrument panel; the Polaroid picture showed a beaming Bea holding Andy, their infant son, swaddled in a teal blue baby blanket. "I'm just anxious to come off the road, Drew. These junkets are just wearing me out."

"I hear you. At least this is the last stopover before we head home. One more night of carousing, and it's back to home and hearth. For you, at least."

"Not a moment too soon," Ourecky said, closing his tired eyes. "And Drew, please do me a favor when we check into lodging. Make sure that you don't draw a VOQ room next to mine."

"Why?"

"Why? You *know* why. Another stop, another girl. Please, Drew, *please* . . . give me a break. Just try really hard to draw a room down the hall or something."

Carson made a radio call to the regional air traffic controller before answering Ourecky. "You never know, Scott. There might not be a girl on this stop. It sure wouldn't be the first time."

"Maybe, but it would sure be the first time since you and I started travelling together. I'm surprised you didn't find one in orbit."

After their harrowing ordeal in Haiti, Ourecky had thought that his most difficult days were behind him. He had been woefully wrong. Although the flight surgeons had not cleared his return to orbit, they had signed his "up" slip to fly with Carson.

The two had spent the past six weeks travelling cross-country, dropping in on base commanders. The overt purpose of their visits was to coordinate emergency landing contingency plans for the Gemini-I, but Ourecky was also very aware that they were conducting a subtle public relations campaign.

At higher levels within the Air Force, the scuttlebutt was quietly circulating about Blue Gemini, which led to considerable speculation and innuendo. Consequently, Wolcott and Tew had launched a protracted campaign of preemptive PR strikes, selectively spreading the word to influential players that the Air Force was in the manned space-flight business.

Ourecky sighed, trying to remember their next destination, only recalling that it was a Strategic Air Command base in the Midwest. The world seemed to swish by in a constant blur.

Although the faces and real estate changed, the drill was effectively the same on every visit. After an incognito landing, they would change from flight gear into their dress uniforms to call on the base commander. Usually, there would be a reception where they would be introduced to the commander's key staff, receive a short briefing about the base's operations, and then be escorted on a tour of the base's facilities. Yet another flight line, row after row of aircraft, runways, hangars, etc., etc., etc.

At this point, he was relatively sure that he had seen just about every airplane in the Air Force inventory, except for possibly the really spooky stuff kept stashed in secret bases in the desert Southwest. And although he flew in what was inarguably one of the most secret programs in the Air Force, even he didn't have clearance to visit some of *those* places.

After their tour, they would provide the base commander with a one-on-one briefing about the Project. The intimate briefing always concluded with Carson solemnly handing the commander a sealed envelope that contained detailed instructions for an emergency landing.

Within the package were the communications and landing procedures, guidance for safely handling the Gemini-I's pyrotechnics and other hazardous materials, security measures and other general instructions for safeguarding the vehicle until recovery specialists arrived.

As an unexpected consequence, after just a few weeks on the road, the pair had evolved into invisible celebrities on a tremendously exclusive circuit, with every base commander vying for one of their secret visits. But what was once a novelty was now just another grind.

Ourecky was not overly bothered by the flying aspect of the junkets. With few exceptions, conscious that his companion was still on the mend from multiple surgeries, Carson kept the T-38 straight and level for most of their cross-country segments. He still managed to accumulate plenty of "aerial combat maneuvering" practice, but typically logged that on short dog fighting hops with the hottest pilots and hottest planes available at the bases they visited.

And being away from home didn't bother Ourecky. While he wasn't fond of being separated from Bea and their new baby, a night's slumber in a VOQ bed was a far sight more tolerable than trying to snooze in the Gemini-I or the Box.

In reality, it wasn't the travel and official business that weighed so heavily on Ourecky; it was the incessant entertainment. Wolcott had insisted that it was just another part of their job, to establish rapport with the generals and other VIPs who might eventually hold sway over the Project's budget and operations.

Within the military's intensely structured environment, even schmoozing was a regimented activity. With few exceptions, *every* stop culminated with an interlude at the base Officers Club. Typically, they squeezed in two bases a day, so Ourecky could anticipate a heavy gut-busting lunch before cramming himself back into the T-38 to whisk off to the next base for the second round.

In the evenings, it was a foregone certainty that they would be feted with platters piled high with hors d'oeuvre, thick steaks and endless rounds of high-octane drinks. Invariably, the Officers Club function was an awkward event, since the base commander usually felt obligated to introduce the mysterious visitors to his more senior officers, even though he couldn't provide any specifics about what they did or why they were visiting. He could only imply—in the vaguest of terms—that their importance far exceeded their relatively meager rank.

Out of a sheer need for self-preservation, Ourecky curtailed his alcohol consumption to the minimum number of drinks necessary to appear sociable. Besides trying to maintain a clear head, he was mindful that his liver was still healing, regardless of the doctors' assurances that he was healthy. Additionally, he had just never developed the same degree of tolerance for liquor that Carson exhibited. So as the evenings drew on, he took it upon himself to be the lucid voice of reason to ensure that Carson applied the appropriate "bottle to throttle" time limitations on imbibing before flight. Of course, the testosterone-infused pilot usually found ample reason to abandon his bender and beat an early retreat to his VOQ room.

Another negative aspect of the travels was that their hectic schedule afforded scarcely any time to keep up with their physical training regimen. Besides, hitting the gym required the self-discipline to climb out of bed to actually go there. As a result of their largely sedentary existence and constant overindulgence, the two men were woefully out of shape. Both had packed on at least ten pounds and just barely fit into their uniforms and flight suits.

So if every evening had its set rituals, every morning had prescribed rites as well. The dawn of their duty day normally saw Ourecky pounding on his fellow traveler's door until Carson unpeeled himself from his bimbo du jour. And every morning, bolstered by a Spartan breakfast of black coffee and plain toast, Carson would soberly make penance, swearing that he would amend his ways and revert to a more austere lifestyle. And his word would stick, at least until the next stop in the itinerary, when the sirens of excess would beckon, and once again he would succumb to the pleasures of the flesh.

Yawning, Ourecky massaged his throbbing shoulders and reminded himself to let out his parachute harness before he put it back on. Months ago, when he had returned to the Project from his brief stint in

California, he could never have pictured the way that things would turn out. Travelling with Carson was like being on a supersonic road trip with Hugh Hefner, where every night was another taxpayer-funded Roman bacchanalia brimming over with rich food and strong drink.

At first he felt that Carson was merely blowing off pent-up steam, particularly after their close call in Haiti, but now he was concerned that the pilot's psyche might have been irreparably damaged by the traumatic interlude in the Caribbean.

Carson was still an excellent pilot, but even Ourecky noticed that he was allowing little details to slip. And while it was likely they would fly Mission Seven in February—provided that Ourecky was cleared in time to participate in pre-launch training—Carson rarely spoke of it, and when he did, it was seldom with any degree of fervor.

Ourecky suspected that Carson had lost confidence in the equipment, even though the Titan II and Gemini-I had both functioned flawlessly during the last two missions. If only Jackson and Sigler had performed to a similar standard, then maybe they would be considered for Mission Seven and subsequent flights, but they had flubbed both Five and Six.

Rampart Air Force Base, Idaho
5:55 p.m.

"And subject to your questions, that's all, General," said Carson, concluding the deskside briefing. He solemnly handed General Dale Astor—a distinguished-looking tall man crowned with a dense mane of silver-gray hair—a sealed envelope bearing the emergency landing protocols.

"*Great* presentation, guys," declared Astor. "Thanks so much for enlightening me."

"There's something else, General," said Carson. "General Wolcott wanted you to have this, with his compliments." Carson proffered a circular embroidered patch to Astor. "It's our mission patch."

Astor examined the souvenir like it was a rare artifact from ancient times. In a sense, thought Ourecky, it was almost that rare.

Unbeknownst to Tew, who would certainly not approve of such a thing, Wolcott had ordered the production of a very limited number of the patches. Carson and Ourecky had never worn the insignia on their uniforms when they rocketed into orbit, and never would; the cloth emblems were intended strictly as "gimme" tokens, to curry favor with high-ranking officers and officials who might eventually influence the future of Blue Gemini. The design of circular patch was intentionally vague; it depicted two lasso-wielding horse-mounted cowboys, super-imposed above the earth, as if chasing wayward cattle over the horizon. Like similar patches produced for classified programs, there was no text or other explanation embroidered on the emblem.

Astor chuckled. "I can definitely see Virgil's hand in this," he said. "It reminds me of that song 'Ghost Riders in the Sky.'"

Ourecky nodded. Astor was right, probably in more ways than he realized. The cowboys of the song and Western legend were doomed to ride forever in the sky, perpetually chasing the Devil's herd. As it was, he and Carson seemed destined to the same fate.

"Hey, look, I'm going to haul you two boys down to the Club and treat you to the hugest corn-fed porterhouse steaks you've ever laid eyes on," said Astor. "But before we do that, I have a little favor to ask of you."

"Certainly, sir. What do you have in mind?" asked Carson.

Picturing a massive slab of beef worthy to be served on Fred Flintstone's plate, Ourecky cringed at the thought of consuming yet another gigantic meal. It was shaping up to be another one of those nights.

Astor opened a red-bordered folder and extracted a black-and-white glossy photograph. Ourecky recognized it immediately. After all, he had snapped the picture himself; it was the infamous image of the brass data plate on Object 2368-B, the objective of their first mission.

"Virgil Wolcott slipped me this a few months ago, when I caught his briefing at the Pentagon," explained Astor. "Of course, I never knew who took it or how it was taken, but now I do. I sure would be honored if you two heroes would grace it with your John Hancocks. Of course,

I'll put it away for safekeeping, but it would sure be a real hoot to have it autographed by the men who were actually responsible for taking it. Would you mind?"

"Certainly, sir," replied Carson, taking the black Skilcraft pen proffered by the general. "We would be honored." Ourecky followed suit.

"Phyllis!" said Astor, returning the photograph to its Top Secret folder and pushing a button on his desktop intercom. "Could you come in here please?"

The general's secretary, a slender attractive blonde in her early thirties, sashayed into the office. She wore a white mini-skirt, a tight-fitting purple blouse, and a matching kerchief tied around her neck. "Sir?" she asked.

"Phyllis, we're headed to the Club for dinner and drinks. Get on the horn and tell Tech Sergeant Cramer to bring my car around." He gestured at the red-bordered folder and emergency landing protocols envelope on his desk before adding, "And stick these jewels in the Top Secret safe."

"The safe with the attack codes, General?" she asked, smiling at Carson. He smiled back.

He nodded. "Yes. Hey, Phyllis, do you still have that Brownie camera stashed in your desk? Why don't you bring it in here and take a picture of me and . . . Carson and Ourecky here."

"I would be glad to, sir," she replied, walking out the door. In a few moments, she returned with the camera. "Just two pictures left, sir. Will that do?"

"Oh, sure." Astor waved the men over to stand beside him behind his huge mahogany desk. "Smile, boys."

"Uh, General," muttered the secretary, pointing at Ourecky. "He's missing his wings."

Astor swiveled to look at Ourecky's chest. "Oh my God!" he exclaimed. "We can thank our lucky stars that Phyllis caught that. We sure can't set foot in the Club with you not wearing your wings. You're my guest, so I would be buying drinks all night if you're out of uniform."

"I think I may have an extra set in my desk," noted Phyllis, gazing at Carson. "And if I don't, I'm sure that Captain Williams is still working down in the Operations section. If I don't have a set, maybe I could borrow his wings for the major?"

"Excellent idea!" blurted Astor. The secretary scurried out the door. Astor nudged Ourecky, winked and quietly said, "I hope she was worth it."

"Sir?" asked Ourecky.

Astor chuckled and quietly said, "Hell, son, don't play dumb with me. I was young once, and I did more than my share of TDY junkets. I know the routine. You wouldn't be the first pilot to employ your wings as a skirt-removal tool in a difficult encounter."

"Uh, sir," stated Carson. "Major Ourecky doesn't wear wings because he's not a pilot."

"He *what?!*" bellowed Astor. "*Please* tell me that you're pulling my damned leg, son. That isn't the least bit amusing."

"Sir, it's true," said Ourecky. "I'm not a pilot. I'm an engineer. I don't rate wings."

"But you did *this?*" sputtered Astor, stabbing his finger at the red-bordered folder that held the photograph.

Ourecky and Carson nodded together. "Major Ourecky actually shot that photograph himself," stated Carson.

"Phyllis, disregard the wings. Just come back in here and take our picture," snapped Astor, jabbing the intercom button. "I cannot *believe* this! With all the pilots who would gratefully give their eyeteeth to fly this thing, Virgil Wolcott had the audacity to send up a non-pilot engineer? You two had better believe that I'm going to be on the horn to Virgil in the morning, and if this is another one of his damned practical jokes, you had better hope that you're already clear of my runways. This is *not* funny."

Seeing Phyllis returning with the camera, Astor pulled the two men close to his flanks and smiled broadly. "Say cheese, boys."

Phyllis snapped the picture and then took a second shot. Advancing the film, she looked towards Carson again and smiled slyly.

"We're leaving now," said Astor, picking up his hat from his desk. Mindful of the silent interaction between the secretary and the pilot, he added, "Phyllis, if you're free this evening, why don't you come down to the Club and have a drink with us?"

"Why, that sounds keen, sir," she cooed, sharing a grin with Carson. "But I need to lock your things in the safe and then tidy up a bit. Can I join you after dinner? Maybe in an hour or so?"

"Splendid," said Astor. "Come on, boys. There's some prime Midwestern beef anxious to make your acquaintance. Let's not keep those steaks waiting too long."

"We're with you, General," said Carson.

Astor frowned and said, "Ourecky, don't think for a minute that you're completely off the hook with me. I *will* call Virgil Wolcott tomorrow, and if you two are yanking my chain about your wings, then I will take it *personally*. I'll be coming after you, and I'll be wearing my golf shoes."

After the three men departed, Phyllis sat in Astor's chair to relax and freshen her makeup. She switched on his small television, waited for it to warm up, and then adjusted its rabbit ear antennas. The evening news was on; most of the stories concerned the day's elections all over the country. Not caring much about political events, she switched the TV off. She applied fresh lipstick, an alluring shade of red not appropriate for the office, and then sprayed her wrists and cleavage with just a slight hint of Arpege perfume.

Rubbing her wrists together, she thought about the scrumptious pilot with the striking blue eyes. Sighing, she hoped that she could latch onto him before one of the O Club regulars managed to sink their feline claws into him. For a moment, she thought about just rushing down to the Club but knew that there were still chores to do, and that they had to be done right if she expected to be paid for her efforts.

She glanced at her Bulova wristwatch, remembering that it had been a gift from Astor before his wife became unduly suspicious of their working relationship, and saw that thirty minutes had elapsed since the general had left. First, she jotted their names—Andrew Carson and

Scott Ourecky—on a scrap of paper, then tucked the note away in her purse.

She made a mental note to have two sets of prints made from the film in the Brownie. Finally, she opened the folder on Astor's desk and examined the photograph inside. She had no idea of what she was looking at, but understood that it had to be important if Astor had asked the two men to autograph it.

She reached into a special pocket sewn into the bottom of her purse and pulled out a camera—a miniature Minox-B manufactured in West Germany—and made several shots of the photograph, taking care to line up everything just so, exactly as she had been taught.

As she tucked away the Minox-B, she looked at the photograph once more before closing the folder and placing it in the Top Secret safe. While she had no idea what the strange writing meant, she guessed that it had to be something important. That sweet old Jewish man would almost certainly be interested in it, just as he was normally always interested in the things that she sent him.

Certainly she felt a twinge of guilt for accepting money for spying, but she had long since grown comfortable with the notion that it really wasn't *spying* if she was doing it for a country that was friendly to the United States. And Israel was their friend, wasn't it?

21

EXORCISM

Rampart Air Force Base, Idaho
9:25 a.m., Thursday, November 5, 1970

Ourecky heaved his B-4 bag into the flight line van and scrambled aboard. It was standing room only, since the boxy passenger compartment already held a somber B-52 crew departing on a twelve-hour alert mission. Like a commuter headed towards a routine day of office work, he grabbed an overhead strap. Carson squeezed into a space on one of the benches, sitting next to the B-52's enlisted tail gunner.

Watching Carson, he sighed; it was business as usual on their cross-country tour. Even though he had sworn a vow of virtual celibacy at breakfast, Carson dutifully transcribed the name and phone number of last night's conquest—*Phyllis?*—from a crumpled cocktail napkin to his little black book. He double-checked the neatly printed entry before exchanging a knowing grin with one of the SAC pilots. After tucking the notebook into his sunglasses pocket, he tore up the napkin and stashed the shreds in the van's ashtray.

After dropping off the B-52 crew, the van stopped to let out Carson and Ourecky where their T-38 waited on the parking apron. Carson deposited his B-4 bag under the starboard wing and then immediately went to speak to an enlisted man who was wheeling a cart-mounted "huffer" unit from a nearby maintenance hangar. Unlike most military aircraft, the T-38 trainer lacked an auxiliary power unit, so it relied on the huffer—technically known as a *palouste*—to provide compressed air to rotate the engines to facilitate starting.

Ourecky crammed their bags into the wing-mounted luggage pod and latched it shut. He normally climbed aboard the T-38 and buttoned in as Carson completed his pre-flight inspection of the aircraft. But today, for whatever reason, he lingered at the base of the ladder, pretending to adjust his parachute harness as he observed the pilot make his walk-around.

He gazed out at their surroundings. The desolate landscape was primarily arid hills of dismal brown earth. The morning sky was dreary gray. He could smell impending snow in the air. A steady cold wind blew in from the north. A B-52 lumbered down the runway and slowly took flight, spewing four parallel plumes of black exhaust as it strained under the weight of the thermonuclear weapons nestled in its metal belly.

He studied Carson. His friend had changed immensely since their last flight in space and the ensuing ordeal in Haiti. Much more lackadaisical than usual, he just didn't seem to be as intently focused as he had been. Even though it was an overcast morning, he wore sunglasses to conceal his bloodshot eyes. His face was puffy and almost without color. And though Carson had once kept himself at the peak of physical fitness, he was developing a noticeable paunch.

Ourecky was concerned that Carson was gradually regressing into a weak shadow of his former self. He was afraid that if Carson wasn't extricated from this perpetual stream of hedonistic distractions, it was only a matter of time before he devolved into a bloated shell of big talk and empty bluster.

Ourecky thought of the scores of former pilots they met while visiting the various O Clubs across the country. Seemingly affixed to their bar stools, only their faces changed from base to base, like an endless carousel where the snarling figures of lions and tigers had been replaced by sullen has-been warriors anxious to pounce on anyone who hadn't yet heard their tales of past glory. He hoped that Carson wasn't destined for the same fate.

Minutes later, Carson announced, "Scott, I've finished my pre-flight. Let's strap in and launch before things get too hectic around here."

Not responding, Ourecky strolled over to the huffer operator. The sergeant was attaching a flexible hose to a manifold port underneath the aircraft. The hose was stiff and unwieldy in the morning cold; it looked as if the man were wrestling a lethargic boa constrictor.

"Come back in ten minutes," said Ourecky.

"But, sir, your pilot told me to start your aircraft as soon as you got aboard."

"Come back in ten minutes," repeated Ourecky. "Go take a break somewhere."

"But, sir . . ."

"Are you questioning my authority, Sergeant?" snapped Ourecky. "Do you *really* want to make the mistake of disobeying my order? Make yourself scarce for ten minutes. That's all I'm asking. Then you can come back out here, turn us over, and we'll be gone. Is that too hard?"

"No, sir," replied the sergeant. Leaving the huffer's hose unconnected, he saluted, sharply pivoted about, and then walked towards the maintenance hangar.

"C'mon, Scott," shouted Carson, clambering up the ladder and swinging a leg into the cockpit. "We're burning daylight. Jump into your seat and strap in."

"Did you say you *finished* your pre-flight?" asked Ourecky casually, as he noticed that the pilot had overlooked at least one serious discrepancy.

"I *did*. Quit stalling and let's move."

Ourecky walked towards the front of the aircraft. He removed a fist-sized locking device from the angle-of-attack vane and held it out accusingly towards Carson. "If you finished your pre-flight, would you care to tell me exactly when you intended to remove this and stow it? Maybe during our roll-out? Perhaps in flight? Or maybe you were just going to wait for the crash investigator to stow it later? Perhaps after our funerals?"

Chagrined, Carson slowly descended the ladder. "I . . . uh . . . uh . . ."

"I uh, *what?*" demanded Ourecky. "Just what the *hell* is going on with you, Drew?"

"I missed that. I'm sorry. It was a mistake."

"I'll say," Ourecky said, frowning. "It could have been one mistake too many. Maybe you've lost your enthusiasm for life and no longer give a shit, but do you recall that I have a wife and child to go home to?"

"Yeah, I *do* remember," retorted Carson angrily, coming to the foot of the ladder. "And you would have never even seen your child if it hadn't been for me! I saved your *life*, Scott."

"You think I don't *know* that? I suppose that makes us almost even, doesn't it? Now, again, tell me just what the hell is going through your mind. *Something* obviously has you rattled, Drew. Are you afraid of going upstairs again? If you are, at least that's something I could understand, because that's the way I feel also. But what I can't understand is how a guy like you can just slowly self-destruct in front of me."

"I . . ."

"I . . . *what*, Drew? Spit it out!" snarled Ourecky, stomping his foot on the pavement. "*Why* are you losing your edge? Are you afraid to fly? If that's why you can't concentrate, then go talk to Virgil and have yourself grounded before you kill *both* of us. Hell, I signed on to fly into orbit, but I didn't sign on to fly with someone who can't keep a clear head."

"That's not it, Scott," replied Carson quietly. "I'm not afraid of flying. I'm not afraid of going upstairs, either. I know it's just a matter of time before we go up again. I'll be ready then. I just need to work through some things."

"*What* things, Drew? Right now I'm much less concerned about going back into orbit than I am making it from here to Ohio. Are you going to have your wits about you for *that* trip?"

Facing each other as if ready to fight, the men were silent as yet another B-52 roared off for a monotonous nuclear patrol. A tear welled in Carson's eye and then streamed down his cheek.

Ourecky had never seen this side of the pilot before, and it disturbed him immensely. But now he felt like a priest called to perform some long-overdue exorcism, compelled to reach deep into Carson's guts to wrench out a wriggling demon. "*What?*" he demanded, thumping his finger into Carson's chest. "What is it? Tell me, Drew! Tell me *now!*"

"I don't want to die alone." Carson's voice quavered and tears poured from his eyes.

"You don't want to die *alone?*" replied Ourecky, with a sarcastic tone in his voice. "Well, Drew, you really shouldn't lose any sleep over *that*. Because it's about a ninety-nine percent certainty that we will die *together*. Of course, it's also a question of whether we're blown to smithereens by a malfunctioning booster or if we're stranded in orbit because our retros won't light or"—Ourecky wagged the wind vane cover in Carson's face—"whether we're killed because you overlook something trivial. So you shouldn't worry about dying alone, because at the rate things are going, we're probably destined to strum the same harp."

"That's not what I mean about dying alone," said Carson, wiping his face with his Nomex flight gloves. "I'll tell you, Scott, Haiti was hard on me. I didn't know what the hell was going on, and I wasn't sure whether we were going to make it out of there or not. You had it easy, because you were unconscious most of the time. Really, I'm not afraid of dying, and I'm not afraid of going back up, but it just terrifies me that I might die without leaving something behind. I guess I didn't realize that until we got back and I saw your little boy. Does that make sense?"

"Sure it does. That's natural. But if that's what you really want, Drew, you need to make some drastic changes in your life. I know the mantra—work hard and play hard—but you have to realize that you can't go on living a life filled with fast cars, fast women, and fancy watches if you really want something of permanence."

"I know that. Scott, I promise I'll change. I will . . ."

"Oh *really?*" snapped Ourecky. "Sorry, but I hear this same litany every morning, and by the time Happy Hour rolls around, you seem to have forgotten your promises. Do you think you can somehow make things better by jumping in and out of bed with an endless string of strangers? In the morning, you leave, and they're just strangers again."

Another B-52 took off and slowly climbed out to the northeast. As the noise abated, Carson said, "You're right, Scott, but zooming around on this damned party circuit isn't helping matters much for me. I've been killing myself for the past three years with no let up, and suddenly I'm pitched into circumstances where I'm able to let my hair down and blow off some steam. And it's not like I brought this upon myself; Virgil *ordered* us to do this."

"Virgil told us to be sociable with these people," Ourecky said, zipping up his nylon flight jacket to ward off the chill. "If that means cozying up to them, going to the local O Club to partake in a drink or two, then *fine*, but he sure as hell didn't order us to *wallow* in it."

"Point taken," said Carson. "When we get back home, we're going to talk to Virgil and ask him to scale back on these junkets. Then we go back into the training routine to prepare for what comes next. Fair enough?"

"Fair enough. Now, do you think you can bring this crate back to Ohio in one piece?"

"I can. I will."

Ourecky lightly punched Carson's shoulder before beckoning the waiting huffer operator with a wave. "That's *my* Carson, *my* brother. Let's strap this thing on and launch."

Disability Claims Office
Veteran's Administration Office, Las Vegas, Nevada
8:25 a.m., Thursday, November 5, 1970

As he waited for his name to be called, Eric Yost watched the clock and tried to ignore the sappy Muzak spilling from a wall-mounted speaker. Listening to his stomach growl noisily, he wished he had some coins for the vending machine in the hallway. He reflected on the recent chain of events that had landed him in the Veterans' Administration office.

In an exceptionally short span of time, his life had undergone a radical series of transformations. Just slightly more than a month ago, he had been sequestered at a snowbound radar site in the wastelands of Greenland. After arriving in Las Vegas and retiring from the Air Force, he took a taxi to the renowned International Hotel. After lugging his cash-jammed duffle bag to the concierge desk, he was swiftly installed in a swanky high roller suite on the thirtieth floor.

His winning streak followed him from Greenland. He almost doubled his stake of over four hundred thousand dollars in the first week, to a high water mark of seven hundred and six thousand dollars. Every amenity was at his fingertips. Anything he desired—expensive meals, booze, cigarettes, clothes, hookers—was comp'ed by the casino. With glitzy women at his sides, Yost occupied a front row seat whenever Elvis took the stage at the International.

He was living high in seventh heaven until right into the middle of the second week, when his winning ways were suddenly reversed. Fourteen days after setting foot in Las Vegas, with his cash entirely depleted, the casino's hospitality abruptly dissipated. Yost learned a harsh lesson of how the casinos classified visitors. All casino guests occupied a specific rung in a three-tiered taxonomy; they were either Winners, Losers, or Soon-to-be-Losers. After he was refused a marker for ten thousand dollars, primarily because he had no collateral and no visible means of income, he was ejected into the streets with his nearly empty duffle bag.

In short order, he hocked everything of value, including his Timex watch, wedding band, and most of his souvenirs from Greenland. He bought a cheap poly-filled sleeping bag at an Army surplus store and presently made his home in a storm drain culvert in an industrial area. He plainly knew that he had to get out of town since there was nothing left for him here, but making his escape was a bit more difficult than he had anticipated. In fact, he had barely enough cash to buy food.

On his retirement paperwork, he had entered his sister's Minneapolis address as his permanent place of residence. He was due a check at the end of the month, but she would have to wire him the funds via Western Union. To make matters more awkward, she no longer accepted his collect calls. He had decided that when he finally got his hands on some cash, he was taking a Greyhound bus to Minnesota to resolve the situation.

In desperation, he recalled something a doctor had told him as he underwent his retirement physical at Nellis. The doc advised Yost to file a claim with the VA for his missing finger. Yost dismissed the idea at the time, but he quickly reconsidered after spending his first night in his dank abode of concrete pipe. After all, it was a cut and dry case, a no-brainer; he had lost his finger courtesy of the Air Force, so he was clearly entitled to the monthly disability stipend—roughly a hundred dollars—as a result.

Moreover, once the compensation was approved, he could specify that the checks be held for him at the VA office. All he needed was one measly check and then he would head north to settle the score with his no-good alcoholic sister. Then, he could resume a normal life, not having to worry about where his next meal would come from or putting a roof over his head.

After filing his claim two weeks ago, Yost dutifully stopped by the VA office every day to see if there might be any progress. Initially, he noticed that people went out of their way not to sit next to him in the waiting room, and as the days passed, many made excuses to wait outside to be called or even come back on another day. Watching them as they scuttled outside with their upturned noses, he laughed to himself; if skipping a bath or two allowed him to advance to the head of the queue that much faster, then so be it.

Barely awake, he heard his name called. "Eric Yost? Eric Yost?" asked the raven-haired young receptionist, as if she didn't already know who he was. "Mr. Yost, Mr. Anderson will see you in Room Six. Right down that hall and to the right, please."

Following her directions, Yost carried his forlorn duffle bag down the hall and entered the specified office.

"Have a seat, please, Mr. Yost," said the claims officer, a heavy-set man in his early fifties. "I'm Seth Anderson. Before we discuss your claim, I would like to thank you for your military service." In front of Anderson was an antique ship's bell clock of gleaming brass, mounted in a walnut cradle bearing a small metal plate commemorating his service in the Navy. The brass and wood were highly polished; Anderson obviously took great pride in the memento.

Yost plopped down into an uncomfortable plastic-backed chair and said, "Uh, thanks, but can we not just get down to business? I filed a claim for my finger, and I want to draw my disability check as quickly as possible."

"In a moment," replied Anderson, wrinkling his nose as he switched on a small desk fan on his credenza. "I hate to be personal, Mr. Yost, but have you bathed recently? Assuming, of course, that you're living somewhere with sanitary facilities. If you're not, I can recommend a shelter for you where . . ."

Yost interrupted him. "Thanks, but I'm fine. I was just in a hurry this morning and didn't have time to jump into the shower." He knew exactly the shelter Anderson referred to; it was a refuge of last resort for transients down on their luck, especially stranded gamblers who lacked the fiscal resources to leave town. It offered a shower, a clean change of clothes and a cot for one night and one night only.

The shelter also endowed each guest with a one-way Greyhound ticket to Los Angeles, courtesy of the Las Vegas Tourism Office and various participating casinos. To this end, it employed a couple of hulking thugs who physically escorted their overnight guests to the bus station to forcibly ensure that the bus tickets were actually used. This pristine desert town held little sympathy for vagrants, wayfaring panhandlers, and their ilk.

"Well, I'm obligated to offer that information. I sincerely hope you don't take offense."

Yost shook his head, leaned to one side, and grunted. Without warning, an extremely noxious odor filled the confined space. As the invisible cloud wafted outside the door, he overheard muffled cursing from the hallway and adjacent offices. "Sorry," he explained, fanning the air. "Pork and beans last night. The cheap store brand gives me a bad case of gas."

A distressed look passed over Anderson's face. On the verge of turning green, he reached into a desk drawer, pulled out a small jar of Vick's VapoRub, and daubed a tiny amount under each nostril. "I'm beginning to catch a cold," he explained. "This will help me stave it off."

"If you say so. I'll just have to take your word for that."

Loosening his tie, Anderson nodded. "Let's discuss your claim, Mr. Yost. We asked the Air Force for your records so we could verify the nature and extent of your injuries, to ultimately determine if they were service-related. We were fortunate that your records were still on file at Nellis. That helped us to expedite your claim."

"Groovy," commented Yost. "That makes sense. Can we get on with this?"

"Now, uh, about your claim," said Anderson, slipping on black-framed reading glasses as he flipped through the pages of Yost's medical records. "According to the paperwork you submitted, you claim to have lost a finger to frostbite while stationed in Greenland."

"That's right," avowed Yost, holding up his left hand and pointing at the void once occupied by his index finger. "This finger right here."

"Interesting. Mr. Yost, before we delve deeper into your claim, I need to discuss something *very* serious with you. It's absolutely *imperative* that you take heed. Okay?"

"I'm all ears. Fire away."

"Mr. Yost, are you aware that filing a false claim with the VA is a *serious* federal crime?" asked Anderson officiously, placing both hands flat on his desk. "That said, would you consider amending your claim or withdrawing it altogether? If you do so now, willingly, we can just let this matter rest with no potential for legal action. No harm, no foul, you might say."

"What the *hell* are you talking about?" demanded Yost angrily, sitting up in his chair. "It's all there in black and white. I'm not trying to *swindle* anyone, least of all the damned government. Hell, I didn't ask to go to Greenland to catch frostbite and have my finger sawed off."

Anderson subtly pushed a small button, like a doorbell, next to his desk, and said, "Here's the problem. According to your medical records, you've never been treated for frostbite nor had your finger amputated. *You* claim that you were stationed in Greenland, but your official personnel file *clearly* states that you have been assigned to Wright-Patterson Air Force Base in Ohio for the past four and a half years. So, as I indicated a moment ago, Mr. Yost, this is your *last* chance to withdraw your claim before we initiate a fraud investigation."

As he massaged the notch where his index finger had once been, Yost felt his pulse pounding in his temples. His teeth gnashed as he tightly clenched his fists. *What is going on here? His missing finger was a legitimate injury, courtesy of the Air Force. What are they trying to pull?* With all of his might, he resisted the pressing urge to stand up and pound some sense into Anderson. "*Wait!*" he uttered. "If I've been at Wright-Patt all this time, why did the Air Force fly me *here* to retire? Why didn't they just discharge me in Ohio?"

"I don't know," replied Anderson smugly. "Frankly speaking, Mr. Yost, I don't *care*. It's none of *my* concern. I'm tasked to evaluate your claim on its merits, and there's no indication that you ever had frostbite or served in Greenland, so . . ."

Yost reached into his duffel bag and tugged out his most prized souvenir, a large figurine of a walrus intricately carved from a whale's tooth. It was the sole remaining artifact of his lucrative sojourn in the Arctic. "But look!" he argued, holding out the talisman. "An Eskimo gave me this in Greenland. Can't you see . . ."

Anderson shook his head and declared, "I don't want to do this, Mr. Yost, but if you insist on pursuing this charade, then I'll be forced to file formal charges."

As his eyes gradually lost focus, Yost no longer heard the bureaucrat's grating voice. Years of frustration and anger welled up within him, boiling and seething, until he could no longer restrain himself. He lurched out of his seat, leaned over the neatly ordered desk, and swung the walrus figurine into Anderson's temple. As blood spattered his desk blotter and meticulously collated stacks of claims paperwork, Anderson slumped unconscious to the floor.

Yost sailed over the desk and continued to pummel Anderson with the ivory sculpture, swinging it like a police baton. In seconds, a security guard burst into the office and wrestled him into submission. Fifteen minutes later, he was fettered in handcuffs and jammed into the back of a Clark County Sheriff's Department patrol car.

After a month to the day that he had arrived in Las Vegas, he found himself in the next and possibly last stage of his metamorphosis. And just as a coin has two sides, so did his current circumstances; on the negative side, his departure from Las Vegas was probably delayed indefinitely, but on the plus side, he no longer had to concern himself with putting a roof over his head or a meal in his gut.

Aerospace Support Project
9:15 a.m., Monday, November 9, 1970

Carson and Ourecky weren't the only ones on the road during the previous week. Tew, Wolcott, and Heydrich convened to share notes on their recent ventures. "Gunter, tell us about your visit to MIT," said Tew. "Any updates on the Block Two computer?"

Heydrich nodded. "*Ja*, Mark, I have an update all right, but probably not the one you care to hear. The Instrumentation Lab people made it emphatically clear that there will be no new computer before Phase Two."

"*Damn* it!" grumbled Wolcott, slamming his coffee mug on the table. "I can't believe that we threw in with those egghead varmints. They promised to deliver this damned machine *months* ago, and all we have to show for it is a bunch of plywood mock-up boxes and wiring diagrams."

"In MIT's defense, Virgil, that's not exactly accurate," noted Heydrich. "They *did* deliver a flight-ready prototype, and it functions exactly to specifications."

"Oh, *yeah*, Gunter," said Wolcott. "Shucks, I plumb forgot that we have that box out there at the HAF. Yeah, it's certified flight-ready, but unfortunately it's as big as a damned *refrigerator,* so it ain't of much use to us if we can't somehow shoehorn it into the spacecraft. But maybe I can ring up some of my buddies at NASA and talk them out of a Saturn V. It's beginnin' to look like NASA will be forced to close out Apollo with some extra hardware available."

"It's a *setback*, Virgil," commented Tew, pouring milk of magnesia into a shot glass. "It's not a show stopper."

"It ain't? Mark, should I remind you that we still have six missions to fly and we only have *two* crews to fly them? And the harsh reality is that if we don't get that new computer as promised, we really only have *one* crew to fly the remaining shots."

Tew downed the shot and nodded. "So be it. Tell me about your visit to San Diego."

Wolcott chuckled and said, "The HAF is operatin' right on schedule. There are three stacks ready to fly, and one in the pipeline. The next flight-ready stack is already encapsulated and loaded on the LST. All we're lacking is a couple of cowboys to saddle up and ride."

"And you also visited ARPS at Edwards while you were out in California?" asked Tew, using a handkerchief to wipe white crust from his lips. "Did you spot any prospective flight personnel?"

"Yeah, pard," replied Wolcott, frowning. "I saw plenty of potential contenders, but ARPS ain't coughin' up any more pilots until our funding is formally approved for Phase Two. I tried to make an end run around them by going directly to the Personnel Branch, but I was shut down cold. The prevailin' attitude is that it will take at least eighteen months to safely spin up any new guys to fly, and the Air Force big hats ain't willin' to lock down a bunch of high-dollar test pilots for that long if it ain't entirely likely they're going to fly. They've already been snakebit with the MOL debacle, and they ain't going to let it happen again, so we fly the remaining six missions with the two crews we have."

"Well, I suppose it's up to me to spread the icing on this cake," said Tew. "I spent most of last week with Kittredge and his staff. Short of a miracle, there *won't* be a Phase Two. Our only hope is if the current Administration remains in the White House."

"So we're dead in the water?" asked Wolcott, frowning as he drummed his fingers on his white Stetson. "Six more missions before the curtain drops?"

Tew nodded.

"But why would they pull us off the trail just as we're hittin' our stride?" asked Wolcott. "Can't they see what we've accomplished?"

Tew replied, "Perhaps, but for them, the box score reads that we played six times. We've taken out three critical Soviet satellites, but we've also missed two and have lost a platform with its crew. Additionally, our original charter was to seek out OBS platforms, and all the intelligence we're now seeing leads us to believe that they're not out there."

"But . . ." sputtered Wolcott.

"Virgil, listen," said Tew softly. "There's something important that I need to share with you."

"What's that, boss?"

"I plan to stick with you for these next six missions, but after that, I'm *done*. Personally, I have no desire to go on to Phase Two, even if it does come to pass. My cardiologist tells me that I'm living on borrowed time as it is, and this incessant stress isn't helping matters much."

"You can't go down to Walter Reed for another surgery? Is there nothing else they can do?"

"Short of a heart transplant, no. Virgil, the docs tell me that my heart muscle is just worn out. There's nothing that they can do about it. I'm sorry, but I would like to spend a few years relaxing and living a normal life before I give up the ghost. Can you understand that, friend?"

"Yeah. You've certainly earned a breather."

"Look, Virgil, if you want to keep Blue Gemini alive, I'll do everything within my powers to make it happen, but you just have to accept that I can't continue on with you. As it is, I'll be damned lucky if I make it through the next six flights. They really wear on me."

The intercom on Wolcott's desk squawked. "Sirs, Major Carson and Major Ourecky are here to see you," said the aide in the outer office.

"Hold them out there a minute, buster," said Wolcott, leaning out of his chair to press the intercom button. "I'll holler at you when we're ready."

"Virgil, Gunter, about my heart, I told you that in confidence," said Tew, adjusting his tie. "I would prefer that you not share it with anyone else."

"I will not speak of it with anyone," vowed Heydrich, clasping his hand over his heart. "You have my word, Mark."

"My lips are sealed, pard," added Wolcott. "You have my word as well."

"Thank you, gentlemen. I know that I can count on you."

"But how about this other fly in the ointment? Ourecky's wings?" asked Wolcott, waving a set of papers towards Tew. "Do I have your blessing to pursue this?"

Glancing at a memorandum stapled to the papers, Tew reluctantly nodded. "You can ask him, Virgil, but ultimately it's his decision. Fair enough?"

"More than fair. I'm sure he'll jump on this. Who wouldn't?" Wolcott leaned back in his chair, jabbed the red intercom button and stated, "Send them in."

As Carson and Ourecky entered, Wolcott whistled and exclaimed, "Well, howdy, it looks like you two have been living *high* on the hog! We send you out on the road for a few weeks, and you come back bustin' out of your britches. What have you gained? Ten pounds? Twenty?"

"Eleven for me," replied Ourecky, frowning as he patted his stomach. "It's difficult not to pack on lard when every meal's a feast and there's no time to exercise."

"I s'pose so," replied Wolcott, gesturing for the two to take seat at the tables. "Does that T-38 still fit you? We're not going to have to let it out at the seams, are we?"

"We can still scrunch into it," answered Carson, easing himself into a chair before adjusting the collar of his Ban-Lon knit shirt. The shirt's pale blue fabric stretched tightly to contain his stomach, like a canvas sail swelling in a stiff wind. "But just barely."

"What's on your mind, gentlemen?" asked Tew. Sour bile and stomach acid surged up in his throat, and he gagged slightly as the acrid taste settled on his tongue. Coughing, he filled a glass from a pitcher of water, and quickly quaffed it.

Leaning forward, Carson placed his hands flat on the table and answered, "Sincerely, sir, we've enjoyed the break, but the fact is that we're losing our edge. We understand the need for these visits, but we want to know if we can scale them back or even curtail them altogether."

"You've gobbled your fill of the fatted calf, pard?" asked Wolcott, grinning.

"We have," replied Carson. Ourecky nodded in agreement.

"Excellent," declared Tew. "Perfect timing." Outside, a pair of fighters screamed off the runway.

"Sir?" asked Carson.

Tew turned to face Ourecky. "The flight surgeons want to take another look at you, just to be absolutely sure, but they assure me that you are physically ready to go back into orbit. The only question that remains is whether *you're* ready to go back up. Are you, young man?"

Hesitating momentarily, Ourecky shot a nervous glance at Carson. "Yes, I'm ready, sir."

"That's what we wanted to hear, pard!" exclaimed Wolcott, slapping Ourecky on the back.

"Unless the flight surgeons stamp you with a bad report, Ourecky, the two of you are going up in Mission Seven. You'll launch in January." Expecting to see a smile or at least some positive acknowledgment from Carson, Tew noticed that the pilot's expression did not change in the slightest. "Are *you* ready to go back up, Carson?" he asked.

Carson paused, looked towards Ourecky, swallowed deeply, and then replied, "I am, sir."

"Good," said Wolcott. "You two gents work with Gunter to hammer together your training program. Whatever you want, within reason, we'll make it happen."

"Our facilities are at your disposal, gentlemen," noted Heydrich. He looked immensely relieved, as if he were thrilled to finally have a break from perpetually coaching Jackson and Sigler.

"I'm delighted to have you two back in the line-up," noted Tew.

"Ourecky, son, I have another pressing issue to discuss with you," said Wolcott, looking anxiously towards Tew. "I've worked out a special deal for you. After you come back in January, I want you to take a few weeks to earn your wings. With the deal we've worked out, you'll be able to nail all the requirements and still be able to cycle back into training for the next mission."

"*Next* mission, sir?"

"Yeah. You two gents will fly Eight in May. And unless things change drastically in the coming months, there's a danged good chance that you will fly all of the *remaining* six missions."

Twisting the wedding band on his finger, Ourecky swallowed nervously. "*All* of them, sir?"

"Probably," replied Wolcott, using a folder to swat a fly on the table. "But we're drifting off the trail here, Ourecky. Did you not just hear me say that we're granting you a *special* opportunity to earn your wings? You'll be able to skip most of the stupid time-wasting crap that most guys go through and focus just on the important stuff. What do you say, pard?"

"I'll pass, sir."

As his face turned red, Wolcott groaned and fanned himself with his Stetson. "You'll *pass?*"

"Yes, sir. I appreciate the opportunity, but I would rather focus my time and energy on preparing to fly these other missions."

"Maybe you're missin' something, hoss," said Wolcott, reaching for a glass of water. "We're offerin' you a chance to earn your *wings* and you won't have to leap through a bunch of stupid hoops to get them. It's all but a damned gimme. I might as well just stick them on your chest right now. We're giving you a golden opportunity, son. Are you really that inclined to fritter it away? Just a few weeks, and then you can have those wings!"

"I'm still going to have to pass, sir."

"Ourecky, do you recollect when we first met, and I asked you why you hadn't become a pilot?" asked Wolcott. "As I recall, you said that you had applied four times."

"Five, sir."

Frowning, Wolcott rolled his eyes and slapped his hand on the table. "Whether you had been turned down four or five or a *hundred* times over, that ain't the danged point, pard. You've more than demonstrated your aptitude to become a pilot. I'm tryin' to slip you through the formalities so we can pin those wings on you. How can you treat this offer so lightly?"

"I'm just not interested, sir. I'm sorry if that offends you."

"*Offends* me? Offends me? No, but let me tell you that your reluctance *would* offend some folks," declared Wolcott loudly. "Hell, I'll give you a case in point, hoss. Do you recall meetin' General Astor last week?"

"Yes, sir," replied Ourecky, nodding his head. "He commanded that SAC base in Idaho."

"Correct. You may not want to believe this, Ourecky, but the fact that you're ridin' a rocket into orbit and you ain't a pilot can be a very sore subject for some folks. You and I both know otherwise, but there are plenty of senior officers who believe those seats should be *exclusively* reserved for guys who wear wings. Pete Astor falls in that crowd, and he threw a danged conniption fit when he found out that we've fired a non-pilot into orbit not just once, but three times."

Wolcott continued. "Needless to say, I didn't particularly relish the ass-chewin' that Astor administered me over the phone. Ordinarily that sort of thing would just slide off me like water off a duck's butt, but it also just so happens that Pete's zoomin' along on a direct course to be the Chief of Staff of the Air Force. He's a bomber guy and has exactly the right pedigree for the job. He might come off as kind of a country bumpkin, but he's a shrewd player with a lot of political clout. There's a very good chance that he'll come in as the next Chief, right after "Three-Finger Jack" Ryan. That could be as early as mid-1972. Any idea why that's significant, pard?"

"No, sir," answered Ourecky, scratching his head. "No idea."

"Because 1972 is roughly the time when we should receive approval to move into Phase Two. And with that sword hangin' over our heads, the *last* damned thing I want is a Chief who has a bone to pick with this Project. Savvy?"

"Yes, sir."

"Of course, on a positive note, Pete Astor's secretary apparently took quite a shine to your compadre here," said Wolcott, winking at Carson. "Any chance that you kept her number?"

A blank look passed over Carson's sallow face, as if he were trying to recall a long-forgotten girlfriend from high school.

"Wednesday, last week," interjected Ourecky, frowning. "That would have been *Phyllis*. Remember?"

"Oh. *Phyllis?*" said Carson. "Oh, sure. I have her number, sir."

"Well, I would greatly appreciate it if you rang her up from time to time," said Wolcott. "Every little bit helps. She followed Pete from his last assignment at PACAF at Hickam Field in Hawaii, so there's a good chance that she'll stick with him all the way to the top rung."

"Will do, sir," replied Carson.

"Back to you, son," said Wolcott, swiveling around to fix a baleful gaze on Ourecky. "Your wings. It'll be just a short TDY stint. You can breeze through the ground phase at your own pace, do your mandatory hops, and be done. Hell, I can't make it any damned easier for you, hoss. Hell, we're practically pinnin' the wings on you for nothing."

"I'm still going to have to pass, sir."

Infuriated, Wolcott jumped out of his seat and slapped his bald crown. "Hell's bells, son, I can't believe that I'm reduced to beggin'. *Please* get your wings. How the *hell* can you be so damned ornery about this?"

"I *can't*, sir," explained Ourecky meekly. "I promised Bea that I wouldn't, and it's probably the only promise that I can actually keep, so *please* understand why I have to decline."

"Bea!" shouted Wolcott. "*Bea?* You're frettin' over a promise to your *wife?* Bea doesn't *have* to know! Hell, you're on the road for weeks at a time already. Beyond that, we've blasted your ass into space three times and she ain't aware of that, *is* she? Just get your damned wings. You don't have to wear them around *her*. If she ever finds them accidently, you can just tell her they came in a Cracker Jack box. Hell, as simple as we're trying to make this, they might as well have."

"But sir, a promise is a promise," replied Ourecky.

"Please, sir," interjected Carson. "I think I see . . ."

As his facial muscles drew taut, Wolcott glared at Carson. "Major Ourecky, are you going to oblige me to ask General Tew to *order* you to earn your wings?"

The room fell silent, awkwardly so. Tew closed his eyes and felt his heart pounding in his chest. He suddenly remembered that Ourecky wasn't the only one who made a promise to Bea.

"Well, pard?" demanded Wolcott.

"Drop it, Virgil," said Tew quietly. "Just *drop* it. We've asked a lot of this young man, and he's consistently delivered, and we *won't* compel him to break a promise to his wife."

22

PAPER DOLLS

Dayton, Ohio
6:30 p.m., Monday, November 9, 1970

Bea used a paring knife to peel and slice carrots for salad. A pan of spaghetti sauce simmered on one eye of the range, while on another red-glowing eye a pot of water gradually bubbled to a rolling boil. As she prepared dinner, Ourecky fed Andy in the living room.

Watching him through the breakfast nook, she scooped the carrots into a bowl, and adjusted her apron. As cool as it was, he wasn't wearing a shirt; his now pudgy abdomen was crisscrossed with pink scars. She had been concerned about him for the past few weeks.

For a man who used to be so fanatical about his health, he had gained a substantial amount of weight and was woefully out of shape. He just seemed miserable most of the time, even though he had healed up from his injuries as well as the surgeries that followed.

"Dinner will be ready in about ten minutes," she announced. "Hopefully you'll have Andy put down by then. Can I bring you a beer? I picked up a six-pack of Schlitz today. Your favorite."

Grimacing, he shook his head as he cradled the baby in his forearm to burp him. "No thanks. And just a light plate for me, baby. Lots of salad, light on the pasta. No meatballs."

"No meatballs?" she replied, raising her eyebrows. *No beer? No meatballs? Was the world coming to an end? Why hadn't she received the memo?* "Scott, you're not sick are you? Are you losing your appetite? I thought you *loved* my meatballs."

"I do, but I just need to start watching what I eat. I need to shed this gut. By the way, I'm getting up early tomorrow. Drew and I are going to start back into our gym routine every day."

"That's great," she replied, sampling the spaghetti sauce with a wooden spoon. Wrinkling her nose, she added a smidge more dried basil. "But can you keep up with that on the road? You said it was nearly impossible to coax Drew out of bed in the mornings, let alone nudge him towards a gym." She watched him as he stretched out on the couch and laid the baby on his chest, cupping his head so that the baby heard his heartbeat. Bea was always amazed with how quickly he could lull the baby to sleep that way.

"We're off the road, at least for a while. Virgil cancelled the rest of the PR trips."

"The trips where you were going base to base to tell them about new equipment?"

Gently patting the baby's back, he nodded.

"Well, babe, that's *great*," she said. "You hated those trips . . ."

"There's more," he said. "And it's not good news."

"Not *good* news? Oh, that's a surprise," she muttered. "And I thought they were finally going to release you from this insane job and let you go back to school like they promised."

"We have to start flight testing again," he replied. "So my schedule will get even crazier."

"When? Are they at least going to wait until after the holidays?"

He slowly got to his feet, padded to the nursery, and gingerly placed the baby in his crib. Returning to the living room, he answered her. "When? The flight testing won't start until January, but we have to start our prep work immediately. I won't be travelling nearly as much, but I won't have very much time off, either."

"Your parents are expecting us at Christmas." She eased the noodles into the boiling water. "They haven't even seen little Andy yet. It would break their hearts if we didn't make it."

"Maybe you can go," he replied. He walked into the kitchen to rinse out the baby bottle. "I really want to go, but it's just not going to happen, Bea."

"So how long is this going to go on *this* time?" she asked. "Until you crash again? Until you're *dead*? Don't they have anyone else? Why must it *always* be you and Drew?"

"It's my job. Drew and I catch most of it because we work well together."

"Okay," she replied, stirring the linguine noodles. "So long as we're going to spoil our dinner with an argument, then I have something to fold into the mix. I want to go back to work."

"Back at the airport? At the gate?"

"No. I want to start flying again. I miss it. The scheduler can adjust my flights so I'll do only regional hops. It will mean going in early and coming home a little later, but no more layovers."

"But why, Bea? We don't need the money. There's no need for you to start flying again."

"I miss it, Scott. It has nothing to do with the money."

"Okay, but what about Andy? If I'm working all hours and can't ever know when I'll be home or even if I'm going to be in town, then who will take care of him?"

"Jill finished that medical transcription course at the junior college. She works at home. They bring her tapes from the doctor's offices, and she types up the records. It's good money, and she can stay home with her little girl. She said she would be happy to keep Andy as well."

He opened the refrigerator, pulled out a sugar-free Tab, opened it, and took a sip. "Sounds like you've already planned this out."

"Can you at least think about it?" she asked, putting a plastic colander in the sink. "You sure expect me to accept a lot for your job. Can you not do this for me?"

"I guess I can, Bea," he replied. "And I suppose that I can eat one of those meatballs, too, if you twist my arm."

Embassy of the Union of Soviet Socialist Republics
Washington, DC
10:25 a.m., Friday, November 13, 1970

In the basement of the Embassy, Morozov shared a cramped workspace and manual typewriter with two other GRU officers. The dingy space was so tiny that all three men had to stand up whenever one came or went. Of course, it was to be expected; once the KGB and custodial staff had picked over the more prime real estate not occupied by Embassy personnel, the GRU staff were jammed into whatever undesirable nooks and crannies that were left over.

As the bureau's designated archivist, Morozov spent much of his time clipping and compiling articles from American newspapers sent from GRU stations across America. While his office mates actually worked sources and conducted surveillance, he was stuck here, with his scissors and a rubber-tipped bottle of mucilage, like a schoolgirl making paper dolls.

Determined to make good at the lackluster task given him, he diligently read each newspaper in detail. For any given paper, much of the printed pages were taken up by national news regurgitated from the AP and UPI wire services. In today's news, a combined force of South Vietnamese and Cambodians had called off a planned five-day offensive after Communist forces had apparently learned of their plans. Jurors had been selected for the trial of alleged war criminal Lieutenant William Calley at Fort Benning, Georgia. The president-elect of Mexico, Luis Echeverria Alvarez, was currently visiting President Nixon in the White House.

Once he had a solid grasp on major current events for the day, he focused on local and regional issues, particularly as they impacted GRU operations. As he skimmed the metro section of a Las Vegas newspaper, a small and seemingly insignificant article caught his eye. It described the arrest of a retired Air Force sergeant for the attempted murder of a Veteran's Administration employee last week. Morozov gulped as he read the offender's name: *Eric Yost*.

An enlarged police mug shot photo accompanied the article. The forlorn subject was a balding, unkempt Caucasian male; his right eye was swollen shut and fresh bruises adorned his face. A placard underneath the image read: *Yost, Eric B., 11-5-70, T312580, Clark County Sheriff's Department*. Sure enough, it was the same Yost he had known from Ohio. *But how could this be?* he thought. *Yost is dead . . . or is he?*

Cutting out the article, he was still absolutely convinced that Wright-Patterson's Hangar Three was a repository for captured UFOs and that Yost's "death" had been part of an elaborate cover-up to safeguard its secrets. If he was just granted some more time and resources, he would eventually penetrate the hangar's veil of secrecy. Maybe then, with one significant coup on his otherwise blank espionage resume, he would be free of snipping out paper dolls, making tea, and scrubbing the samovar.

As his heart pounded in his chest, Morozov loosened his collar and jotted down notes. He would contact the Las Vegas GRU station to determine if they could acquire a copy of the official police report from the incident, as well as any other information that might be relevant. Then, he would compile his facts and produce a comprehensive report. He looked up at a stack of cassette tapes. The small library of tapes contained recorded lessons on conversational Vietnamese. He had intended to spend the weekend immersed in his language studies, but this new wrinkle would require his undivided attention.

10:25 a.m., Thursday, November 26, 1970

Almost two weeks had elapsed since Morozov had submitted his report on Yost's arrest in Las Vegas, but he had heard nothing in response. Surely, his bosses had to comprehend why this information was so relevant. He pushed Yost and Hangar Three from his thoughts as he winnowed through the stack of newspapers that had arrived this morning.

Trying to stay abreast of developments in Vietnam, Morozov perused an article concerning an American POW rescue attempt at Son Tay earlier in the week. The POWs had previously been moved from the camp, so the raid was notionally a failure, but the operation had created quite an uproar with their North Vietnamese allies. They feared that more such raids were imminent, leading to extensive discussions on how to effectively manage the hundreds of POWs they currently held. The GRU recommended consolidating the POWs into larger camps, protected by air defense sites to preclude the arrival of more American helicopters. *Isn't that obvious?* thought Morozov, lighting a Winston cigarette. He inhaled deeply, savoring the taste of fine tobacco that only the Americans seemed able to grow. He heard a phone jangle nearby.

"*Anatoly Nikolayevich!*" barked a cipher clerk. "Upstairs! The Resident wants to see you."

"Perhaps the Crippler wants to install you as a Hero of the Soviet Union," sneered one of his office mates. "Maybe the Kremlin is so impressed by your proficiency at clipping out paper dolls that they have finally mailed your Gold Star."

Listening to the other GRU officers snicker, Morozov stubbed out the cigarette and replaced it in the half-empty pack. He shuffled around chairs to clear a path and then quickly made his way up the back stairs to the Resident's office.

Colonel Federov's office wasn't large, but it was comfortable and well-furnished by GRU standards, with real wood paneling and furniture. Studying reports, the red-haired officer looked up as Morozov furtively tapped on the doorframe.

"Come in," said Federov brusquely, closing a folder. "Have a seat."

The double-breasted coat of Federov's American-styled business suit was draped across a valet stand in the corner. The coat, as well as his oxford cloth shirts, had to be custom-tailored to accommodate his massive shoulders. Without a doubt, he was the most physically intimidating man that Morozov had ever encountered. If his enormous size and physical prowess weren't enough, he was reputed to be a voracious reader and genius as well.

Morozov stepped forward and began to sit down in a sturdy mahogany chair magnificently upholstered in plush red velvet.

"Not that one, idiot." Federov gestured towards a straight-backed wooden chair. "*That* one."

Morozov sat down in the wobbly chair, nervously cleared his throat, and said, "Happy Thanksgiving, sir."

Scratching his square chin, Federov glowered at Morozov as if deciding to kill him now or whether the loathsome chore could wait until after lunch. "*What?*" he blurted.

"It's Thanksgiving, sir," muttered Morozov, immediately conscious that he had committed a grievous error. "It's a traditional holiday in America, from when the Pilgrims . . ."

"I *know* what Thanksgiving is, Anatoly Nikolayevich. I'm an *intelligence* officer, you buffoon, so I know my enemy's holidays. That doesn't mean that I *celebrate* them."

Morozov noticed a *spetsnaz* hatchet mounted in a frame behind Federov's desk. He surmised that it wasn't a ceremonial weapon, since the blade's edge bore deep nicks and was marked by ominous dark stains. "Sir, did you want to hear about the Americans' failed POW raid?" he asked, hoping to calm the formidable Resident. "I just reviewed the report, and—"

"*Hush!*" barked Federov, opening another folder. "I was just looking at your report concerning this retired American sergeant in custody in Nevada."

"*Da,*" blurted Morozov anxiously. "Sir, I would gladly apprise you of any—"

"Do you not think I have enough to do?" growled Federov. "Otherwise, why would you waste my time with such mundane matters as an American pensioner trying to bludgeon some sense into a government bureaucrat? Am I not burdened *enough*?"

"But the circumstances of the subject—Eric Yost—are directly related to what I had reported about the Americans' UFO studies at Wright-Patterson Air Force Base!" declared Morozov emphatically.

"*Da.* As I recall, just a few months ago, you claimed that Yost had been murdered to cover up what the Americans were doing, and now you're suddenly changing your tune? And after you insisted that your mysterious hangar was used to store captured UFOs, did you not later report that it was merely a workshop where Soviet aircraft were studied? And now this fellow Yost has been reincarnated in Nevada? Can you not find some consistent story and stick to it?"

"But, Comrade Colonel . . ."

"*Enough!*" declared Federov. "Since you can't seem to focus on matters at hand, Anatoly Nikolayevich, let's discuss this case at this Wright-Patterson Air Force Base. I want to put this matter to bed. Permanently."

"As you wish, Comrade Colonel. For the sake of chronology, I'll start . . ."

Federov scowled and held up a hand sufficiently large to swat an airplane from the sky. "Be *quiet*, you insolent nitwit. I don't need a damned chronology. I have reviewed your reports. My time is limited, so there is no need for you to regurgitate trivia and drivel. Anatoly Nikolayevich, the fact is that you severely bungled this case and squandered precious resources."

"Bungled, sir?" replied Morozov, shuddering with the notion that he might have failed.

Federov sniffed. "Yost fed you a line of shit, moron. Good field operatives must be intuitive. You should have caught on much earlier before we spent so much time and money on this escapade. Just so you know, Anatoly Nikolayevich, there were *never* any alien corpses or flying saucers in that hangar of yours. We know *exactly* what the Americans were doing in there."

"No flying saucers? Then how about the claims of reverse-engineering, where the Americans were studying our aircraft?"

Federov shook his head. He opened another folder and slid a photograph in front of Morozov. "*Nyet.* No flying saucers. No hijacked

MIGs, either. Your man Yost wasn't the only one taking pictures at that hangar. In January of last year, we caught wind that a space mission simulation system was being transferred from NASA to the Air Force. It was from NASA's Gemini program, so NASA obviously considered it obsolete. Guess where it ended up?"

Perplexed, Morozov shook his head.

Federov took a sip from a small glass of hot tea, smacked his lips, and then continued. "Anatoly Nikolayevich, it was installed in *your* mysterious hangar in Ohio. We followed the shipment and positioned a man to surreptitiously photograph it. The Americans are stupid about their security, so it wasn't difficult to piece this puzzle together. Our analysts were able to tell the complete story just on the basis of this single photograph."

"They were?" asked Morozov sheepishly, examining the image. It showed computer cabinets and other large pieces of equipment being unloaded from a flatbed trailer.

"See this man there?" asked Federov, pointing at a figure standing near the entrance. The man had dark hair and wore a heavy parka. Even though the photograph was obviously taken through a telescopic lens, the man's features were distinguishable. Pointing to a piece of equipment as if directing the workers, he appeared to be in charge of the operation. "Our analysts were able to identify that man as Lieutenant Colonel Edward Russo of the US Air Force."

"Colonel, the analysts were able to determine the purpose of this facility by a single man?"

Obviously pleased with himself, Federov guffawed. "*Da!* You see, although it had not been officially announced, Russo was slated to be in the next group of military astronauts assigned to the Americans' Manned Orbiting Laboratory program! So, once you fit all the pieces of this puzzle together, this hangar obviously housed a training facility for the MOL. Why else would the Air Force want a mock-up for an outmoded NASA spacecraft?"

"But the MOL program was cancelled last year," stammered Morozov. "In June."

"Correct," said Federov, smashing his ham-sized fist on the table. "For once, donkeyhead, you've done your homework! And now the final two pieces of the puzzle, the things that confirm our suspicions about this place. We have an informant who has been able to drive by the hangar periodically, and she states that it is rarely used anymore. We suspect it's just being used for storage."

"You implied that there was another piece?" asked Morozov.

Federov nodded. "*Da.* The man I showed you earlier . . . Russo? After the MOL program was cancelled, he was placed on a liaison assignment with the US Navy. We know *precisely* where he is at this very moment, and his duties have absolutely nothing to do with space flight, flying saucers, or aliens."

"Where is he, sir?"

"He is presently enrolled at the US Navy's nuclear power school at Bainbridge, Maryland. After he completes their training course, he is going to a temporary assignment aboard one of the Americans' nuclear submarines."

"This all has been very enlightening, sir," said Morozov. "But I assure you that I was only following orders when I met with Yost. I know that we spent a considerable amount of time, money, and resources on the Ohio effort, but there's no way that I could have known . . ."

Interrupting him, Federov smiled. "Of course. You know, Anatoly Nikolayevich, when I was first stationed here, my predecessor used to regale me with stories about your administrative expertise. He would rattle on and on about your affinity for paperwork and details."

"That's true," stammered Morozov, dreading the notion that he might be relegated into an even less significant clerical role within the station. "But a good operative must not only know tradecraft, he must also be proficient with administration as well. The ability to write accurate reports is contingent on it, so a field operative . . ."

Interrupting him, Federov nodded and said, "Funny you should mention that, because he insisted that you just weren't cut out to be a field operative. He was curious to see how this assignment in Ohio would play out, and now we know the outcome."

"I've been studying Vietnamese," offered Morozov in desperation. "In my free time, of course. I listen to the conversational tapes and do the exercises. It's a difficult tongue, but I . . ."

Federov ignored him. "It would be a shame if the GRU didn't adequately exploit your abilities, particularly at this late stage of your career."

"I agree wholeheartedly," blurted Morozov. "I think that I could better serve the Soviet Union if I was stationed with our Socialist brothers in Hanoi!"

"So you wish to be reassigned?" asked Federov. "This is your desire?"

"I would, sir." *Do I wish to be reassigned?* thought Morozov. *Do I wish to be reassigned? Perhaps that's why I have tendered so many transfer requests!*

Federov opened an envelope and slid an Aeroflot ticket across his desk. "Then your wish is granted forthwith," he decreed. "You will fly to Moscow on Monday. You will be delighted to know that I have discussed your situation with my superiors at the Aquarium, and we have decided on the perfect place to exploit your abilities. Your next assignment, and probably your last, will be at the Encyclopedia. It is an absolutely *perfect* posting for someone who thrives on minutiae."

The Encyclopedia? As sultry visions of Hanoi evaporated from his thoughts, Morozov's heart sank. *Nothing* could be worse. The Encyclopedia was the nickname for the GRU's Department of Archives and Operational Research. Occupying the two lowest levels of the Aquarium's basement and other facilities, it was the repository for the immense volumes of intelligence accumulated by the GRU's vast espionage enterprises throughout the world.

Federov wasn't being the least bit facetious when he said that the Encyclopedia would likely be Morozov's last assignment within the GRU. Archivists and researchers at the Encyclopedia were routinely exposed to so much sensitive information that they could never be allowed to venture outside of the Soviet Union. In fact, Morozov had heard grim rumors that when Encyclopedia workers reached the end of their careers, they were moved into forced retirement at a remote location.

He had been there once, during his initial training, and that was enough to convince him that he didn't ever want to return. Within the GRU, there were plenty of jokes about the wretched place, most cautionary in nature. As an example: *Two long-time Encyclopedia archivists fell in love and married; their offspring looked like moles, although not quite as attractive.* Perhaps the most telling joke was one that bore a certain degree of dire truth: *The Encyclopedia wasn't Hell, but you could certainly see Hell from there; all you need do is look up.*

Scowling, Federov asked, "The Encyclopedia does not suit you? Should I remind you that the Aquarium's basement has *another* floor? Perhaps you might be more comfortable there. After all, I'm sure that you've heard the old joke about the Encyclopedia. How does it go? The Encyclopedia isn't Hell, but you can see it from there. All you have to do . . ."

". . . is look up," interjected Morozov quietly. He swallowed deeply and added. "No, I think that the Encyclopedia suits me just fine. I look forward to serving the Soviet Union there, sir."

Federov waved his hand like he was shooing away an obnoxious child. "That is all, Anatoly Nikolayevich. Dismissed. Don't miss your plane. And don't stuff yourself with turkey and dressing today. Gluttony is a sin, or so I've heard."

Morozov stood up, saluted, and pivoted about. He was almost out of the office when he heard Federov's voice behind him. He turned around slowly to face the Resident's desk.

"Nice shoes, Anatoly Nikolayevich. They look new. Florsheims?"

Morozov swallowed. "*Da.* They are Florsheims, sir. You have a good eye. But they're not new. I bought them second-hand in Ohio. I just take very good care of them."

"If you say so, Anatoly Nikolayevich, but I caution you to remember that there is so much more to the Aquarium's basement than just the Encyclopedia. If you insist on standing out too much from your comrades, you may find yourself visiting another floor."

23

ENCYCLOPEDIA

Pacific Departure Facility, Johnston Island
3:12 a.m., Friday, January 15, 1971

Almost two years after the tragic accident that killed Howard and Riddle, and after they had been into orbit three times themselves, Carson and Ourecky were poised to go yet again. They listened to the groans, clicks, and other noises of the Titan II, knowing that the booster's turbo pumps would soon spin to life, initiating the massive exothermic chemical reaction that would blast them clear of the confines of Earth. At this point there was little to do but wait.

"You awake over there?" asked Carson.

"Barely," replied Ourecky. "I guess I should be a tad more excited. If there was room I would do some jumping jacks to raise my pulse a few notches, just so the docs don't think my heart has stopped."

Just a few minutes later, they were beyond the point of no return. "Launch vehicle is transferring to internal power," stated the CAPCOM. "Stand by for engine gimballing."

"On internal," replied Carson. "Waiting for gimbals."

"T minus one minute and counting," stated the CAPCOM.

"One minute and counting," answered Carson. "Thanks, guys. Have a safe flight home. Don't stuff yourself on the luau pig in Honolulu."

"Thanks. Have a safe trip yourselves," said the CAPCOM. "Stage Two Fuel valves opening in five seconds."

"Minus thirty seconds," stated the CAPCOM.

"Once more into the breach?" asked Carson, placing his right hand on the center console.

"Once more, dear friend," replied Ourecky, tapping Carson's hand.

"Minus twenty seconds," stated the CAPCOM. "See you later, alligator. And ten, nine, eight, seven, six, five, four, Ignition, three, two, one, Zero. Hold-down bolts are fired. Lift off!"

"Lift off and the clock is started!" called Carson. "Scepter Seven departing."

**Headquarters of the *Glavnoye Razvedyvatel'noye Upravleniye (GRU)*
Khodinka Airfield, Moscow, USSR
4:18 p.m., Friday, January 29, 1971**

Momentarily looking up from his notes, which were dimly illuminated by a forty-watt bulb suspended over his worktable, Morozov removed his reading glasses and rubbed his irritated eyes. He had labored in the dismal bowels of the Aquarium for the past two months. Although the Encyclopedia was certainly not an assignment that he relished, and one from which he would likely never emerge, he diligently applied himself to his tasks. His industriousness paid off; he had already been promoted, graduating from entry-level archivist to Third-Class Analyst in record time. In that capacity, he now supervised a Third-Class Analysis Section consisting of himself and four archivists—three "retrievers" and one "filer"—who assisted him in his research.

The Aquarium's basement contained three massive subfloors; the Encyclopedia occupied the lowest two. Apart from the worktables for the Analysis Sections, most of the space was filled by row after row of index card files, much like those found in any library. The Encyclopedia existed solely to placate the GRU's voracious appetite for information. More specifically, since the GRU already possessed the information in raw format, Encyclopedia workers toiled day and night to satisfy the GRU's incessant craving for collated and pre-digested information.

Although poorly lit, the Encyclopedia was at least comfortably warm. Besides being well below ground, where the temperature remained fairly constant throughout the year, banks of humming dehumidifiers—installed to safeguard the paper holdings from dampness and mold—provided warmth as well. The windowless walls were painted an earthy shade of taupe, perhaps to remind the workers of their subterranean setting.

As he reported to work every morning, a Second-Class Analyst issued Morozov a stack of cards, each of which contained one or more questions. Some of the questions were mundane, requiring scarcely any effort on his part: "*Who is this man? Who does he interact with?*"

Other queries—the ones that Morozov particularly enjoyed—were considerably more esoteric, and required more extensive research and sleuthing. "*What is manufactured at this facility? How are these two sites related?*" Morozov prioritized the questions, developed a research plan, and then deployed his retrievers to fetch the raw information required to answer the questions. In a sense, his task was like weaving a rope out of tiny bits of fiber, carefully splicing threads together. In his short time here, he had determined that an analyst's most important skill was intuitively knowing which threads to chase and which ones to ignore.

Although a single question might entail that his retrievers amass a collection of hundreds or even thousands of cards, only a minute fraction of the Encyclopedia's vast information holdings were to be found in the index card files. The rest was stored in adjunct facilities—massive warehouses that contained photographs, movie film, books, tape recordings, source documents, and such—in the vicinity of the Aquarium.

The retrievers retrieved and the analysts analyzed, but probably the most thankless job fell to the filers, who were responsible for ensuring that the cards found their way back to their appropriate slots in the cabinets. By far, the filers worked the longest hours, often staying until midnight or later, long after the others had left for the day.

In a practical sense, the Encyclopedia could be likened to a gigantic brain. An army of intake archivists processed the raw data as it arrived, collating it and annotating it to the index cards. They were like the five senses, gathering the information and storing it into memories. If the intake archivists were the senses, then the Analysis Sections were the neurons; when a question required an answer, they were energized to winnow through accumulated memories, sparking through their unique collection of axons and synapses. And like the neurons of a biologic brain, the Analysis Sections were layered by hierarchy, with the First Class sections responsible for higher level processing and conceptual thought, down to the dreary Third-Class sections—like Morozov's—that performed little more than limbic functions.

The Encyclopedia was a clunky and inelegant solution to an abundantly complex problem. But like so many other clunky and inelegant Soviet solutions, it *worked*. The venerable Automat Kalashnikov assault rifle wasn't chic or attractive, but it was robust and inherently functional. When its trigger was pulled—no matter whether its stamped metal receiver was immaculately clean, jammed with the grittiest desert sand or packed with frigid Arctic ice—it *fired*. Morozov was sure that the American CIA probably collated its intelligence gatherings in massive computers, as the KGB likely did as well, and that the GRU would eventually do likewise. Until such time, the Encyclopedia would have to suffice. It *worked*, and he was an integral part of it.

Since the Encyclopedia would soon outgrow its current environs, there was talk of a new facility. The whimsical architectural sketches showed a modernistic building as spacious as a tsar's summer palace, warmly lit and climate-controlled. If it was ever realized, it would be equipped with a semiautomatic retrieval system, and an expansive network of pneumatic tubes to swiftly convey index cards and documents.

Examining a color-coded card, Morozov looked forward to leaving in a few hours. He jotted some notes on a slip of paper and then handed the note to one of his retrievers, a dainty and kind-faced woman in her early forties. They shared something; she had been in Stalingrad during the Siege. They often chatted about their experiences over tea at lunch, but Morozov had come to suspect that she—as a teenage girl—had survived the ordeal by consorting with Nazi soldiers.

As he placed the card in a careful array on the table, the solitude was broken by the muted sounds of a man pleading for his life. Morozov and the retriever anxiously gazed up towards the whitewashed ceiling. The floor directly above them was a massive labyrinth of cells, a veritable factory of misery, where hapless GRU prisoners were held, questioned and often executed.

Unlike the KGB, which could arbitrarily snatch ordinary citizens off the street and detain them indefinitely, the GRU's charter almost exclusively limited its arrest powers to members of the Soviet military forces. At any given moment, every branch of the military forces was represented in the GRU's prisoner rolls, including generals, admirals, and a substantial number of GRU personnel suspected of treason or espionage.

Hearing the man yelp again, Morozov held his breath and cringed. There were at least forty-five centimeters—roughly eighteen inches by the English system—of reinforced concrete overhead, so he was certain that most of noises were muffled, so that he was only hearing the cries and screams of the tormented when they were at their absolute peak of agony. If the bloodcurdling noises weren't enough, the ceiling was marked with ominous stains and blotches, where blood and gore had seeped through minute crevices and seams in the concrete.

The unseen man shrieked like someone being slowly disemboweled by a giant hook, which could likely be his exact predicament at this precise moment. The retriever appeared to be on the verge of tears. The note fluttered in her hand as she attempted to read it. "How do they expect us to work in these conditions?" she asked meekly. "I don't sleep anymore. I just go to my apartment and cry all night. My husband doesn't know what to make of me."

Morozov's hands shuddered as he cautioned, "*Focus*. Don't dawdle. Give thanks that we are down here and not up there." Remembering that it was Friday, he looked at his watch, anticipating what would happen shortly.

Suddenly there was a gunshot, followed by a heavy thud. Apparently, the interrogator who "owned" the cell upstairs was a creature of long-established habit. Morozov could set a clock by the sounds that emanated through the concrete at different intervals. In his mind, he pictured the methodical inquisitor as a working class cook with a clearly set regimen of daily specials to prepare during the week. Simple but brutal beatings were the stuff of Mondays. Tuesday's menu featured hammers, saws, and other hand tools. Electricity and flame were Wednesday's fare. Some specialized torture tools—which Morozov could not yet picture—were served up on Thursdays. Friday's offerings were clearly the most grue-some and yielded the most dreadful screams; he imagined that it likely involved amputation and evisceration. And then on Friday afternoon at five, or roughly thereabouts, came the *pièce de résistance*: a merciful bullet to the back of his prisoner's head, served up just in time for the torturer to catch the subway home to a meal of buckwheat *kasha* washed down with a tumbler of *Moskovskaya*.

There was a faint scraping sound, obviously the noise of a cadaver being dragged across a rough-textured floor, and then blissful silence. Morozov believed that there was a reason that this particular table was set aside for fledgling Third-Class Analysts: to indelibly etch on their thoughts the consequences that could befall the overly inquisitive. Curiosity may have killed the cat, but the feline's demise was likely swift, not stretched over the course of five agonizing days.

Yet again, Morozov resolved himself to toil as hard as necessary to achieve the ranking of Second-Class Analyst, if for no other reason than he and his archivists would move to the floor below. But besides a respite from the sounds of the tormented, there was another benefit in advancing—or perhaps *descending*—to Second-Class. Second-Class Analysts had much greater leeway concerning the information they requested and reviewed. Third-Class Analysts such as himself were required to maintain a detailed ledger that accounted for every scrap of information that they saw, and there were dire consequences for anyone caught with something that did not apply to a particular question that they had been called to answer.

Amongst the Third-Class, freelance research was expressly forbidden. Vigilant appraisers, themselves veteran analysts, roamed the floors and audited the ledgers. It was a deadly serious business. If an analyst couldn't satisfactorily explain his transgressions, then he would be sent upstairs to occupy a chamber until such time as he could.

Besides moving downstairs and having more freedom, what Morozov really desired was to solve the mystery of Hangar Three. The Americans had obviously gone to tremendous lengths to deceive him; he wanted to know *why*, and he didn't buy Federov's conclusion that the Ohio hangar simply housed a now defunct training facility for the Air Force's MOL program.

Before he left America, he had spent hours reviewing his files and memorizing key names. He strongly suspected that the ultimate key to unlocking the mystery was to gather information about the test pilots assigned to the Aerospace Support Project. He wagered that they had been flying *something*—if not captured flying saucers, at least something of great significance—and he was determined to discover what that was. Now, it was just a matter of patiently biding his time until he acquired access to the necessary information. And time was something he had, in great abundance; although it was tempting to go to the stacks to just yank out the appropriate cards, he knew that he had to wait until the answers arrived at his table.

24

THE BITTER LEGACY
OF SOYUZ "YANTAR"

Burya Test Facility
Kapustin Yar Cosmodrome, Astrakhan Oblast, USSR
9:15 a.m., Monday, October 4, 1971

Together with Gogol and Travkin, Vasilyev was jammed into the narrow rear seat of a *Zhiguli* sedan. The crew had been summoned to the *Krepost* headquarters to personally meet with Lieutenant General Abdirov, the leader of the project.

Much had happened since their "*Kochevnik*" *Soyuz* mission last year. In May, Vasilyev had lost his entire family in a Moscow automobile accident. Vasilyev had been participating in a survival training exercise at the time, and the exercise directors hadn't even pulled him out of the field to inform him that his family was killed. Scarcely returning home in time for the funerals of his beloved wife, Irina, and the two young daughters that they adored, he was still bitter about the incident and its aftermath.

He had barely been granted time to mourn. Because he was assigned to the prime crew for a critical strategic mission, Vasilyev was effectively expected to set aside his grief, summon his fortitude, and dutifully trudge on with their training. It was the soulless Soviet system at its worst.

The past several months had been a miserable ordeal. He and Travkin despised Gogol, but could do nothing to escape his clutches; they were destined to fly with him, whether they liked it or not. Most of their routine, like working in the simulators for hours on end, was just painfully monotonous. They had their procedures down cold but were essentially marking time until the long-anticipated *Krepost* station was completed.

To make matters worse, at least whenever they were at Kapustin Yar and not training at another site, Vasilyev returned to a cold and empty apartment at the end of every day. To their credit, Travkin and Ulyana did their best to console him, but his heart was shattered. As much as they strived to convince him that he would be healed by the passage of time, he ached for Irina and their toddlers and wished that he could see them just one more time.

In late June, Vasilyev was impacted—although indirectly—by another horrific tragedy, a terrible accident that brought all Soviet manned spaceflight to an absolute halt. The incident occurred during the historic *Yantar*—"Amber"—*Soyuz* mission, in which a three-man crew—Vladislav Volkov, Georgy Dobrovolsky, and Viktor Patsayev—docked with and occupied the *Salyut-1* space station. After spending twenty-three days in orbit, the three cosmonauts succumbed to asphyxiation when a faulty vent malfunctioned in their Descent Module just prior to leaving orbit.

Although the rest of the *Soyuz* spacecraft's automatic systems functioned correctly during reentry, the three men were found dead, strapped into their contoured couches. The disaster was the second fatal incident involving the *Soyuz* spacecraft; on its inaugural manned flight in 1967, cosmonaut Vladimir Komarov had been killed when his parachute failed to deploy after reentry.

The *Yantar* tragedy caused great consternation in the *Krepost* project, since the effort was so dependent on the *Soyuz*. Although the designated crews were slated to continue orientation and training missions until the *Krepost* was ready for flight, those sorties were grounded until further notice. Consequently, only two crews—Gogol's and one other—had actually flown in orbit. More importantly, even if the *Krepost* was delivered on schedule, it could not be occupied if the *Soyuz* was still grounded.

9:50 a.m.

An aide escorted the cosmonauts into Abdirov's office, where they occupied leather-upholstered chairs in front of the general's desk. Try as he might, Vasilyev would never be comfortable in the presence of Abdirov. The general's grotesque appearance was more than disconcerting. If he had suffered such ghastly injuries, Vasilyev would be strongly tempted to chew on the muzzle of his Makarov.

Assisted by the aide, Abdirov deliberately and painfully rose from his chair, slowly walked around his desk, and stood next to Vasilyev. The general's posture was stiff, like a pine plank. Placing his scarred hand on Vasilyev's shoulder, he softly said, "I am truly sorry for your loss, Pavel Dmitriyevich. I apologize for not personally offering my condolences before today."

"Thank you, sir," answered Vasilyev, trying mightily to maintain his composure instead of yielding to anger. "Losing my family has been terribly difficult for me, but I am truly fortunate that I am able to focus my energies at such a truly worthy effort as the *Krepost*."

"Then I am happy for you," said Abdirov. "Speaking from experience, I know just how easy it is to become mired in pain. When calamities strike, it is crucial for us to go on living and continue our service to the Motherland, since our daily sacrifices are the best memorial we can offer for those who have been lost."

As strange as it seemed, Vasilyev sensed that Abdirov was truly sincere in the sentiments he offered. As Abdirov gradually returned to his seat, his secretary circulated through the room, cordially pouring tea and serving spiced *pryaniki* biscuits from a silver tray.

After his aide helped Abdirov back into his chair, the general announced, "Gentlemen, Major General Yohzin will brief you concerning some significant developments with the program." He motioned towards a corner, where Yohzin stood. Although Vasilyev had often heard his name, this was the first time that he had actually laid eyes on the Yohzin. Wearing a meticulously tailored RSVN uniform, Yohzin was tall and of medium build, with dark brown hair combed back neatly across his crown. His face was broad, with very typical Russian features; his skin was dry, with a reddish cast, as if he had spent much of his life outdoors.

"As you are no doubt aware, the *Yantar* incident has dealt us a tremendous setback," said Yohzin. "The Korolev bureau is reluctant to allow further flights of the *Soyuz* until an exhaustive investigation is completed and these discrepancies are adequately resolved."

Vasilyev stifled an urge to laugh. *Reluctant?* After two fatal accidents, the Korolev bureau *certainly* had ample reason to be reluctant. They were heavily invested in the *Soyuz,* and couldn't risk the grim prospect of even more fatalities. After all, the *Soyuz* was envisioned to become the mainstay of Soviet spaceflight. Like the ubiquitous American jeep in the Great Patriotic War, the *Soyuz* could be adapted to perform in a multitude of roles. It could be employed as a taxi to ferry crews to space stations and had been a key component of a planned lunar expedition to beat the Americans to the moon, even though that effort had fallen by the wayside.

Designed from the outset with automatic features that managed all aspects of flight, the *Soyuz* could be modified to fly as a crewless resupply vehicle or tanker, as well as a platform for various types of other unmanned missions. There was even talk that special military variants might eventually be produced, perhaps including interceptor and reconnaissance versions of the spacecraft. So, with so much at stake, the Korolev bureau clearly wasn't willing to gamble. Moreover, the bureau wasn't the sole competitor in the race to produce the next generation of utility spacecraft; other aerospace design bureaus were anxious for an opportunity to finally shove the vaunted Korolev bureau from their high pedestal. Definitely the leading contender, the Chelomei bureau had already made significant progress on a prototype—the TKS transport supply spacecraft—that could readily supplant the *Soyuz*. To make matters worse, the Korolev bureau had lost its visionary leader, the famed "Chief Designer" Sergei Pavlovich Korolev, in 1966. Now led by Vasily Pavlovich Mishin, contending with budget shortfalls and a plethora of technical complications, the bureau struggled to maintain its preeminence in Soviet manned spaceflight.

"Since unexpected depressurization is the greatest immediate concern with the *Soyuz*," said Yohzin, "an entirely new type of pressure suit is being designed. It's called the *Sokol,* and it is intended strictly as a precautionary measure, to protect the crew inside the spacecraft in the event of catastrophic depressurization. Until the engineering issues are entirely resolved and the *Soyuz* is properly recertified, flight crews will wear *Sokol* emergency suits for the ascent and descent phases of the mission."

"But the *Sokol* suits are not ready yet, Comrade General?" asked Gogol, stirring a tablespoon of sugar into his glass of hot tea. He turned up the glass, drank most of the steaming tea in one draw, and smacked his lips afterwards. Abdirov smiled faintly at the gesture.

Yohzin nodded. "*Da*. That is correct. But even when the *Sokol* suits do become available, there is still a much greater issue for us. Because of weight and space issues with the *Soyuz* Descent Module, only two cosmonauts will be able to fly at a time if they are wearing pressure suits. As we speak, the Korolev bureau is working to modify the vehicle to allow this, but it may be at least eighteen months or even two years, before the *Sokol* suits are ready and the Descent Module is adapted accordingly."

Good, thought Vasilyev. It sounded as if the circumstances might grant him and Travkin at least temporary reprieve from flying again with Gogol. And maybe, given some extra time and with any luck, Gogol's aberrant behavior might be revealed in some other way, and the miscreant would be summarily excised from the program.

"As you are aware, the first *Krepost* station will be ready for launch early next year," revealed Yohzin. "Because of the pressing need to make it operational as swiftly as possible, the General Staff has authorized us to fly the *Soyuz* for our missions, on an interim basis, to deliver crews to man the station."

Yohzin continued. "The Korolev bureau has agreed to support this, with certain stipulations. Until the *Sokol* emergency suits become available, *Krepost* crews will wear SK-1 suits, just like the one you wore on your Vostok missions, Gogol, during the ascent and descent phases of their mission."

Gogol laughed. "Now, I regret burning my old pumpkin suit in China. It fit very comfortably."

"And because of the requirement to install some temporary fittings for oxygen flow and to rapidly pressurize the SK-1 suits in an emergency," said Yohzin, "only two cosmonauts may fly in the Descent Module, just as with the *Sokol* suits once they are delivered."

The foolhardy scheme did little to instill confidence in Vasilyev. He felt like they were gladiators being ordered to ride on a chariot with loose wheels and an axle on the verge of breaking. If the glaring technical glitches with the *Soyuz* didn't yield sufficient cause for concern, the SK-1 pressure suits were based on outdated technology. The heavy orange suits were awkward and unwieldy at best.

Abdirov cleared his throat and spoke. "Which brings us to the most pressing question, gentlemen: Which two of you will fly the first mission to occupy the *Krepost*?"

As the three cosmonauts leaned forward in anticipation, Abdirov wasted no time in delivering the punchline. "After reviewing your qualifications, here is my plan: Lieutenant Colonel Gogol, you will fly with Major Vasilyev on the first mission. Major Travkin will continue to train with this crew as a back-up."

Since they occupied three chairs arranged roughly in a semi-circle in front of Abdirov's desk, Vasilyev clearly saw the reactions of the other two men. The corners of Gogol's thick lips turned up slightly, in an almost imperceptible smirk. Obviously relieved that he was spared from flying with Gogol, Travkin flashed an involuntary smile, which he quickly—and quite consciously—turned into a frown. As for himself,

Vasilyev didn't need a mirror to know that he had probably scowled at the unwelcome news.

Furtively, Vasilyev turned to glance at Abdirov. Although it was virtually impossible to read the general's horrifically scarred face, it was clearly obvious that Abdirov was puzzled by their responses.

"On to another matter," said Yohzin. "As you might imagine, this turn of events has compelled us to make changes with other *Krepost* systems. Obviously, the deployment control system for the warhead will have to modified, so that the deployment sequence can be activated by just two men and their keys, instead of three. Once these changes have been incorporated into the simulator mock-up, you will begin training with the new procedures immediately."

Imagining what it was going to be like to fly by himself with Gogol, Vasilyev struggled to focus on Yohzin's words.

"There's yet another modification to the weapon deployment system," said Yohzin. "We have to anticipate circumstances where the two of you may be required to deploy the warhead without a prior authorization from the ground, in the event that hostilities have already commenced and communications links have been neutralized. To address that contingency, your code safe will contain a set of autonomous action codes, one for each of you. Between the two of you, you will have to agree to autonomous action, and both of you would have to enter your personal codes and turn your key within the allotted time, so the procedure still precludes the possibility of one of you overpowering the other."

"And lastly," said Abdirov quietly. "We have to contemplate one more contingency, and that's the very unlikely possibility that one of you might become incapacitated, either due to illness or injury, during a mission. Since we still must ensure the capacity for autonomous action, should it be necessary, we will radio up a special independent action code that would allow one of you to deploy the weapon by yourself. The independent action code would override the second key station, so that turning one key would suffice to deploy the weapon."

One key? Vasilyev could not comprehend that one man might be granted so much power. To possess a fifty-megaton nuclear warhead that could be directed on any spot on Earth? And that *Gogol* might be granted such authority? It was just too much to conceive.

"Lieutenant Colonel Gogol," said Abdirov. "I have to ask, just for my own edification, would you be willing to deploy your warhead using the autonomous action codes or the independent action code, if the circumstances warranted such action?"

"Of course, Comrade General," vowed Gogol, answering without the slightest hesitation. "You can be assured of this, sir."

Certainly, Gogol would drop the Egg without even a moment's hesitation, thought Vasilyev. In fact, Gogol probably wouldn't be able to restrain himself if he was issued an independent action code.

"Good," said Abdirov. "*Very* good. You gentlemen are released back to your duties."

As the three of them walked into the outer office, Gogol grinned. "So the two us will be in orbit *together*," he said quietly, punching Vasilyev's shoulder. "For *weeks*, no less! What a truly *delightful* turn of events."

Flinching from the blow, Vasilyev shuddered at the dire thought of spending sixty days in orbit with the despicable Gogol. As his stomach plummeted, he heard the voice of Abdirov's secretary behind him. "Comrade Major Vasilyev," she said.

In a single motion, he stopped and spun around. "*Da?*"

"The general wants to speak with you. Alone."

"We'll wait for you in the car," said Gogol, leaning towards him and speaking quietly. "And be mindful of what you chat about, Pavel. A thimbleful of caution goes a long way."

Vasilyev walked back into Abdirov's office and reported. Yohzin had left; he and Abdirov were entirely alone. He heard the door click shut behind him as the general gestured for him to take a seat.

Abdirov sipped from a glass of water, cleared his throat, and asked, "Tell me, Major Vasilyev, and be frank. Is there some reason that you are reluctant to fly with Lieutenant Colonel Gogol?"

Vasilyev felt like the general was reading his thoughts. *But what could he say?* It was like walking a tightrope over a deep pit brimming with burning coals. This was his chance to finally break from Gogol, and while it was very tempting to finally reveal the truth, the consequences of a misstep could be catastrophic. If Abdirov didn't accept his explanation, then his career would be over if he refused to fly with Gogol. Moreover, if he didn't fly with Gogol, then Travkin would be obligated to fill the second seat, and he could not sentence his friend to such a fate.

"I have your permission to speak openly?" asked Vasilyev. "Without fear of repercussion?"

"*Da.* Of course."

"Some of his personal habits concern me," answered Vasilyev matter-of-factly. "And frankly, sir, he sometimes displays a cavalier attitude towards his duties."

The corners of Abdirov's damaged lips turned up in his unique half-smile and half-sneer. "I suspected as such," he said. "I've heard that he can be very abrasive when talking to Control, if the conversation concerns a matter that doesn't suit him, and I have to imagine that he exhibits the same sort of behavior with you and Travkin. And his *habits* offend you, Pavel Dmitriyevich?"

"Maybe I was speaking out of turn, Comrade General," answered Vasilyev sheepishly.

"Then would it surprise you to know that I know all about Gogol's shenanigans and obnoxious habits?"

Vasilyev was shocked; if Abdirov had such intimate knowledge of Gogol's behavior, then why on earth would he let him fly again? "You do, Comrade General?" he asked.

"Sure. Suffice it to say that your commander did not adequately clean up after himself. After your Descent Module was brought back here last year, our technicians found two cigarette butts inside. I was made aware that you and Travkin did not smoke, and when I confronted Gogol about the butts, he admitted that they were his."

"I didn't know that you were aware, Comrade General," said Vasilyev.

"I was. Trust me, Pavel, you will not be bothered by his obnoxious habit again, at least not on mission. I cannot compel him not to smoke on Earth, but he has promised that he will leave his cigarettes here when he departs for the next mission. Moreover, I will ensure that his personal kit is scrutinized to make sure that he cannot smuggle any smokes to orbit. Sincerely, I am sorry that he placed you at such great risk."

"Pardon me, Comrade General, but I didn't think I was in great danger. To his credit, Gogol demonstrated to me that there was no danger of fire in weightlessness. While we weren't fond of breathing his smoke, Travkin and I did not feel that we were in danger."

"You didn't think there was a danger?" scoffed Abdirov. "You might not know it yet, but the *Yantar* crew had to fight a fire aboard their *Salyut* while they are up there, so I am not so prone as you to share in Gogol's naiveté concerning the threat of fire in space. By the way, the *Yantar* crew taught us another lesson, and that is that crew personnel *must* be compatible. Two of them were almost constantly at odds with each other up there, and their squabbling almost prematurely ended the mission. So, Pavel, besides Gogol's smoking and coarse behavior, is *any* other reason that you two can't occupy the *Krepost* together for a few weeks?"

For a fleeting instant, Vasilyev felt the need to be candid, but then realized that his honesty would likely come to naught, if not worse. "*Nyet*, Comrade General."

"Good, then on to other questions."

"Other questions, Comrade General?"

"*Da*. I'll tell you, Pavel, there are many attributes that I value in a subordinate, but for this mission, reliability is the quality that I value above all others."

"I agree, Comrade General. Considering the mission, you're exactly right."

"Then can I trust you to be reliable, Pavel?" implored Abdirov. "Can I trust you to do your duty?"

"*Da*, Comrade General," replied Vasilyev. "Without question. I could not possibly shirk my duties as a Soviet officer."

"Then if you were instructed to deploy the warhead, you would do so without hesitation?"

"*Da*, Comrade General."

Abdirov nodded his head. "That's good, but it is absolutely crucial that you be resolute in executing your duties, even if the situation is not completely clear. For instance, what if the circumstances were such that Gogol was incapacitated and *you* were issued an independent action code? Would you still be so enthusiastic about deploying the warhead?"

Choosing his words carefully, Vasilyev replied, "Comrade General, truthfully, I could *never* be enthusiastic about deploying the warhead, but if that was my task, then I would do so without hesitation."

"Good." For whatever reason, Abdirov's demeanor seemed to change. His tone became less like a superior officer counseling a subordinate, and more like a father speaking to a child, advising him on the true nature of the world. "You know, Pavel, as military officers, we are often faced with unique challenges that compel us to make some very difficult decisions. Sometimes, in order to make those decisions, it is vital to understand the *intent* of the mission, and not just the mission itself as it is expressed in a formal order. Does that make sense to you?"

"*Da*, Comrade General."

"Good. Pavel, just so it's absolutely clear: Do you realize that I am the *only* one granted the authority to release an independent action code to a cosmonaut in orbit on the *Krepost*?"

"*Da*, Comrade General."

"And you do understand that I would never release an independent action code to you except in the most dire of circumstances, don't you? I could only undertake such action if it was clearly obvious that we were in danger of losing all communications and that hostilities were imminent. And when I say hostilities, I am not talking about some petty regional conflict, but an all-out nuclear exchange between the Motherland and our enemies. You do understand this, don't you, Pavel? You do understand why I cannot possibly take these duties lightly? It is quite a burden that the General Staff has entrusted to me, don't you think?"

"I agree completely, Comrade General," replied Vasilyev. Focusing on a patriotic painting hung on the wall behind Abdirov's desk, he swallowed, trying to anticipate where this conversation was leading.

"Good. I'll tell you, Pavel, I have great disdain for the indecisive. When the circumstances warrant, you must be fully prepared to exercise initiative."

"*Da*, Comrade General."

"Good. So that you might clearly understand *my* intent, Pavel, in the future, if the circumstances are such that I release an independent action code to you, then it's logical for you to assume that you are being directed to take the initiative and *act*. Do you understand?"

As his thoughts gelled, Vasilyev realized what Abdirov was conveying to him. Once in orbit, if Gogol became incapacitated, he was to interpret receipt of independent action code as a tacit order to deploy the warhead. Still, he could not believe that Abdirov was crossing this line; merely discussing such matters with a junior officer, even in the most theoretical of terms, was a tremendous gamble.

"Did you hear me, Pavel Dmitriyevich?" asked Abdirov. "Do you understand?"

"Comrade General, are you instructing me to deploy the warhead if I ever receive an independent action code?" asked Vasilyev softly.

"That's *not* what I said," replied Abdirov. "To be clear, I want you to *listen* to me, Pavel. When the circumstances warrant, you *must* execute sound judgment."

Suddenly, Vasilyev could not draw a breath. He felt as if he was being hypnotized; Abdirov's one-eyed stare pierced deep into his soul, like a gimlet boring into a tenacious stone. He tried to speak, but could not.

"Did you hear me, Pavel Dmitriyevich? When the circumstances warrant, you *must* execute sound judgment."

Vasilyev struggled for words, but found only Abdirov's. "Comrade General, when the circumstances warrant, I *will* execute sound judgment."

"*Good.*"

Burya Test Facility
Kapustin Yar Cosmodrome, Astrakhan Oblast, USSR
5:45 p.m., Friday, December 31, 1971

It was late in the evening, and the *Krepost* headquarters were largely vacant; most of the workers were already at home with their families, preparing to celebrate New Year's Eve, the biggest holiday of the year in the Soviet Union. As they remembered the year and its momentous—and tragic—events, Abdirov and Yohzin shared a bottle of *Stolichnaya*.

Yohzin heard a light tapping at the door; it was his driver, reminding him that the car was warmed up and waiting outside.

"I suppose it's high time for you to head home to the family, eh?" asked Abdirov, pouring them another round.

"You know that Luba and I would be delighted to have you, Rustam," answered Yohzin, downing the shot. He wiped his lips with the back of his hand. "There's ample food and drink to share, and she would be thrilled to set you a place at our table again. Come with me, friend, and celebrate tonight with my family."

"*Hah!*" snorted Abdirov. "Listen, Gregor, I would be delighted, but even if your lovely wife can still pretend to tolerate my loathsome appearance, I'm very aware that I scare the dickens out of your sons. Why should I ruin this special night for them? Besides, I *am* celebrating with my family. *You're* my family, Gregor."

Yohzin suddenly felt tremendous sadness for his friend. "But you know that you are always welcome," he said, slightly slurring his words.

"You are very kind, Gregor, but there is still work to be done tonight."

"Work?" asked Yohzin. "Even on this holy night?"

Abdirov frowned at him. "Holy night? Listen, I know that you and your family still celebrate Christmas, for the boys if nothing else, but you might be a little more cautious about mentioning that it in the future. Some might not be so amused."

"Sorry. So, what is so pressing that it demands your attention on New Year's Eve?" asked Yohzin.

Half-drunk, Abdirov awkwardly stood up and stumbled towards the conference table. The table's gleaming surface was completely covered with blueprints and schematics. "Join me here," he said. "And bring that bottle."

Yohzin did as he was asked and poured them another round. Saluting their fallen comrades, far too many to name as individuals, they downed the shots. "So, what's all this?" he asked, gesturing at the mass of paperwork.

"I received all this yesterday," explained Abdirov, scratching a scaly patch of scarred skin on his neck. "They were delivered by special courier, flying directly from Moscow. You remember that I told you that the *Krepost* will eventually be connected to *Perimetr*?"

"*Da*," answered Yohzin, studying the diagrams. *Perimetr* was a vast secret network of detectors and controllers that would automatically initiate a retaliatory nuclear strike in the event that other command and control networks had been neutralized. To those who were aware of it, the theoretically foolproof network was also colloquially known as the Dead Hand.

"To accommodate the *Perimetr* system, these are the proposed modifications to the existing weapons deployment system on the *Krepost*. Obviously, none of this can be installed in time before the first station is launched, but all future *Krepost* stations will be controlled by *Perimetr*. Moreover, special hardware will be produced so that the first *Krepost* can be retrofitted with the *Perimetr* equipment."

"Then the *Krepost* will be unmanned, Rustam?" asked Yohzin. "If it will be controlled by *Perimetr*, will there still be a need for a crew?"

"Initially, for the sake of redundancy, our men will still occupy *Krepost* as a backup to the automatic deployment system," answered Abdirov. "The most significant change is that we would lose our ability to issue an independent action code."

No independent action code? That's a huge relief, thought Yohzin. Dispensing with the independent action code was a truly fortuitous development. He had always been deathly afraid of the ramifications inherent in granting such immense power to a single individual. If the

Perimetr network was frightening, then the notion of the independent action code was even more so. Unfortunately, Yohzin was conscious that Abdirov was a diehard believer in the special code, and would be loath to surrender such power to a coterie of machines.

"I don't like this," muttered Abdirov. "I think you're aware that I am not a fan of *Perimetr*. I think that the entire concept is insane, and it will accomplish nothing but could potentially jeopardize a warhead deployment in a crisis. Robots should not make decisions concerning the deployment of nuclear weapons; there should *always* be a human hand at the controls."

"I agree, Rustam," noted Yohzin. "Absolutely."

As if they were conspiring to kill an offending Party official, Abdirov leaned towards Yohzin and said quietly, "I am delighted that we're in accord, Gregor, because I want you to study these schematics and devise some means to bypass *Perimetr*."

Bypass Perimetr? What exactly did he mean? Confounded, seeking clarity, Yohzin apprehensively asked, "I don't understand, Rustam. You wish me to circumvent *Perimetr* so that the warhead will *not* be deployed if *Perimetr* is triggered?"

Abdirov chuckled. "Very amusing, Gregor. You think that I *don't* want the weapon deployed? Quite the contrary. I *do* want to drop the Egg, but once we relinquish control to *Perimetr*, that opportunity will be snatched from our hands. The independent action code is our only means of doing so, and we are on the verge of losing that capability."

"But you *want* to drop the Egg, Rustam?"

"Certainly. I'll tell you, Gregor, I am an old man, and I am weary of this incessant standoff between the East and West. It will eventually come to a head, but *when*? Why must it drag on so long? It's just a matter of chemistry. There needs to be a catalyst to start this reaction."

Cringing, Yohzin resisted the urge to gasp. Struggling to maintain his composure, he became physically ill.

"I'll tell you, brother, if ever the opportunity presents itself, even ever so briefly, we should strike while the iron is hot," declared Abdirov with fervor. "I am absolutely convinced that we can defeat the

Americans if we attack first. Once the Egg is dropped, those damned fat politicians in the *Politboro* will have no option but to immediately unleash the rest of our arsenal!"

"Then you are willing to deploy the warhead without American provocation?"

"Of course!" howled Abdirov.

As his heart raced and veins pounded in his temples, Yohzin heard a faint tapping at the door again. "It's my driver," he explained. "I think he's impatient to get back to his comrades in the barracks, and he can't join in the revelry until I release him from his duties."

"Then be on your way, Gregor. I'll be fine here."

"I'll have Luba fix you a plate and some goodies, and I'll send my driver back with them."

"*Spasiba*," replied Abdirov. "You are very kind."

With that, the two men embraced and Yohzin went on his way. Tugging his overcoat around him to ward off the cold, he climbed into the back seat of his *Moskvitch* sedan. Patting Magnus on the head, he considered the chilling exchange with Abdirov. Ironically, he once believed that the American nuclear arsenal was the greatest menace in the world, but now he knew better, since the Americans were at least practical enough to implement sufficient checks and balances to ensure that their weapons were not arbitrarily or accidently fired. He was convinced that Abdirov had completely lost his equilibrium, and that it was now his task to save the world—his sons' world—from a good man gone mad.

He knew that he must act, and act swiftly; there was no time to spare. Although he had promised himself that he would never betray Abdirov and reveal the *Krepost* to the West, this development indelibly altered the relationship between him and his old friend. It would be a tremendously risky undertaking, but he vowed to sneak his Minox camera into the *Burya* facility to capture all the intelligence that he could. He would not just direct Smith and his Americans to the keyhole, he would shove them through the damned door if that's what it took to postpone Armageddon.

25

GIFT HORSE

Aerospace Support Project
9:45 a.m., Monday, March 6, 1972

As Ourecky waited for an intelligence briefing to start, he reflected on the events of the past year. Last year had passed without incident, at least as far as Blue Gemini was concerned. The hardware had functioned flawlessly, and he and Carson had flown four perfect missions in 1971. While they had yet to see an OBS—and now questioned whether the nuclear nemesis had ever actually existed—they had successfully knocked down two Soviet recon platforms and two military communications relays in the past fourteen months, for a grand total of seven Soviet space vehicles destroyed since their first flight. Intelligence reports indicated that the Soviets were scrambling to determine why their most critical satellites seemed to function normally—at least for a while—only to suddenly fail or fall inexplicably out of the sky.

At this juncture, the two men had completed seven missions in orbit, more than any other American. Although they held the record for number of flights—unpublished, of course—they couldn't lay claim to the mark for overall endurance; that auspicious record belonged to NASA spacefarer James Lovell, who had logged over 715 hours during his four flights into space.

Blue Gemini's leadership dynamics had changed immensely over the course of the past year. Virgil Wolcott had gradually drifted towards greater authority as General Tew's health continued to decline. Tew still called the shots—at least notionally—but most often he just rubber-stamped whatever Wolcott offered.

After they flew their last mission in January, Blue Gemini had lapsed into something of a hiatus. Although there were no missions currently scheduled, Ourecky was painfully aware that two more complete stacks—each consisting of a Titan II and Gemini-I combination—were held in readiness at the San Diego HAF. The two last arrows in the quiver were obviously earmarked for special targets, which Wolcott had yet to reveal.

Although Wolcott wasn't forthcoming about their potential mission or the prospects for the future after Mission Twelve, it wasn't any secret that he was aggressively lobbying to have Blue Gemini extended for at least another six missions. *Six more missions?* Ourecky was close to the breaking point and wasn't sure that he could make it through six more flights. Wolcott repeatedly promised that more pilots would join Blue Gemini when and *if* they transitioned to the next phase, and he insisted that Carson and Ourecky would migrate to a training role. As it stood, though, no one seemed inclined to interfere with a machine that consistently worked, so he and Carson were obviously destined to fly the next two missions, regardless of the targets.

Besides the pair, only Wolcott and Colonel Seibert were present in the office; General Tew was supposed to be here but apparently had checked into the base hospital for an extended series of tests. Seibert wore a blue knit shirt over khaki trousers; he looked as if he was destined for the golf links as soon as the briefing was over. As a counterpoint, Wolcott looked as if he was on his way to compete in a rodeo;

then again, Wolcott *always* looked as if he was on his way to compete in a rodeo.

"Gents, Colonel Seibert has something momentous to share with us," announced Wolcott gleefully. "Mark Tew and I have already seen this, and all I can tell you is to prepare to be amazed."

"I'm sure that you're aware that the Soviets launched a small space station last year," stated Seibert, wielding a slightly warped yardstick to gesture at a diagram on a briefing chart. "They call it *Salyut*. It's very similar to the Skylab station that NASA plans to launch next year."

Seibert sneezed, wiped his nose with a handkerchief, and then continued. "The overall *Salyut* mission profile is almost identical to Skylab. The station is launched unmanned on a UR-500 booster rocket, which the Soviets now also refer to as the Proton launcher, and then a crew is launched later to occupy it. *Salyut-1* was launched last year in April, and the first crew was sent up in a *Soyuz* three days later."

"They docked, but weren't able to enter the station," stated Seibert. "Another crew went up in June. They docked successfully and occupied the station for twenty-three days. They lost cabin pressure during reentry and all three cosmonauts died. According to NORAD, the *Salyut* station succumbed to orbital decay last October.

"Here's the next item that the Soviets have up their sleeve," said Seibert, flipping over the briefing chart to reveal another image. "What you're looking at here is a military equivalent of *Salyut*. It's called *Almaz*, which means 'diamond' in Russian." He tapped his finger on the drawing and glanced towards Wolcott. "We've been aware of this jewel for over a year now. It shares some major components with the *Salyut*, but the two stations are produced by competing design bureaus."

Noticing that Carson was having difficulties reading the chart because of a glare coming through the window, Seibert shifted the chart stand slightly and continued. "*Almaz* is intended to remain in orbit for a year or longer. The Soviets plan to swap out crews roughly every ninety days and restock consumables with unmanned supply ferries. It was developed by the Chelomei design bureau, primarily for military reconnaissance and surveillance missions.

"Here's the kicker," he noted, pointing at a stick-like object in the drawing. "The *Almaz* is armed with a 23-millimeter automatic cannon."

"*Armed?*" exclaimed Carson, pushing himself up out of his chair and stepping forward to scrutinize the diagram. "It has a gun? *Ouch!* That's an ominous development. Do we have any reason to believe that they might be on to us? Why else would they mount an automatic cannon on a manned space station?"

Wolcott cleared his throat and interjected, "That was exactly my concern also, pard. The intel folks have checked into it. According to what they're tellin' us, there's no reason to suspect that the Russkies are aware of Blue Gemini. It's more likely in response to the work done on unmanned satellite interceptors like the SAINT."

Seibert put up another chart. Ourecky noticed that while the previous charts contained mostly vague drawings, the new chart contained an intricately detailed diagram of a space station very similar to the *Almaz* they had just seen.

"And this, gentlemen, is what we have been searching for," noted Seibert, popping the chart with his pointer. "This is *it*, at long last, the Holy Grail of sorts. It is the Soviet platform for their nuclear orbital bombardment system."

"So it's real?" asked Ourecky.

"Very much so. It's codenamed *Krepost*. Just so you know, *krepost* is a Russian term for a fortified outpost, like a citadel or fortress. The *Krepost* is essentially a conglomeration of components from several different Soviet spacecraft and competing design bureaus. The core block, particularly the crew living space and control module, is fabricated mostly from components of the *Almaz* military manned space station that we discussed previously. The *Krepost* will rely on modified *Soyuz* spacecraft to execute the crew transfer and cargo functions."

Seibert clicked on an overhead projector, and used his rubber-tipped pointer to gesture at a diagram on the screen. "I'll describe the station's configuration from bottom to top. As this drawing shows, the main hull is cylindrically shaped, approximately sixty feet long and thirteen feet wide, and weighs approximately twenty tons."

Pointing at the bottom of the cylinder, Seibert continued. "The nuclear warhead is located in the aft end of the *Krepost*. Assuming that our intelligence is correct, the warhead is a much larger version of the Soviet's RDS-37 two-stage design. It's encased in its own self-contained reentry vehicle. The reentry vehicle—warhead, retro rockets, and associated equipment— comprises over a third of the *Krepost* station's overall mass."

"Just forward of this is a service module, which contains electronics, the maneuvering system, and fuel tanks. Forward of the service module is a control area, which houses the instrumentation for the platform and the warhead. Moving forward from the control area is the crew's living space, which contains a galley and individual sleeping compartments."

"Your diagram shows three individual bays," stated Ourecky, gesturing with a Skilcraft ballpoint pen. "Is that accurate? Are we to believe that it's manned by a three-man crew?"

"Looks like a mighty tight fit for three folks," noted Wolcott. "You sure they can bunk three hands in there?"

"No, sir," replied Seibert, looking towards Wolcott. "It appears that the Krepost was originally designed for a three-man crew, but our current intelligence indicates that they will staff it with only two, quite possibly to reduce the requirements for food, water, and other consumables. Additionally, we're aware that they have been recently forced to make some significant safety modifications to their *Soyuz*, in the aftermath of the fatal accident last year, and we speculate that they can only fly two personnel on a *Soyuz* now, rather than three."

Wolcott nodded.

"Just past the living space is their docking hub, which was apparently purpose-built for the this station. It has three docking ports, set at one-hundred-twenty-degree angles, perpendicular to the long axis of the station, arranged like the spokes on a wheel. Each port can accommodate a *Soyuz* vehicle, in either the crew configuration or cargo configuration. There's also an inflatable airlock, here," stated Seibert, tapping the pointer on the screen. "Between two of the docking ports. We suspect that the airlock is only for contingency purposes, since there doesn't appear to be any logical requirements for extravehicular activities."

"The docking hub is equipped with the Soviet's *Igla* automatic docking system. The docking hub incorporates one additional docking port, located at the stern end. It does not have any hatch or other provision for transferring personnel or cargo. It's there strictly to accommodate yet another variant of the *Soyuz*, which would be an unmanned propulsion module. The propulsion module would be used to provide additional horsepower, in the form of thrust, to make adjustments to the *Krepost's* orbit. This *Soyuz* propulsion module variant is still on the drawing boards; we anticipate that the Soviets plan to produce and launch it in sufficient time to prevent the *Krepost* from falling out of orbit."

Seibert gestured a pod that protruded from the side of *Krepost* station. "Unfortunately, gentlemen, the *Krepost* is also armed with the same automatic cannon that we saw previously on the *Almaz*."

Switching off the projector, Seibert concluded by saying, "This is going to be a mighty tough nut to crack, even if we do get a chance. Although we're still confident that the Soviets aren't aware of Blue Gemini, they seem to be obsessed with the notion that we have an operational unmanned satellite interceptor, hence the automatic cannon. Of course, the gun is a moot issue if their station doesn't have a crew aboard. And that leads to our second problem.

"This *Krepost* is an ingenious design," observed Seibert. "Very diabolical. The three docking ports enable them to efficiently rotate crew ships and cargo ferries, so it can likely remain operational indefinitely, with continuous manning. If that's the case, it's very likely that we may *never* have a shot at it while it's unmanned."

"Gents, if you ain't figured it out yet, you're gazin' at your next target," revealed Wolcott. "This will culminate Phase One of the Blue Gemini and simultaneously validate some of the things we'll do during Phase Two, including EVA operations, if Phase Two is approved."

Wolcott paused to spit his chewing tobacco into a trashcan and replace it with another lump. "Our objective is to launch as soon as the *Krepost* has been inserted into a stable orbit," he stated, tucking his Red

Man pouch into his pocket. "We aim to sneak you up there before they launch a crew to occupy the station."

Wolcott continued. "Once you're upstairs, we have three critical tasks for you. First, just like your previous missions, we want you to execute a detailed fly-around inspection. Second, we want you to disable their docking ports to prevent the *Krepost* station from being occupied. Third, we want you to bring back physical proof that this monster is armed. That would be in violation of the Outer Space Treaty that they signed, as we did also, five years ago."

"Virgil, I hate to be obstinate," said Carson, returning to his seat. "I sure don't like the thought of the Soviets arming their spacecraft, and I'm aware that it's a violation of the treaty, just as you said, but isn't this a bit like the pot calling the kettle black? After all, we've been knocking their satellites out of the sky for about three years now. Begging your pardon, but wouldn't you think that planting explosive charges on their satellites could also be construed as a violation of the same treaty?"

"There's a *tremendous* difference, pard," declared Wolcott adamantly. Turning slightly redder than normal, he loosened his bolo tie, undid his top collar button, and fanned himself with his white Stetson. "True, you gents *did* emplace explosive charges on satellites, but there was only one single instance where a charge was actually detonated, and in that single instance a Navy *admiral* personally pushed the button. The rest of those satellites just eventually became unstable, perhaps of their own accord and perhaps not. But even more importantly, unless someone plans to talk out of school, no one can prove that we *ever* intentionally destroyed any satellites that didn't belong to us. Now, Major Carson, you ain't inclined to blab to anyone about what happened upstairs, are you?"

Carson shook his head and sipped from his coffee.

"All of this seems very contingent on timing," said Ourecky. "I don't quite understand how we can expect to prepare and then launch in sufficient time in order to beat their crews to orbit."

"We have a very reliable intelligence source over there," explained Seibert. "I can't delve into particulars, but this Russian is well placed within their Strategic Rocket Forces, and he's been funneling us detailed information about other programs for over a year. He's given us substantial information on their mission plans for the *Krepost*, including specific details on their intended orbital inclination, apogee, perigee and the like. Right now, according to information that he has provided, we have at least a month's lead time to start preparing."

Seibert continued. "Although we have at least a month's leeway, another source assures us that there will be another UR-500 launch before November. This source works at their Tyuratam launch site, so although he's sure a UR-500 Proton is being prepared, he can't absolutely state what the next payload will be, but he says that it's roughly a sixty percent certainty that it's going to be a *Krepost*. As for the time-sensitive issues, he claims he can provide us with at least twenty-four hours advance warning before a launch, and possibly as much as a month's notice."

"That should alleviate your concerns about precise timin', pard," said Wolcott, grinning.

"Sir, it does, but I'm still slightly confused on one issue," said Ourecky. "What happens if we climb upstairs and find a civilian *Salyut* instead of an *Krepost*?"

"Well, pardner, I s'pose we'll just cross that bridge when we come to it."

"And what if we don't beat the Russians to orbit?" asked Carson. "What if it's occupied when we close on it?"

"Yet *another* bridge to cross, pard," replied Wolcott. "Anyway, you've heard all the *good* news, so now we have to bestow the part that probably ain't goin' to set too well with you. We've studied this from every conceivable angle, and the only practical way that we can pull this off, given the criticality of the timing, is to forward-base until the launch."

"Forward-base, sir?" asked Ourecky. "What does that mean exactly?"

"Hoss, the time-sensitive circumstances compel us to lean well forward in the saddle. We intend to stage the next stack at the PDF at Johnston Island indefinitely, at least until our window shuts in November."

"That's great, Virgil," stated Carson. "But even if the hardware was permanently on pad alert out there, I don't understand how you could deliver us and the launch crew to the PDF in time to launch. With forty-eight hours, maybe, but with *twenty-four* hours? It wouldn't work."

Wolcott nodded. "Correct. And that's why we're stagin' everything— lock, stock, kit and caboodle—at the PDF, once we receive word that the Soviets are making preparations to launch a UR-500. So the whole package—the stack, the launch crew, and you two stalwarts—will be forward-based."

"Until November?" asked Ourecky, chagrined.

"Not necessarily," answered Wolcott. "You'll be there until you *launch*. That could be in November, but it could be even a month or even a week after you set up camp at the PDF."

Closing his eyes, Ourecky tried not to groan. He pondered about how he could possibly explain this new wrinkle to Bea.

"You don't look overly thrilled, hoss," observed Wolcott. "Let me set this in context. When the time comes, we're sending you to a tropical island in the Pacific where all you have to do is bide your time and twiddle your thumbs until you receive the word to strap on a rocket and go to work. On the other hand, at this very moment, hundreds of your contemporaries are receivin' orders to go to much more desolate locales, missile silos, SAC bases, Vietnam, and the like. You savvy, Ourecky? You need to look upon this more as a paid vacation."

10:50 a.m.

After Carson and Ourecky had departed, Seibert packed away his charts and collected the rest of his classified material. He started to leave, thought better of it, and turned back. "A word, Virgil?" he asked.

"Sure, hoss," answered Wolcott. "Pull up a chair and take a load off. What's on your mind?"

Seibert took a seat at the table. "I've got access to all the raw material on this *Krepost,* although all the source information has been redacted," he said. "So, while I don't know who he is or where he is, he had to be fairly high up in the Soviet food chain."

"And?"

"This whole business bothers me," explained Seibert. "After years of whispers and rumors about a military space program, it took us years to acquire anything substantial about their *Almaz* platform. Now, this damned *Krepost* is just falling into our lap. It just doesn't make any sense. I just don't like this. It's almost *too* perfect, like a pretty little package all tied up in a bow."

"Ted, ain't you ever heard the expression that you never look a gift horse in the mouth?"

"I have, but in the intelligence realm you learn that when something *looks* suspect, it's because it *is* suspect."

"So, pard, you think the Russians are tryin' to lure us into a trap?" asked Wolcott.

"Perhaps," answered Seibert, unwrapping a cough drop. "But to be honest, Virgil, I have to entertain other possibilities as well."

"Other possibilities? Such as?"

"Virg, you know damned well how reluctant Mark Tew is about moving on to Phase Two. For years, he's been absolutely fixated on this damned orbital bombardment system. Finally, everyone is coming to the realization that it was all a big pipe dream, and probably never existed in the first place. So just when Mark is ready to concede the fight and go home to retire, all of this *Krepost* stuff appears out of thin air."

"Just what exactly are you implyin', Colonel?" demanded Wolcott.

"I am implying that the timing is *extremely* suspect," answered Seibert. "Previously, we haven't gotten so much as a trickle about this thing, and suddenly we're deluged with this torrent? A complete package with intricate diagrams? Like I said, Virg, it's *suspect*."

"Then you're implyin' that it might not be the Soviets at all?" asked Wolcott.

"I don't know, Virgil," answered Seibert. "I really just don't know."

Department of Archives and Operational Research
GRU Headquarters, Khodinka Airfield, Moscow, USSR
3:45 p.m., Friday, April 14, 1972

Over the course of the past year, Morozov had cultivated his Analysis Section into a highly disciplined team. They had garnered a brilliant reputation for their efficiency and generally were issued only the toughest questions to answer. Their diligence and hard work yielded dividends. Once they completed their tasks for the day, at least until other assignments were forthcoming, they were allowed to relax.

Of course, if they weren't actively pursuing an answer to an inquiry, they couldn't leave the Encyclopedia, just in case another pressing question came up. And even though they could have a break, they were still effectively confined to their assigned table. Actually, in recent weeks the rules had been eased slightly; if more than one Analysis Section was inactive at any given moment, it was acceptable for the idled workers to visit each other's tables and quietly socialize, but it was still forbidden for anyone to arbitrarily meander around in the stacks.

So this afternoon, while the other Analysis Sections still toiled, Morozov and his expeditious workers basked in their leisure. As two retrievers knitted woolen shawls, Morozov played chess with the third retriever; a young man newly arrived from his home in the Ukraine.

With his elbows on the table and his chin cupped in his hand, Morozov skimmed a summary of international headlines as he awaited his opponent's next move. The Americans were preparing to launch yet another mission—Apollo 16—to the moon on Sunday. IRA terrorists had detonated thirty bombs in Northern Ireland in the past two days. The fighting in Vietnam had reached a new level of intensity as the North Vietnamese had stepped up their fighting throughout South Vietnam.

Sighing as he read the last item, conscious that he would never set foot in Vietnam, he set the paper aside and reflected on how quiet it was. Although he was still a year away from being promoted to Second Class, he had at least earned his section an opportunity to move to a more tranquil table.

He still overheard enough to remember what happened upstairs, but it was not nearly as terrifying as the sounds of a year ago. Now, instead of muffled screams, he just heard soft snoring; he looked to his right to see his filer—a portly woman in her mid-thirties—slumped over with her head down on the table, fast asleep.

He smiled. His section had become like a close-knit little family, content in themselves and comfortable in his leadership and protection. If this were America, this charming scene could be a subject for a *Saturday Evening Post* cover, whimsically painted by Norman Rockwell.

The cozy image vanished from his mind as he heard the sharp squeaks of boot soles on the hardwood floors. He looked up to witness heavily armed *spetsnaz* soldiers pouring out of the elevator and the stairwell. Clad in loose-fitting camouflage coveralls, the blank-faced soldiers brandished Kalashnikovs and were festooned with bandoliers of ammunition and grenades.

As the soldiers scattered throughout the stacks and amongst the tables, a *spetsnaz* officer tried to speak through a handheld loudspeaker. Morozov covered his ears and winced as the loudspeaker emitted squealing feedback. The officer cursed, made some adjustments, and announced, "*Everyone!* Remain seated with your hands in view. Do not touch any of the reference materials on your tables. Wait at your tables until soldiers come to escort you."

Expressing no outward sign of emotion, even though his heart pounded furiously in his chest, Morozov watched as the soldiers went from table to table. The Encyclopedia workers were ordered to collect their personal belongings, and then they were abruptly ushered—as integral sections—to the stairwell. Some of the women wept openly as they were led away.

Morozov could not imagine what had precipitated this wholesale action; it reminded him of the Stalinist purges when entire villages were rumored to have been marched away into bleak oblivion. While he could not fathom how so many could be simultaneously guilty of an infraction, he was gravely concerned for them. But the pragmatic man that he was, he could not picture how the cells upstairs could possibly accommodate all of them. Soon, there were only ten tables still occupied, but the soldiers still continued with their chores.

And then suddenly it dawned on him, a painful revelation that rattled through his mind like the aftermath of a thunderclap. It was something so *blatantly* obvious that he could not believe that he had not previously grasped its significance.

Shuddering with dread, he realized that since all of the dismissed workers were allowed to collect their coats and belongings as they left, they were most likely headed for home, *not* the grisly cells upstairs. Whatever this witch hunt entailed, the other Analysis Sections—now being hastily herded towards the stairs—were *not* the targets.

"I have chocolate," whispered one of his retrievers excitedly, furtively showing Morozov a miniature Hershey's Kiss candy concealed in her palm. "It's an American candy you gave me several weeks ago. I was saving it." The chocolate treat was hardly recognizable; she had hoarded it so long that it was a misshapen lump wrapped in dull silver foil.

"*Eat* it," replied Morozov quietly.

"But they might see me unwrap it," she replied. Despondent, she was on the verge of tears.

"Eat it *with* the wrapper," he hissed. "*Now.* Stuff it in your face and gobble it down before you buy Makarov bullets for *all* of us."

Fifteen minutes after the *spetsnaz* soldiers had appeared, only five Analysis Sections—twenty-five workers in all—remained at their stations. Standing next to Morozov's table, a First Class Analyst addressed them. "Leave your work and gather over here." As the other Analysis Sections made their way to the table, the First Class leaned towards Morozov and said, "Sorry to interrupt your chess match. Looks like you were making very good headway."

Politely thanking him, Morozov smiled nervously.

The First Class spoke to the assembled Encyclopedia workers. "I apologize for disrupting your afternoon, but we have an extremely awkward situation brewing in America. A well-placed source close to a high-ranking American general has been compromised. She is presently in the custody of the Office of Special Investigations of the US Air Force, so we must assume that she is undergoing extensive interrogation."

Morozov silently breathed a sigh of relief and then mouthed the word "Sorry" to the whimpering retriever struggling to swallow the foil-wrapped candy.

The First Class continued. "It is a foregone conclusion that the general will be forced into retirement if he is fortunate enough not to land in prison. This source—his personal secretary—has been a veritable fount of valuable intelligence." Grinning broadly, he added, "Of course, in her defense, she believed that she was helping the Israelis. Such is the power of suggestion . . ."

"With all that said, I'm sure that you are speculating why you are here and why your comrades have been directed to leave the premises," said the First Class. "We have been directed to undertake an immediate high priority effort to mitigate any potential damages. Your sections have proven yourselves to be trustworthy and efficient, so you will participate in this effort. We will not leave here until it is concluded. Because we must operate under an abbreviated timeframe, we will suspend many of our normal operating and security practices for the duration of this project."

11:25 p.m.

Bleary-eyed and groggy, Morozov strained to remain conscious as he slowly winnowed through the burgeoning pile of index cards before him. At the opposite end of the table, his retrievers dozed as he jotted notes regarding the next group of cards to be collected. After he issued them their new guidance, he resolved himself to put his own head down for a quick catnap.

He marveled at the sheer volume of information that the American woman had provided to the GRU. Clearly, her boss was a man of profound importance, since his fingers were poked in so many different pies. And the extensive breadth of information made Morozov's task that much more difficult.

His job—as well as that of the other four sections that remained at work in the Encyclopedia—was containment. While he pitied the wayward American secretary, her fate was effectively sealed, and there was little that could be done to lessen her misery, even if the GRU felt compelled to do so. She was probably still convinced that she had collected information on behalf of the Jews, and it might be weeks or months before she could be persuaded otherwise.

Since the secretary had been written off, the overall objective of the containment process was to safeguard other sources and agents. It was extremely rare for the GRU to lend absolute trust in intelligence data from any given source; consequently, another source or agent was typically tasked to gain access to the information from another perspective.

Thus, for every piece of information provided by the secretary, Morozov and his counterparts had to methodically ferret through the stacks for similar information that might have been provided by another source. It didn't matter whether the source had been tasked to furnish the information or whether they just happened upon it, because it was a certainty that as the American counterintelligence agents worked through their own protocols of damage control, once they identified that a piece of information had been compromised, they would eventually determine who else might have had access to the same information. Right now they just had the secretary within their grasp, but literally scores of other sources and agents could eventually be compromised as a result.

Consequently, the containment process was not unlike the watertight compartments on a ship. If a breach was discovered in one compartment, its watertight doors were sealed, ideally after the people in the compartment were evacuated or otherwise protected. Of course, as Morozov well knew, plenty of the same watertight doors would be hastily slammed in the faces of sources that they could not evacuate or otherwise protect, but that was just the brutal nature of the business.

Arriving at the last card in the present stack, he nudged the closest retriever and directed her to wake the others so they could be ready to go to the stacks again. He sipped from a glass of water as he read the remaining card. His heart literally skipped a beat when he recognized a name of a test pilot associated with the Aerospace Support Project at Wright-Patterson. Staring at the card, he dipped his fingers in the glass, splashed water in his face, and then rubbed his tired eyes.

He was not mistaken. The name—*Drew Carson*—was there, in neatly typewritten letters, on the index card. The cross-reference card indicated that the secretary—*Phyllis*—had submitted two photographs, on the same day, that were either of Carson or in some way related to him. Morozov's temples throbbed as his pulse quickened.

He placed the card face down and then issued his new instructions to the three retrievers. As they parceled the tasks amongst themselves, he flipped over the card with Carson's name.

Feigning disinterest, he said, "Oh. Dig through the stacks to see if there's a dossier on this fellow, Drew Carson. And ring the Number Two Archives to request these two photographs. Probably nothing we should squander much time on, but we do need to be thorough."

3:24 a.m., Saturday, April 15, 1972

Morozov was cautious not to look at the two photographs until he had dispatched his retrievers on yet another foray of gathering cards and documents. As they prowled in the stacks, he studied the images.

The first—apparently taken by Phyllis herself—showed two US Air Force officers—both majors—standing beside the general—Astor—who had been the secretary's boss. The two men looked quite ordinary. Although their faces were plastered with seemingly forced smiles, they appeared to be extremely uncomfortable in the general's presence. Unlike the general, neither man wore any of the distinctive ribbons that denoted heroism or wartime service. Morozov could clearly read the nameplates on their right breast pockets; the one on the left—Carson—wore pilot's wings while the other—Ourecky—did not. The image looked like a quickly posed snapshot, apparently taken in the general's office.

The second image was far more perplexing. It depicted a metal data plate with Cyrillic characters that described a space vehicle's physical attributes and manufacturing information. Even more puzzling was that the image had apparently been autographed by the two men— Carson and Ourecky—seen with the general in the first photograph. Morozov immediately surmised that the two majors were somehow responsible for the second photograph, although that made little sense to him.

Carson's dossier revealed very little as well. Ironically, the information had been gleaned from personnel records at Morozov's request months ago, but it was the first time he had seen most of it. According to his file, Carson was an exceptional pilot who had graduated with top honors from the various flying courses that he had attended, including the prestigious test pilot school at Edwards Air Force Base in California.

Carson had been assigned to the Aerospace Support Project for nearly four years. Morozov could not envision why a pilot of Carson's caliber had not been granted an opportunity to prove his mettle in the ongoing war in Southeast Asia. He was obviously involved in something quite secret, but what could it be?

Morozov looked across the room at an obese, slovenly man sitting alone at a nearby table. He was Dmitry Anatolyevich Popov, recently assigned to the Encyclopedia after working three years in the GRU's Directorate of Cosmic Intelligence.

"Cosmic Intelligence" sounded rather ethereal, like a group of analysts who conducted séances, fussed over astronomical charts, or perhaps investigated UFO sightings. In actuality, the Directorate of Cosmic Intelligence concerned itself with tracking the thousands of man-made objects in the heavens, as well as collecting intelligence on the space-related activities like launching facilities and related technology. Their function was not unlike that of the American's satellite tracking facility operated by NORAD—North American Aerospace Defense Command—at Cheyenne Mountain in Colorado.

Popov was brilliant but eccentric, with extensive academic training in astrophysics and mathematics. Like so many others, he had landed in the Encyclopedia after apparently committing some grievous error. Morozov had heard a rumor that Popov had miscalculated the orbital path of an American spy satellite, which led to the compromise of several sensitive facilities before his blunder had been detected. But if anyone had an answer about the mysterious photograph of the data plate, then surely Popov would know.

Pausing to fill a glass at a silver samovar, Morozov strolled towards Popov's table. He placed a folder before him and held out the steaming glass. "Dmitry Anatolyevich," he asked, "could you help me with this puzzle? With your background, surely you can help me unravel this. This is a photograph provided by the American secretary. I'm certainly not an authority, but to the best of my analysis, this appears to be an artifact from one of *our* space vehicles."

Popov opened the folder and peered at the photograph through thick spectacles. "Oh, *sure*," he said, accepting the tea offered by Morozov. "I recognize this. It's from a Type Four reconnaissance satellite. This is the data plate, with all of the Type Four's specifications."

"So it *is* one of ours?"

"*Da.* But where the hell did this come from?" asked Popov nervously. He removed his smudged spectacles, wiped them on his shirt, replaced them, and stared at the photograph. "How could it possibly be related to this American woman?"

"Honestly, I have no idea," replied Morozov, shaking his head solemnly "But since she gave it to us, it obviously means something."

"If you insist, Anatoly Nikolayevich, but I cannot comprehend how she could have gotten her hands on this. This particular satellite was designed and built under the highest secrecy. The security on that plant was airtight. It would have been impossible for an assembly worker to take the picture and spirit it out, and I cannot conceive of any way in which the Americans could have infiltrated the plant."

"So this picture must have been taken *inside* the assembly plant?" asked Morozov.

"Of course!" averred Popov. "There could have been *no* other way."

"But could it not have been taken at the launch site?"

"*Nyet!*" sniffed Popov. "Immediately after this satellite was assembled, an aerodynamic shroud was installed even before it left the plant. It was mated to its booster after it arrived at the launching site. The *only* possible place this picture could have been taken was at the plant."

"So what happened to this Type Four satellite?" asked Morozov. "Was it ever launched?"

"*Da.* It was fired into orbit, in April of 1969. I remember it well. It functioned perfectly for about six weeks, and then it started behaving erratically. The controllers claimed that the design was fundamentally defective, and the designers argued that the controllers were sending the satellite erroneous instructions."

Morozov nodded.

"You know, we experienced similar failures with some other satellites," mused Popov, staring at the ceiling as he scratched his ear. "*Six* others, to be exact, at least while I was working at Cosmic Intelligence. It caused quite some consternation with the design bureaus. It was almost as if . . ." Suddenly, for whatever reason, his broad face turned pale, as if he recalled something particularly horrific.

"Dmitry, what is it?" implored Morozov. "You look like you've seen a ghost."

"I did. *Scores* of them," said Popov quietly. "You should sneak this photograph back into the stacks and forget that you ever laid eyes on it. I know that they told us that the rules are temporarily suspended, but I warn you—*nothing* good can come of this. This is outside our realm, and the less we know, the better off we'll be."

"But this photograph was in the archives," said Morozov quietly. "Perhaps no one has realized the significance of it or how it must have been captured. This could be quite a coup. I would gladly share the credit with you."

"*Credit?* This is the *GRU,*" hissed Popov. "Surely, you are not so *ignorant* to believe that no one has yet grasped this photograph's significance.

Don't delude yourself. You can rest assured that they know of this image and how it was taken."

"But. . . ."

Popov leaned across the table. Turning the photograph facedown, he said quietly, "Let me enlighten you, Anatoly Nikolayevich, concerning how much the GRU knows. The facility that built this satellite no longer *exists*. More importantly, the *people* who worked at that facility no longer exist. Do you understand? Do you?"

Morozov swallowed and nodded.

"Unless you're in a rush to trot after them, you should drop this matter, particularly before you drag me along with you," warned Popov. He closed the folder and thrust it towards Morozov. "Walk *away* from my table, Anatoly Nikolayevich, *now*, and forget that we ever spoke."

Saying no more, Morozov slinked back to his section. Popov's dire admonitions did little to dispel his curiosity. Placing the intriguing images side by side, he stared at them, as if he might be able to meld them in his mind to draw out the missing clue. It was simple to infer that these men had somehow taken the photograph of the data plate, but how was that possible? Were they espionage agents? Did they fly spy planes? The Americans possessed spy aircraft that flew incredibly high and fast; was it possible that they had somehow developed a system that could collect images through walls or ceilings?

Scanning between the photos, he studied the names—Carson and Ourecky—yet again, and burned them into his memory. Someday, no matter how long it took, even if he had to wait patiently until he became a Second Class Analyst with almost unfettered access to the stacks, he would unravel the mystery of these two men and how they managed to capture this image.

26

SOJOURN

Petro-Dive Training Institute, New Orleans, Louisiana
2:15 p.m., Monday, April 24, 1972

At this moment, Ourecky desired little more than a tall glass of ice water and an opportunity to scratch his incessantly itching nose, but he would have to suffer through at least another miserable hour before he could indulge in either. Encased in a space suit, he was submerged in an enormous water tank to simulate working in the weightless environment of space.

With three extra layers of protective fabric, his new suit—designed specifically for Extravehicular Activity, more commonly known as EVA or spacewalking—was even more restrictive and rigid than the lighter model he wore during his first two flights into orbit. To realistically practice spacewalking, his suit was fully pressurized, just as it would be in a vacuum, so it was even more stiff and awkward. To compensate for its buoyancy, it was ballasted with small pouches of lead pellets, carefully distributed in strategically placed pockets.

Wishing that he could somehow massage his aching shoulders, he paused to reflect on the recent events. Since the *Krepost* briefing last month, their schedule had evolved into a relentless regimen that alternated between two arduous cycles: "Box Weeks" and "Tank Weeks." Box Weeks had them in the Simulator Facility at Wright-Patt, slogging in the Box—the procedures simulator—or the paraglider trainer. At this point, they were so proficient that the Box was more of a monotonous nuisance than a challenge. It was exceptionally rare when the simulation staff could contrive a glitch or combination of glitches that they couldn't resolve in short order.

In their past simulations and actual missions—in even the worst cases—they generally had at least a week to study and absorb the sequence of planned maneuvers necessary to close with a target satellite. With the new scenarios, they were often locked into the Box with a day's preparation or less, in some cases executing an intercept from a virtual cold start. The rapid-paced simulations were a prelude to a sprint to orbit to beat the Soviet *Krepost* crews.

But as much as Ourecky had come to despise the Box, he dreaded Tank Weeks even more. Tank Weeks were spent here, at the Petro-Dive Training Institute, a defunct commercial diving school in New Orleans. Petro-Dive, which had suffered bankruptcy last year, was home to a massive indoor pool built to train hardhat divers to service offshore oil platforms. The deep tank was sufficiently large to immerse a complete Gemini spacecraft mock-up, as well as other mock-ups used as practice targets. So with very few exceptions, if they weren't in the Box, they were underwater in the Tank.

Every day in the Tank was a painful ordeal. He was thirsty and had an excruciating headache as a result of his dehydration. Even though it was made exactly to his specifications, the suit constantly chafed at his hips, shoulders, and other locations. Because pure oxygen—with no moisture added—was piped down through his umbilical, the mucous membranes in his mouth and nose were habitually dry and his lips were constantly chapped. The suit's coolant loops often failed to keep pace with his exertions, so he had to carefully pace himself to prevent his visor from fogging up.

He paused for a moment to mentally work through his next assigned task, a seemingly simple drill of deploying an experimental work platform for later chores. The sturdy platform was mounted at the end of the boom previously used to extend the Disruptor hoop. It was intended to alleviate many of the problems experienced during NASA's EVA missions. Once the boom was extended, Ourecky would make his way to a pedestal where he would lock his boots into secure footholds. A chest-high titanium T-bar extended up from the pedestal's base; by shifting his feet to lean into or against the T-bar, he could use it as a fulcrum to apply a significant amount of leverage when using various tools.

In a sense, the platform was not unlike a homeowner's stepladder. The apparatus was an expedient solution, hastily devised from scrounged components. The secure footing and solid base certainly offered him greater mechanical advantage than floating free, but there was also a considerable amount of risk involved. He and Carson had to work in close concert, orchestrating every act, to ensure that Ourecky wasn't accidently smashed against their target.

After opening the hatch and standing up, a tedious process that required roughly thirty minutes to accomplish, he had to slowly pivot around to face the adapter end of the spacecraft to deploy the work platform. So now, although he theoretically stood upright in his seat, the spacecraft mock-up was oriented so that he was effectively lying on his right side. That probably would not be too bad, except he had been in this *same* position for the past hour. Although the exercise planners could pretend otherwise, the laws of gravity were still very much in effect, and they were wreaking havoc with his muscles and circulation. His right shoulder ached as if it had been pounded with a sledgehammer, and his right arm was partially numb.

His cumbersome gloves significantly hampered his dexterity; he wiggled his knotted fingers in an attempt to improve blood flow. Every action required a concerted effort, even simple acts like grasping tools, so he was mindful of the need to think ahead to maximize efficiency and conserve energy.

For every action, Newton's third law of motion stipulated, there was an opposite and equal reaction, except that the reactions seemed greatly exaggerated in a weightless environment. Just the simple act of turning a wrench could send him in an uncontrollable spin. To compensate, he paced himself, moving cautiously and deliberately, in virtual slow motion.

Ignoring his various aches and pains, he mustered his strength to begin the process of deploying the boom-mounted platform. Once it was locked in place, he would clip onto a safety cable and work his way along a handrail to shimmy to the top of the platform. Once there, his first job was to wield a "snake stick" to cut an inch-thick metal rod that simulated the barrel of the *Krepost* station's automatic cannon.

The snake stick was a ten-foot pole with a two-foot crossbar at the end. A noose protruded from either end of the crossbar, much like the pole-mounted nooses that snake handlers used to snare rattlers. Once he positioned the nooses on the target, he switched on an electric motor to cinch them on the rod. The first noose was a flexible braided metal tube containing an incendiary material called thermite; once ignited, the burning thermite sliced through the rod. The second noose was simply an anchor cable that allowed him to retrieve the rod once it was sheared.

When and if they intercepted the *Krepost*, retrieving the gun barrel was a secondary task. Tew and Wolcott wanted it essentially as a souvenir, for sentimental reasons, but the more pragmatic reason for grabbing it was to prove to the world—if need be—that the Soviets were putting weapons in orbit, in defiance of the 1967 Outer Space Treaty. It was an ace in the hole that they wanted in the unlikely event that the Blue Gemini project was ever revealed, so as to prove why such an aggressive action was justified.

Ourecky's primary EVA task was to use the snake stick and other tools to disable the docking mechanisms and related *Igla* radar equipment, in order to prevent the *Krepost* from being occupied. The physical layout of the *Krepost*, particularly its diameter and the positioning of several antennas, precluded employment of the Disruptor, but other means were being studied to destroy the station.

As Ourecky practiced his spacewalking chores, Carson rehearsed various emergency procedures—disconnecting the umbilical and closing the right-side hatch—that would become necessary if Ourecky was disabled and unable to return to the snug sanctuary of the cockpit. He also maintained a sharp lookout for his right-seater, acting as an extra set of eyes to spot potential hazards and to ensure that the two vehicles remained close but not too close.

As miserable as Ourecky felt, he knew that the Tank was also taking its toll on Carson. Working in the suits sapped their strength. At the end of the day, they were physically spent. The routine had also put a significant damper on Carson's social life. After a day spent underwater, they had barely enough energy to take a shower or eat dinner. As the days dragged into the week, they had to exert substantial willpower just to climb out of bed and go to work.

As Ourecky watched the boom extend, a SCUBA-equipped diver drifted nearby, casually snapping pictures with a Nikonos underwater camera. Ourecky heard the exercise controller's voice through his earphones: "Scott, this is Topside. How's it going down there?"

"Just fine," he replied. "Be aware my right shoulder is cramping."

"Okay," replied Topside. "Look, cease what you're doing. I'm sending down the safety divers to bring you up."

"No thanks," replied Ourecky. "This is the last evolution for the day. We'll have this finished in an hour. We're in the homestretch, so I want to finish it before I come out of here."

"I admire your tenacity, Scott, but you're done in the Tank. Virgil just called. You and Carson are flying out to the PDF tomorrow morning. It's on."

Dayton, Ohio
6:48 p.m., Monday, April 24, 1972

Ourecky was packing in the bedroom when Bea got home. Squealing with glee, Andy scampered to him and hugged him around the waist. It still surprised him that Andy was no longer crawling. It was as if he had just blinked and somehow missed several crucial stages of his son's development. In a sense, he had.

"Going somewhere?" she asked. She was still in her blue Delta stewardess uniform; occasionally, she changed at Jill's place when she picked up Andy, but must not have had time. Andy sat down on the floor to play with some wooden alphabet blocks.

"Uh, *yeah*," he replied. "And I have to leave tomorrow."

"I know *that* look," she said, unbuttoning her jacket. "Let's not beat around the bush, Scott. This is obviously not going to be just an overnighter. How long this time? A week? Two weeks? That's usually what one of these flight tests take."

"Bea, I may not be back for a few months," he replied. He regretted that he had not broached this issue before, to at least get her somewhat accustomed to the notion of his extended absence. "I could be gone until November."

Fuming, she plopped down on the bed and kicked off her shoes. "You're leaving *tomorrow* and you won't be back until *November*?" she asked angrily.

"Bea, you weren't listening. I *may* not be back until November. I'll probably be back much earlier than that."

"Okay," she answered. "It's not as if we can ever make any long-term plans, anyway. Where will you be?"

"Can't say."

"Can't say or don't know?"

"You *know* the answer to that," he replied, cramming a pair of sneakers in his bag. "Why do you even ask? You know *damned* well I can't tell you, Bea."

He thought she was going to pout the rest of the evening, but her face grew strangely calm. "When I go to the airport in the morning, I'm going to turn in my two-week notice," she stated matter-of-factly.

It took him aback, so much so that he couldn't speak for a moment. "Bea, that's great, but I'm not asking you to stop flying," he finally replied, sticking an extra tube of toothpaste in his shaving kit. "I'm thrilled that you'll be able to spend more time with Andy, but I was gone almost all of last year, and you were still able to fly."

"I'm not quitting for you, Scott," she replied, removing her jacket. "Jill is sick. The doctors don't know what's wrong with her, but it looks bad. She's still working at home, but she can barely do that now. She's helped me so much by taking care of Andy while I'm flying, so I figure that helping her with Rebecca is the least I can do. If you're going to be gone, I might as well just stay with her until she gets better."

Ourecky nodded. "She's lucky to have a friend like you. And I'm lucky to have you, too, Bea. I love you."

She stood up, undid her scarf, and hung it on a peg inside the closet. "Look, Scott," she said softly. "I love you, too. I don't understand whatever it is that you're mixed up in, but I'm not in the mood to argue. I just hope that someday you figure out what's important, and I hope that you figure it out before it's too late."

But I do know what's important, he thought. But how could he explain to her that he had inadvertently made a pact with the Devil and that his choices were no longer his own? Zipping closed his kit bag, he wondered if there was a clear path out of this predicament. But if nothing else, he had learned something that was probably unknown to any other human being. With every day that passed, it became increasingly clear that the Devil wore cowboy boots.

Pacific Departure Facility, Johnston Island
6:15 p.m., Thursday, July 27, 1972

Ourecky reclined in a chaise lounge in the shade of their "veranda," a section of tent canvas that they had rigged to the front of the suit-up trailer. The rickety chair's pale green webbing was sun-dried and frayed; every other strap was missing or on the verge of breaking. The faded canvas flapped and rustled in the evening breeze.

Scrutinizing a topographic map of a Contingency Recovery Zone in New Guinea, he struggled to remain awake in the sultry heat of late afternoon. He had reached the point at which further study was futile;

his brain was practically saturated with technical data, mission details and the like.

Virgil Wolcott had definitely hit the nail square on the head; in the three months they had been holed up on the Island, they had an overwhelming abundance of leisure time. They had bided that time until calendars and clocks were meaningless, and they had twiddled their thumbs until the digits were cramped and bleeding.

There was literally little to do but wait. While the pace was considerably less hectic than the past year's, the monotony was all but excruciating. It was incredibly hard to maintain any degree of fervor about anything. They were almost delirious with boredom.

He contemplated calling home, but quickly abandoned the thought. The operations shed had a shortwave radio that could reach a MARS—Military Affiliate Radio Service—network station back in Ohio, where a participating HAM radio operator would patch the radio call into the local phone network.

As much as he missed Bea and their son Andy, Ourecky hated making the calls on MARS. Amongst other things, the MARS protocols required that they say "over" after every sentence, so a call seemed less like an intimate conversation and more like an official transaction. And while he enjoyed hearing Bea's voice, there was just very little that they could talk about.

He could say nothing of what he was doing or where he was, and since she spent almost every waking moment with her friend Jill, there was little she could say except to report—as always—that Jill was growing progressively sicker. Worse, every time he called Bea, he was painfully aware that it might be the last time they spoke, since this mission might be the time when the law of averages finally caught up to him.

Since it had never been intended to be fully occupied for extended periods, the PDF had evolved extensively to support the new mission. Before this latest sojourn, Ourecky had never lingered here for more than four days prior to a launch; during the rapid spate of launches last year, the launch crew had refined the procedures to the extent where he and Carson rarely even stayed overnight before launching.

Usually, both LSTs remained on site until just prior to the launch, but now only the LST that ferried the stack was still here. To accommodate the launch site workers who were normally quartered on the ship, a small tent city had sprung up. Gasoline-fired barrels burned human waste, and generators whined night and day.

After the first month, the launch support crew had been reinforced by twenty-five percent, which allowed for a quarter of the personnel to regularly rotate to Hawaii for R & R. But since there was no one to augment them, Carson and Ourecky remained on the Island for the duration.

The spacecraft mock-ups from the Tank had been flown in by C-141 from New Orleans, and a makeshift weightless training facility had been improvised in the dredged deepwater area beside the docks. Even though it allowed them to continue practicing EVA skills, they could use the facility no more than twice a week. The corrosive salt-water was incredibly harsh on the suits' aluminum fittings and other hardware. For every day they trained underwater, the suit technicians required three days to rinse the suits in fresh water, meticulously clean the hardware, and then carefully air-dry the garments. But as much as Ourecky hated climbing into the suit, the EVA training was at least a break from the perpetual boredom.

Even as the Tank training suits required almost constant mainte-nance by the fastidious technicians, the two men donned their mission suits on a daily basis. Because of the oppressive heat and their lack of appetite, they constantly struggled to keep on enough weight to ensure that the suits fit correctly. They jogged daily and lifted weights regularly in the workout tent.

They had to be careful even with their exercise regimen, remaining vigilant not to bulk up with excess muscle that would alter the fit of their suits.

After watching the sun fall below the horizon, Ourecky went into the trailer to relax and read. He sat down on his cot and opened a dog-eared paperback. Realizing that he had read the book three times already, he set it on the wooden box that served as his nightstand.

The PDF director, Lieutenant Colonel Ted Cook, entered the trailer and sat down. Like most of the men on the launch support crew, he wore only khaki Navy dive shorts and flip-flops. Like Carson, Ourecky, and virtually every other man at the launch site, he was deeply tanned. He wiped sweat from his brow with a faded blue baseball cap and scratched his sharp-pointed nose. "Evening, gents," he said, closing the door behind him.

"Beer, Ted?" asked Carson. He reached into a small refrigerator, extracted two glistening silver cans, and held one out.

"No thanks," replied Cook. "And don't bother opening yours, Drew."

"What's up, boss?" asked Ourecky, kneading a tennis ball. He had squeezed it so much that he had worn off the fuzz from the ball.

"Guys," announced Cook, "we just got a hot cable from Wright-Patt. The Soviets are making final preparations to launch a UR-500. It's going up within the next twenty-four hours. As soon as I leave here, I'll give the order to break the stack out of encapsulation and erect it onto the pad. We're launching, gents. You'll be headed upstairs within the next forty-eight hours."

Pacific Departure Facility, Johnston Island
8:15 a.m., Friday, July 28, 1972

With their launch imminent, Carson and Ourecky lingered in the communications section's open air shed, anxiously awaiting more detailed instructions. Ourecky sipped coffee and worked a crossword puzzle as Carson read a week-old Honolulu newspaper. Trying to summon a three-letter word for "flightless bird," his thoughts were abruptly interrupted by the loud chattering of a Teletype machine.

"That's it," noted Carson, setting aside his newspaper. "Ready for this hop?"

"I'm ready to be *done* with it," replied Ourecky, thinking of Bea. "Should we call the blockhouse so they can start spinning up?"

"We'll give it a minute. Really, Scott, there's no sense in getting in a huge rush. It will still be at least a few hours before we can launch."

The Teletype rattled on for a minute, spewing out a continuous print-out, and then suddenly stopped. A communications technician tore off the paper, verified some authentication information in the message header, logged the transaction, and then handed it to Carson.

Grinning, Carson read the print-out; in seconds, his expression changed to a frown and he hammered the table with his fist. "We're scrubbed," he declared angrily. "Wolcott's ordered us to stand down. *Damn* it!"

"We're *scrubbed?*" asked Ourecky. "For today?"

"For a month, at least," replied Carson, handing the paper to Ourecky. "The Soviets launched a UR-500 about three hours ago, but its second stage malfunctioned. Our target went to the bottom of the Pacific Ocean. So much for Virgil's big shindig."

"So what happens now?"

"Our stack will be shipped back to the HAF for its ninety-day work-up, so Ted and the PDF crew are going to catch an extended breather," said Carson. "As for you and me, we've been summoned back to Wright-Patt. After that, it's anybody's guess."

Ourecky smiled. Although he wanted to fly just to put this mission behind him, the timing was fortuitous. According to Blue Gemini's safety mandates, an assembled stack had to undergo a detailed maintenance overhaul every ninety days. The overhaul procedure compelled that the Titan II be returned to San Diego, where it would be partially disassembled to facilitate intensive inspections of critical components. So, although their PDF ordeal was not yet entirely over, the overhaul granted them at least a thirty-day respite. Hopefully, it would be an opportunity to catch up with Bea and Andy, and return to normalcy, if only temporarily.

Aerospace Support Project
9:25 a.m., Monday, July 31, 1972

Ourecky tapped lightly on the door before entering the office. Only Tew was present, seated at his desk and immersed in paperwork. He had not seen Tew since departing for Johnston Island three months ago;

appalled, he now barely recognized the ailing general. Tew was drawn and gray, like a shadow on a late afternoon in winter. Tew had once confided to him that the immense stress of each mission aged him five years, and now Ourecky definitely believed him.

Announcing his presence, he softly cleared his throat. "You wanted to see me, sir?" he asked politely.

Tew signed a few more papers, then looked up and smiled. "Yes, I did. Come on in, Ourecky. This will only take a minute. Have a seat over there, son."

Taking a seat at the conference table, Ourecky looked towards Wolcott's empty desk and asked, "Is Virgil not here today, sir?"

"He's up flying. He took one of the T-38s on a cross-country hop," answered Tew, joining him at the table. "But what we have to speak of doesn't concern him. I suppose that you're aware we're seeking an extension for Blue Gemini?"

Ourecky cringed at the thought of more flights. He swallowed deeply and apprehensively replied, "I am, sir. Six more missions?"

Tew gravely nodded his head. "Ourecky, I ask that you hold this in strictest confidence, but I want you to know that when and if the Project is extended, I don't intend to continue on with it. I think that I've contributed all that I can contribute, so I'm planning to file my retirement paperwork immediately after we finish our last mission in the first phase."

Ourecky was speechless for a moment, since this was the first indication he'd seen that Tew intended to retire. Finally, he exclaimed, "That's excellent, sir. Congratulations. We'll sure miss you here, but I'm happy for you."

Tew smiled feebly. "Thanks. I'm more than ready to retire, but I'm not leaving until I've accomplished all that I've set out to do, and there's still some business yet unfinished. Moreover, it's a loose end that concerns you."

"How so, sir? Have I done something wrong? Does Virgil still want me to earn my wings?"

"No. None of those things. Look, you've done your part, Ourecky, and we are incredibly indebted to you. I'm sure you know that we're not receiving the new computer before Mission Twelve, so we definitely need you to fly the next two missions. I'm sorry for that, but there's no other way, since we've grown so reliant on you and Carson. But after those two missions, you're *done*. I *promise* you that I will not submit my retirement paperwork until your transfer is complete and you've departed from here, preferably to earn your doctorate at MIT, assuming that's what you still desire."

"It is, sir. Very much so, but you shouldn't be so concerned with me."

Shaking his head, Tew said emphatically, "Ourecky, on my honor, I am obligated to make sure that you get completely clear of this place and that you *never* return." He looked askance towards the blocked white Stetson cowboy hat on Wolcott's desk, grimaced, and added, "Unfortunately, since I cannot trust *anyone* else to act so altruistically on your behalf, if it takes me a few extra weeks to close that transaction and run you out of here, then so be it. The golf courses will just have to wait."

"Thank you, sir. I greatly appreciate it."

"You're more than welcome." Changing the subject, Tew asked, "So, how is Bea? How is your son? Andy, isn't it? Isn't he about two years old by now?"

Forcing a smile, Ourecky replied, "Oh, they're doing great, sir. I just got back into town yesterday, but they're doing super. We have a lot of catching up to do, but . . ."

Tew leaned forward and looked directly in Ourecky's eyes. "Son, don't try to lead me down the primrose path. We're placing you under a huge amount of strain, and I've seen plenty of very strong marriages disintegrate under far less trying circumstances. So, tell me: How are things *really* going between you two?"

Hanging his head as he stared at the tabletop, Ourecky truthfully confided, "Not great, sir. Not even good. I thought that Bea would be thrilled that I was home again, but I think that she just expects the phone to ring and for me to leave again. I'm not sure how much more of this we can take before things just spin out of control. I'm just not sure that I can keep making things work."

Tew put his hand on Ourecky's shoulder and said, "I know it's just a distant glow right now, but we're starting to see the light at the end of tunnel. Please try to hang on, son. Don't lose what you have."

Tears welled in Ourecky's eyes, and he fought hard to contain them.

"Son, how much leave time do you have saved up?" asked Tew, placing his hands flat on tabletop. His desktop intercom buzzed, but he ignored it.

"Uh, right now I've accrued about three weeks' worth," replied Ourecky. "I burned a lot of it last Christmas, when Bea and I went out to Nebraska to see my parents. We try to do that every year. Pardon my curiosity, sir, but why do you ask?"

"Because I want you to go home and spend some time with your wife and child," declared Tew. "Do you understand me, Major Ourecky? Take Bea and go *far* away from here. Take a trip to California or Hawaii or Florida or *somewhere*, but take advantage of this time. And I don't want you hanging around with Carson. I know you two are friends, but I'm confident that Carson can fend for himself for a few weeks."

"A few *weeks*, sir?"

Tew nodded. "Your stack is en route to San Diego for maintenance overhaul. Consequently, since we're at a lull, I'm ordering an operational stand-down for all non-essential personnel. We're going to keep some of the key players here, but otherwise we want everyone to take advantage of this break. I know that you have been under considerably more stress than anyone else in this Project, so I wanted to tell you personally."

"That's great, sir, I'm sure that everyone . . ."

Holding up his hand, Tew interrupted. "We're going to be on hiatus for at least a month, so I don't want to see you for another thirty days. Don't worry about your leave time. I'll handle it with the personnel people so your furlough is not charged to your account. Just *go*, Ourecky. Do we understand each other? Am I making myself sufficiently clear?"

"Yes, sir. Thank you, sir."

"Ourecky, we may still have two more missions to fly, but I want you to put all of that out of your mind until you return. Focus on your family. That's what's really important. Don't make the mistake of losing them.

27

STORM

On Orbit
7:25 a.m., Friday, August 4, 1972 (GET: 16:04:23:00)

As the result of a rather strange chain of coincidences, Lieutenant Colonel Ed Russo was one of two human beings orbiting the Earth on the morning of August 4, 1972. Ironically, considering how long he had aspired to fly in space, his extraterrestrial experience hadn't been particularly pleasant; more accurately, he *hated* it and was desperately anxious to return to the surly bonds of Earth. And although he anticipated that this would be yet another painfully long and agonizing day, he had no way of knowing that it would also be one of the longest days in his life.

Pondering his circumstances, he glanced at his mission clock. He sighed; he had been in orbit for sixteen days, four hours, twenty-three minutes, and six seconds. If the mission proceeded as planned—which was tremendously doubtful at this juncture—he would remain up here for *twelve* more days. He really wasn't sure that he could bear twelve more days aloft.

The novelty of weightlessness had evaporated early on his first day. Seemingly always in motion, his stomach did more somersaults than a troupe of circus tumblers. He had no appetite and was constantly nauseous. The disorienting sensation of floating really unnerved him, so he strived to keep a handhold on the nearest solid object as he compelled his mind to establish a firm "up" and "down" even when there were none. Despite his efforts to adapt, his confused vestibular senses could never reconcile themselves with the paradoxical reality.

Although he had no means to weigh himself, he was woefully aware that he was losing considerable muscle mass and growing progressively weaker by the day. He rarely got any decent rest; his sleep was often disturbed by sporadic flashes in his eyes, which were apparently caused by stray electrons jostled by cosmic rays. To make matters even worse, the environmental heat exchanger wasn't working to capacity, so the cabin's sweltering heat kept him perpetually drenched in sweat and did little to help his sleep situation or lack of appetite.

Life in space was far less glamorous than he had bargained for. He longed for gravity's familiar pull. He wanted things to fall when he dropped them. He wanted objects to stay put where he laid them. Despite an abundance of Velcro to anchor loose items in the cabin, things constantly drifted away and became lost. He lost his sunglasses the first day and hadn't seen his toothbrush in over a week. Strangely enough, more so than anything else, he had an intense desire to pour water in a glass and watch it remain there.

Russo inhabited an enormous metal cylinder—the Manned Orbiting Laboratory—that whirled around the world every ninety minutes in a polar orbit. As he reminisced about how he came to dwell in these miserable circumstances, he was painfully conscious that the MOL's sides were emblazoned with bold white letters that spelled out "US NAVY" above the familiar stars and stripes of the American flag.

Years prior, five interlocking variables contributed to the MOL's radical transformation from a marginally public Air Force program to a highly secretive Navy endeavor. First, the Space Task Group had successfully argued that the MOL was little more than a duplication of NASA's manned space efforts. Moreover, they contended that most unique Air Force requirements could be undertaken by unmanned satellites currently in development. These arguments, among others, led to the public cancellation of the Air Force's MOL program on June 10, 1969.

The second variable involved the difficulties of performing surveillance on ocean-going targets versus land-based targets. A Soviet armored division couldn't rush out of its barracks to arbitrarily arrive at the Fulda Gap the next day. Its deployment was limited by terrestrial transportation networks—highways and railroads—which greatly simplified the task of tracking its progress from garrison to the battlefield. On the other hand, a Soviet fleet was not so constrained, since the seas literally provided a fluid medium on which to chart a course.

The third factor was that the United States was on the cusp of losing the controversial war in Vietnam. More accurately, senior leaders had concluded that investing more American lives and treasure was a losing proposition, so prosecution of the war was gradually being handed over to the South Vietnamese, making it *their* war to lose. Equally inevitable was the notion that they *would* lose the war, probably sooner than later. And when this defeat occurred, the Soviets—the patient benefactors of the North Vietnamese—would finally gain access to an immensely valuable strategic resource: an all-weather, deep draft port—Cam Ranh Bay—from which they could readily project power into the Pacific.

The potential consequences of a Soviet warm water port on the Pacific were staggering, easily rivaling the massive Japanese naval buildup in the years prior to Pearl Harbor. A space-based ocean surveillance sentinel was essential in expectation of the time when—not if—the Soviet Navy could slip quietly into the trackless waters of the Pacific to threaten the Philippines and other previously invulnerable strategic locations.

Fourth, the Air Force envisioned an MOL launched atop a Titan IIIM booster, manned by two military astronauts who would return to Earth in a Gemini-B reentry vehicle. The missions were expected to be no more than thirty days in duration, after which the MOL would be discarded. The Navy's ocean surveillance MOL would remain in orbit indefinitely, which would entail swapping out crews, as well as replenishment of oxygen, water, food, and other staples.

A power-hungry synthetic aperture radar—SAR—was the keystone of the ocean surveillance systems. A nuclear reactor was the only viable option for providing reliable power to the SAR. While NASA used small reactors to power experiment packages flown on Apollo lunar missions, their plutonium-fueled generators weren't of sufficient magnitude to sustain the SAR and other functions of the MOL.

Without question, the Navy had more experience handling nuclear reactors than any other entity in the free world. Moreover, they just happened to possess a mission-validated device—the diminutive reactor carried about the NR-1 research submersible—that could be readily adapted to the ocean surveillance MOL.

Finally, the fifth factor that led to the MOL's evolution was that the White House was occupied by a president who had served as a naval officer during World War II. While he had not risen to great rank or prominence in the Navy, and had not participated in any heroic actions, he did comprehend the strategic necessity to maintain dominance over the seas. So it didn't take much for Admiral Tarbox and his coterie of determined confederates to convince him of the pressing need to keep an unblinking eye on the world's oceans. Consequently, control of the MOL program was quietly transitioned from the Air Force to the Navy, and with its well-publicized cancellation, it disappeared from the public consciousness altogether.

And so it was that the MOL came to be in orbit on that August day, but Russo was aboard more or less by accident. The MOL's first flight should have been a purely Navy effort. Two Navy officers—Commanders Chris Cowin and Jeff McKnight—had been slated to fly it, but in the week prior to the flight, McKnight had broken an ankle during a launch pad evacuation exercise.

Russo was the sole Air Force officer still assigned as an MOL astronaut; the remaining five were Naval aviators handpicked by Admiral Tarbox. Russo was designated as the right-seater on the mission's back-up crew. At the time, he and McKnight were the only men on the roster who had successfully completed the Navy's Nuclear Propulsion School; as such, with McKnight grounded, Russo was the only MOL astronaut fully qualified to manage the station's nuclear reactor.

Although the Navy wasn't thrilled with the prospect of an Air Force officer rising into space on their hallowed chariot, they could either send him up, delay the mission for several months, or fly one of their own guys with minimal reactor training.

Despite their reluctance, it was entirely logical to send Russo; he was intimately familiar with the reactor's workings, and no one could argue his competence with the ocean surveillance systems. Besides, he was arguably the most proficient Gemini pilot in the entire MOL program. Like the Air Force's Gemini-I, the Navy's Gemini-B reentry vehicle was equipped with a paraglider rather than a parachute. As a result of his Blue Gemini liaison stint at Wright-Patterson, Russo had more time in the paraglider simulator and more hours flying the actual paraglider than all of the Navy astronauts combined.

Cowin petitioned adamantly to avoid flying with Russo. Despite his prowess with the reactor and other systems, Russo wasn't well received in the Navy clique; the other men resented him as an opportunistic inter-loper who exploited his relationship with Tarbox. Over Cowin's strin-gent objections, the Ancient Mariner decreed that Russo would occupy the right seat to orbit.

But after a perfect night launch from Vandenberg Air Force Base, the MOL's maiden voyage had been far from uneventful. In addition to the malfunctioning heat exchanger, the synthetic aperture radar proved to be chronically temperamental. Their radios had been failing inter-mittently; as of this morning, it had been over four days since they had communicated with their mission controllers. While Cowin had consist-ently tried to cast Russo as the scapegoat for all their technical woes, the nuclear reactor—Russo's exclusive domain—was the only system aboard that had consistently operated without any difficulties.

To say that the two men were incompatible would be a tremendous understatement. Russo despised Cowin, and the feeling was certainly mutual. From the very outset of the mission, they quarreled constantly, over even the most seemingly insignificant details. Besides a cramped spacecraft and helium-oxygen atmosphere that grew increasingly more foul with every passing day, the only thing they shared was a festering hatred of one another.

The communications outage drew them ever nearer to the breaking point. Russo argued that they should cut their losses and abandon the mission. Cowin countered that while the failure was a distraction, it wasn't a mission-stopper, and that they should dutifully remain on station—collecting essential data—for the planned duration of the flight. As the mission commander, Cowin had the final say in the debate. Besides, while Russo desired little but to pack up and head for home, Cowin thrived up here and would remain indefinitely if granted an opportunity.

Three days ago, their animosity and lingering distrust came to a violent head. Cowin became enraged with Russo's vocal insistence on cutting the mission short. The Navy astronaut took a swing at the Air Force pilot, landing the only solid blow in the scuffle, and then the confrontation quickly deteriorated into an awkward wrestling match in zero gravity. Shortly after the fray, nursing painfully tender ribs and a badly swollen black eye, Russo gathered his personal belongings and withdrew to the far aft of the outpost, where he hung his cocoon-like sleeping restraint next to the reactor control console.

Cowin kept his quarters in the wardroom at the stern end, just aft of the Gemini-B reentry vehicle, where their sleeping compartments were located. When the circumstances dictated, they grudgingly met in the operations module just forward of the reactor control area. The operations module was home to the monitoring station for the ocean surveillance system, as well as the overall controls for the MOL.

Yesterday, even the stoic Cowin had to admit that the comms outage was more than a passing nuisance. But instead of curtailing the mission, he decided to resolve the situation. He spent the entire day traversing the station from one end to the other, meticulously tracing cables and painstakingly examining every communications component that was accessible from the MOL's interior. When he failed to locate anything out of proper kilter, he impulsively announced that he planned to don an EVA suit to venture outside today to physically check the antennas and antenna connections.

Waiting for Cowin to appear, hoping that he had ditched his EVA scheme since yesterday, Russo loitered in the galley. The galley space was just aft of the wardroom; the two spaces, approximately a third of the MOL's pressurized compartment, comprised the living quarters for the astronauts. To remain stationary and keep his bearings relatively fixed, he wrapped his legs tightly around the metal post of a table-like fixture bolted to the "floor" of the galley and then wedged himself against the adjacent bulkhead. The MOL's designers had apparently gone to great lengths in a futile effort to render the station as homey and as familiar as possible, but the "furniture" and similar appointments became little more than annoying obstacles once gravity fell away.

He watched the sharp-pointed northern coast of Madagascar drift by in a viewport. Although he wasn't hungry, he sampled a food bar made of dehydrated apricots; although it sounded appetizing, the wafer had the consistency and taste of compressed sawdust. He forced himself to nibble half of it and then stuffed the remainder in a waste bag. He had no sooner swallowed the last bit before he felt an unsettling wave of spasms in his gut. He snatched a waste bag and vomited into it. *So much for a quick bite,* he thought, stashing the bag in a trash receptacle.

He heard a noise from the stern end and glanced that way to see Cowin casually soaring from the wardroom into the galley. Cowin had already donned the white suit liner that was worn underneath the bulky EVA suit. He nimbly floated into the galley and foraged through a food locker until he located one of his favorite entrees: roasted chicken and enriched rice.

Using a pair of blunt-tipped scissors, which he had thoughtfully tethered with a strand of nylon cord, he snipped off a corner of the meal's plastic envelope. Holding the transparent envelope to his mouth, he carefully kneaded it to squeeze out chunks of rice and chicken. The

Navy MOL rations were experimental ready-to-eat "wet-packs" which didn't require reconstitution or special preparation. Ironically, the two men had initially squabbled over the choicest morsels, at least until Russo's appetite had completely abandoned him.

"Yum, that's *mighty* tasty," declared Cowin, smacking his lips with gusto. "I don't know if I'm going to be content with my wife's cooking after munching all this good chow. Too damned bad you can't enjoy this." He intentionally squirted some globs of rice into the air; jeering at Russo's obvious discomfort, he snorted and laughed as he caught the white clumps between his teeth and quickly gulped them down.

"Chris, are you *absolutely* sure that you want to go outside?" asked Russo, desperately hoping to reason with his argumentative cohort. "You don't think it's too dangerous to go EVA if we can't make contact with the ground?"

If the truth be known, Russo was far less concerned about Cowin's safety than he was about his own personal and professional survival. Although it was highly unlikely that Cowin might become injured or otherwise incapable of reentering the station, it was still certainly a possibility, and Russo wasn't absolutely sure that he could successfully return to Earth by himself. Besides that, even if he did somehow make it home, he knew that he would face months and possibly years of classified inquiries, rumors and innuendo. It would be bad enough if Cowin survived, but if he was inexplicably lost overboard, Russo would be compelled to account for the MOL mission's failure. His military career would be *over*, since the Navy would be swift to latch onto him as a scapegoat, and he had already torched most of the bridges that might otherwise return him safely to his Air Force roots.

"I think we've discussed this topic ad infinitum, Ed," stated Cowin, scowling as he dug a large chunk of chicken from the plastic packet. He jammed the scrap in his mouth, chewed noisily, swallowed, and licked his fingers. "And there is *nothing* left to discuss. Besides, I'm the mission commander, and since there's no one on Earth or in space with the authority to override me, I call the ball. I'm going outside. Period. You stand watch on the airlock, just like we rehearsed. Everything had better go smoothly on the lock-out and lock-in. If it doesn't, you'll be due to draw another pummeling. Got it?"

"Aye aye," grumbled Russo, furtively rubbing his puffy eye. "Got it." Watching Cowin eat, Russo's stomach seemed even more queasy than normal.

"Good. I'm glad that we could reach an understanding." Cowin finished the ration and discarded the bag into a waste receptacle. "Hey, what's that damned sour smell?" For dramatic effect, he poked a finger down his throat and feigned retching. "Did you puke *again*, flyboy?"

Gagging as acrid bile surged up in his throat, Russo groped in his thigh pocket for another airsickness bag, but was too late. It was yet another mess to clean up.

3:05 p.m.

After sponging splotches of vomit from the galley bulkheads with a damp cloth, Russo assisted Cowin with the time-consuming process of suiting up. Cowin then entered the airlock for pre-breathing. The MOL's atmosphere was a fire retardant mixture of helium and oxygen, but the EVA suit provided oxygen only, which necessitated the lengthy ordeal of purging residual helium from the body.

Right now, Cowin was outside. In accordance with their EVA procedures, Russo remained at the stern end of the spacecraft, functioning somewhat like a topside tender for a hardhat diver. He vigilantly monitored Cowin's vital statistics on the life support panel next to the airlock, and kept in fairly constant communications. Cowin had been outside for over two hours, and had successfully removed a series of protective panels that allowed access to the suspect antenna connections.

The EVA session was proceeding so smoothly that Russo was on the verge of falling asleep from boredom. Almost hypnotized by the pulsing light that represented Cowin's respiratory rate, he half-wanted the foray to succeed just so they could converse with someone on Earth. On the other, hand, he wouldn't mind if the abusive hothead failed to fix the radios, since even Cowin had conceded the need to scuttle the mission if they couldn't make communications.

Russo briefly went to the operations module to verify—once again—that the synthetic aperture radar was powered down; located on the "bottom" of the MOL, the SAR emitted so much energy that it could be extremely hazardous to Cowin.

Returning to the airlock, Russo was taken aback by a glimmering dazzle in his eyes. It was almost exactly like the transient sparkles that frequently disturbed his sleep, except considerably more severe. The bright flashes lasted only a few seconds, but he was suddenly over-whelmed with an inexplicable feeling of dread.

Momentarily paralyzed with fear, he heard a warbling sound and knew immediately that it was the distinctive alarm associated with the MOL's radiation detectors. Spurred to action, he instinctively keyed his headset and spoke to Cowin. "Chris, I've got radiation alarms."

"No kidding," replied Cowin through his earpiece. "They're blaring in my left ear out here. Can you not squelch that damned alarm tone?"

Russo turned around and threw a switch that disabled the shrill alarm that fed into the intercom and communications circuit. "Better?" he asked.

"Yeah," answered Cowin. "So what the hell happened? Did you acci-dently trip something?" Even through the intercom loop, there was a skeptical hint to Cowin's voice, as if he insinuated Russo was intention-ally sabotaging the mission.

The cabin alarm still sounded; the warbling was growing progres-sively louder. "I didn't trip anything," declared Russo defiantly. "I think the reactor may be going south."

"*Damn* it!" snarled Cowin. "Well, we have to assume the worst case. Hustle aft and check it. If it's crapping out, go ahead and scram it. I'm not done out here, but I have to replace these covers. The sun will fry these components if I leave them exposed to the elements."

"Got it. I'm headed aft." Russo's heart pounded as he unplugged the jack for his headset. Normally, he would work his way to the reactor control station at a virtual snail's pace, maintaining a constant handhold. Instead, he planted his feet on the forward bulkhead next to the life support panel, gripped a nearby rail, and pulled his body into a deep squat.

As he had watched Cowin do at least a thousand times, he tilted his head up and picked an unobstructed trajectory that would take him all the way through to the aft compartment. Unimpeded by gravity, he kicked hard and launched through the air like a comic book hero.

Almost as soon as he had sprung off, he realized his error; just a gentle push would have been plenty sufficient, but now he was moving much too quickly. Instinctively preparing for impact, he unconsciously extended his right hand. As he smashed into the aft bulkhead, he heard an ominous snap and felt immediate pain in his wrist. Crying out in pain, he seized his hand as he careened off the bulkhead and rebounded back into the operations module.

Clasping his injured hand close to his chest, he slowly worked his way back to the reactor console. He examined the radiation alarm display; the indicator needle registered a dose-rate of ten rems per hour, which was commensurate with a serious reactor incident.

The "rems per hour" was a shorthand notation on how much radiation they *might* absorb. The actual dosage depended on several variables, such as where they were physically located in the MOL's cabin. The greatest potential exposure risk was at the point closest to the reactor, where Russo was presently located. Even then, over half of the MOL—a service module that contained water, fuel, radar equipment and other items—shielded him from the reactor.

Since roughly five hundred rems was considered a lethal dose, he would have to physically remain at this spot for fifty hours to soak up that much radiation, if the reactor continued to emit at a constant rate. If he moved forward, especially if he climbed into the Gemini-B and closed the circular hatch cut through its heat shield, his exposure would be significantly less.

If the detectors indicated ten rems per hour *here*, it was a sure bet that the situation was vastly different on the far side of the service module. An indicator placed at the reactor would likely peg out at a couple of hundred rems per hour. He didn't think that they were experiencing a catastrophic meltdown, but the situation could be swiftly deteriorating in that direction.

The corrective action for a serious incident was an emergency reactor shutdown, also known as a "scram." Locking his leg around his sleeping restraint, Russo flipped up a safety cover and poised his finger over the SCRAM switch. When he threw the switch, a series of electromagnets would be de-energized, freeing several spring-loaded control rods to slide into the reactor core to smother the fission process. Simultaneously, a dense liquid solution containing boron would also be squirted into the reactor to further impede the nuclear reaction. The process was not unlike closing the damper on a chimney.

Even after the faulty reactor was scrammed, it would still generate heat for several hours, if not days, until the fission process finally abated. The residual heat was plenty sufficient to continue driving the steam turbine that generated electricity, so they would still have plenty of juice for the MOL's life support and other essential functions long after the reactor had sputtered out.

In even the worst case scenario, where the reactor core might generate sufficient heat to be swiftly rendered into a glob of liquid metal and highly radioactive steam, he had the option—if he reacted swiftly enough—to push a red plunger labeled EMER JETT. When he mashed the big plunger, several explosive bolts would simultaneously fire, followed shortly by the ignition of a small solid rocket motor, which would physically jettison the reactor vessel from the MOL.

That contingency was available *if* he reacted quickly enough. Of course, if he didn't respond swiftly enough, their only option was to pile into the Gemini-B and hope that they were able to power up and flee before the MOL's cabin was breached by the disintegrating reactor.

Virtually every scenario—short of a catastrophic meltdown—called for a measured response. Depending on the magnitude of the emergency, once they scrammed the reactor, they still had roughly forty-eight hours to execute an orderly retreat. That entailed a very laborious process of physically transferring tape cartridges—which contained the raw data from the SAR and signal intercepts—from the operations module to either the Gemini-B or a data reentry pod. In any event, even though they would likely sop up a lifetime's dosage of radiation in the process, they would not wantonly abandon ship and leave the invaluable findings behind.

Transfixed on the gauges, Russo was about to throw the switches to scram the reactor, but hesitated. In larger reactors, like those on submarines and ships, the scramming process could be reversed. But because of its unique failsafe design, once the MOL reactor was scrammed, it could not be restarted. They would still have the residual power and batteries, but the mission was *over* as soon as he threw the switch.

Clutching his throbbing right hand to his chest, he forced himself to stop. Up to this point, he had been reacting out of panic, and one lesson the Navy's Nuclear Propulsion School had hammered into him was that while a malfunctioning nuclear reactor might eventually kill you, panic would do the job *much* faster. Panic and nuclear reactors just didn't mix. While it was important to act quickly, the key to resolving their dilemma was incisive and deliberate action.

He painstakingly examined the reactor controls, but found no discrepancies. The reactor's heat and pressure levels were well within normal limits. Every critical gauge was mirrored with a trend monitor that displayed the highest and lowest readings within the past twenty-four hour operating cycle. The trend monitors also showed that the reactor had been operating within normal limits; there had been no abnormal spikes in any of the critical indicators.

Russo knew that scramming the reactor was a swift ticket home. Because of the communications outage, there was no telemetry, so no one would ever know. Once the deed was done, he would be back on Earth in less than forty-eight hours, possibly much sooner.

Strangely, as he considered his circumstances, he remembered the praise heaped upon Ourecky for saving the paraglider trainer in Alaska. Ourecky had also been an outsider, but the close-knit Blue Gemini pilots readily accepted him after his decisive actions saved the ship. *Would the same fate be his if he remained calm and saved the Navy's vaunted MOL mission?*

He slowly blinked and then focused on the radiation indicators. They still registered a dose rate of almost ten rems per hour. *This did not make sense.* All of the other instruments showed that the reactor was perfectly healthy. He was elated that the reactor was not malfunctioning, but he was also perplexed. It was abundantly clear that they were being bombarded by radiation, but the barrage apparently was *not* emanating from the reactor.

He audited the reactor's vital signs yet again. Everything was well within acceptable operating boundaries. *The reactor is healthy*, he told himself. *Where is this radiation coming from?* Perhaps there is *no* radiation. *Could the radiation detectors be defective?*

Although the alarms were activated by the detectors next to the reactor console, two redundant detectors were located in the MOL cabin. He gingerly navigated his way forward to verify the mid-deck detector. *Ten rems per hour. Good, the aft detectors were working correctly.*

Next, he gradually drifted towards the forward end of the cabin to check that detector. *Ten rems per hour.* He breathed a sigh of relief; the aft detectors were definitely working correctly. But the readings still didn't make sense; while the needles fluctuated slightly, the samplings were uniformly consistent throughout the MOL cabin, even though they should be theoretically weaker at mid-deck and weaker still at the stern end. Then suddenly it dawned on him. The readings made *perfect* sense. Despite all evidence to the contrary, they *weren't* erroneous, and that indicated a problem much more serious than a crippled reactor.

Aghast at the potentially dire implications, fumbling with his one good hand, Russo scampered towards the airlock station. Shuddering with dread, he clumsily jammed his headset jack into the communications panel and blurted, "Chris! *Chris!* Jump back into the airlock *now!*"

"*Huh?*" grunted Cowin. He was breathing hard in his exertions. "Hey, don't get your panties in a wad, flyboy. I'm still buttoning up this cover. I should be done in a few minutes."

"*No*," ordered Russo. "Drop what you're doing and climb back in *now*. I'll explain after you lock in."

Aerospace Support Project
7:32 a.m., Sunday, August 6, 1972

Strumming his fingertips on the table, Virgil Wolcott gritted his teeth as he waited impatiently. He was absolutely confounded; even though they were scarcely a week into the month-long operational stand-down, a much-needed breather for all hands, Mark Tew had phoned him less than an hour ago to initiate an emergency recall of all key staff personnel.

Wolcott didn't know if this was some sort of quick-reaction exercise, perhaps hastily conceived by Kittredge's staff at the Pentagon, but he damned sure wasn't in the mood for games or any other sort of asinine tomfoolery on a Sunday morning. He had planned to spend the entire day fly-fishing, a recently acquired distraction he found almost as relaxing and therapeutic as horseback riding. In fact, he would be probably casting on the lake right now if Tew had called only a few minutes later.

Resolving himself to set a good example for the others, Wolcott kept his mouth clamped shut and his temper tightly in check. After enduring long months in a pressure cooker environment, several of the men stumbled in looking as if they had spent a long night drinking before dutifully struggling out of bed to answer Tew's impromptu reveille. No one looked too thrilled to be here.

Unshaven and obviously not yet entirely sober, Gunter Heydrich was the last straggler to appear. Severely disheveled, he looked to be suffering from the aftermath of serious bender. His greasy black hair was uncombed, his face was puffy and his eyes were swollen and bloodshot. His shoes were mismatched; one was black and the other brown. His haphazardly donned shirt was at least two buttons out of alignment. Reeking of schnapps and *Jägermeister*, he flopped into a chair, reached for a pitcher, and poured a tumbler. He dug two Alka-Seltzer tablets from his shirt pocket and plopped them into the water. The effervescent tablets fizzed and bubbled at the bottom of the glass. Rubbing his temples, he nodded towards Wolcott and grunted.

Wolcott returned Heydrich's greeting, swiveled to face Tew, and announced, "The troops are assembled, boss. So what's this all about?"

"I apologize for summoning you all on a Sunday morning, especially since we're on operational stand-down, but we've received a warning order," explained Tew. His faltering voice was weak, almost feeble, and his hands trembled slightly. His pale forehead glistened with a faint sheen of perspiration. "Gentlemen, we have been instructed to immediately develop a contingency plan to support a Navy mission."

"*Tarnations*," mumbled Wolcott, rolling his eyes. "Navy? Our friend Tarbox, *again*? I should have known. What celestial object does the Ancient Mariner want wrangled out of the sky *this* time? The moon? Pluto?"

Tew shook his head as he slipped a TELEX print-out towards Wolcott.

Wolcott's eyes opened progressively larger as he scanned the flimsy paper. Flipping it upside down, he swallowed deeply and said, "*Oh*. Well now, ain't this just a doggone corker? I wasn't even aware they were up."

"Do you see why I wanted *everybody* this morning, Virgil?" asked Tew.

"Yup, but is it safe to assume that everyone here has been cleared to clap eyes on this?"

"They have, Virgil."

Wolcott slid the flimsy paper to Heydrich, who read it and gasped aloud. "*Schiesse*," muttered the German engineer, passing the paper to the next man at the table. He gulped down his Alka-Seltzer cocktail, looked askance at Tew, and accusingly asked, "So, Mark, *this* is the secret you were keeping from me?"

Virtually all of the staff officers reacted in the same manner as Heydrich, outwardly expressing at least some degree of shock. And like Heydrich, some clearly resented the fact that they had been deliberately left out of the loop. But without exception, all of them snapped back to immediately focus on the problem as presented; without any prodding from Tew or Wolcott, they began assessing the situation from their unique perspectives. To a man, they made notes, sketched out simple diagrams, worked on preliminary calculations, and listed potential courses of action. Colonel Ted Seibert, Blue Gemini's intelligence officer, scurried out of the room, returning almost immediately with a ponderous sheath of papers.

After everyone had digested the initial information, Tew cleared his throat. Looking up, the staff officers set aside their pencils and notepads and the room fell deathly silent. He softly spoke. "We need to produce an initial feasibility assessment for General Kittredge within the hour. It doesn't have to be detailed, mainly just a broad brush yes or no on whether we can conceivably execute the mission, and a rough timeline to show the earliest possible date we can execute. By the end of the day, we are to submit a more comprehensive plan. Understood?"

As most of the men nodded in affirmation, Wolcott spoke. "We understand, boss. Now, are you sure you're up to this? You're lookin' a mite peaked."

"Virgil, you're right. I am feeling under the weather, so I would appreciate if you would lead the rest of this discussion."

"As you wish, Mark," replied Wolcott. "Okay, gents, I'm woefully sorry that we kept this little nugget from you, but we're all in cahoots now. Since there obviously ain't no time to dawdle, let's settle down to the brass tacks, startin' with the most fundamental stuff. How about flight hardware? Can we even support this contingency?" He turned towards Grady Rhodes, Blue Gemini's chief logistics planner.

"It would be a *big* stretch," replied Rhodes. Once grossly overweight, the colonel was now a gaunt shadow of his former self; constant stress, combined with a strict diet, had caused him to shed pounds like a sick bird molting feathers. Apparently, he had no time to shop for a new wardrobe to accommodate his altered physique; his oversized clothes hung loosely off his skinny frame, almost comedically so. "We have the flight hardware, but the problem is *where* it's at."

"How so, pard?" asked Wolcott.

Rhodes explained. "After the *Krepost* mission was scrubbed, the PDF crew encapsulated and loaded the Mission Eleven stack to be returned to the HAF. Right now, it's at sea, on the LST. It should be just a hair over 600 miles out of San Diego, as we speak. This is an older generation landing ship, so it can only make twenty knots headway at full steam. That puts them back in port in two days. It will take at least a day to safely unload the Mission Eleven stack, and then another day to load the Mission Twelve stack, then ten days at sea to transit back to the PDF, plus two days to offload, break encapsulation, refuel and conduct critical component testing. That's *sixteen* days minimum."

"And if we just turned the LST around right now?" asked Wolcott. "Instead of swappin' out stacks, could we not just send Eleven back to the PDF?"

Rhodes shook his head as he replied, "I hadn't even considered that as an option." He jotted a few notes, checked some figures, frowned, and noted, "Theoretically, Virgil, we could launch in *ten* days, but the Eleven stack was already scheduled for its mandatory ninety-day maintenance overhaul. It's still in date, but only barely. If we go in ten days, its maintenance paperwork will expire the day before launch."

Scratching his nose, Wolcott leaned to his right, glanced at the logistic officer's figures, and then pivoted towards Tew. "Your call, Mark, but I would turn the LST."

"Considering the circumstances, that's a risk we're going to have to accept," noted Tew. "Virgil, after we're done here, pass word to have that ship turned about."

Rhodes nodded. "Virgil, we also can't forget about the servicing LST. It's still loaded with fuel, and is presently anchored off Hawaii. The crew is on liberty call in Honolulu."

"Then tell them to round up their hands, saddle up and skedaddle towards the PDF as well. Is that copacetic with you, Mark?"

Lightly clutching his abdomen, Tew tacitly acknowledged by weakly nodding his head.

"Done," vowed Rhodes.

"There's another issue," stated Seibert, the dapper intelligence officer, quickly shuffling through a batch of current weather forecasts.

"There's an ugly tropical disturbance brewing about six hundred miles off Mexico. It appears to be moving west and gaining strength. They anticipate bumping its status up to tropical storm by the end of the day. They already have a name for it: *Celeste*. Our ship's course should keep well north of it, but it might cause problems later."

"Well, Ted, thanks, but obviously we hold no sway over Mother Nature," Wolcott said. "Instruct your weather folks to keep an eye on it and keep us well posted. Unless things take a significant turn for the worse, we ain't changing our plans. Savvy?"

"We'll continue to monitor," replied Seibert. "There's something else, though. We always have a security picket encircling Johnston Island when we're conducting pre-launch operations. I'm sure that you're aware that the Soviets have been sniffing around for a while, and they've intensified their surveillance operations. We're fairly sure that they're aware of the Project 437 Thor launches, but I don't think that they've caught on to what we're doing."

Seibert continued. "If we're seriously looking to launch in ten days, we won't have sufficient time for the Navy to shift their forces to establish their perimeter. At one end of the scale, we may end up with an intelligence trawler in the neighborhood, but in the worst case, a Soviet submarine loaded with frogmen commandos could wreak havoc on the island."

"True," replied Wolcott. "But since this is an effort requested by the Navy, I would sincerely hope that they would move heaven and earth to do what has to be done to post that danged picket."

Heydrich belched loudly, excused himself, and then observed, "We'll also need to set the tracking and communications assets."

"We should be able to work with the Navy to integrate with their network," answered Tew. "They're strictly using ground-based and ship-based relay sites. I suppose we should make a plan to incorporate some of the ARIA aircraft as well, to cover dead spaces."

"We will also need to contact General Fels at Eglin," said Heydrich. "So he can begin making preparations to deploy his teams to all of the equatorial Contingency Recovery Sites."

"Isaac's teams are *already* deployed to those sites, Gunter," interjected Tew.

"Oh. *Ja*, obviously," noted Heydrich, rubbing his eyes. "I should have guessed that."

"Virgil, I think that everyone has the sufficient information to work up the feasibility assessment," stated Tew. "We only have an hour, so let's focus on that."

"Will do, but there's still a loose end," replied Wolcott. "The stack is already configured for Carson and Ourecky. Their seats and gear are on board, and the vehicle has already been swung for weight and balance. Do you want me to call them in here?"

Tew shook his head, clenched his fists, closed his eyes and gritted his teeth. Wolcott was concerned at his friend's uncharacteristic behavior; Tew seemed to have lapsed into some sort of unresponsive daze.

A painfully long and awkward moment passed before Wolcott asked, "Uh, Mark, do you want to call them or no?"

Tew opened his eyes, drew in a deep breath and replied, "Carson's in town, but Ourecky is on leave in . . . Nebraska?"

"Yup," replied Wolcott, nodding.

"Well, since this is just a feasibility study, let's not jump ahead of ourselves. Those two deserve a break. When and if we receive the execute order from Kittredge, we'll recall them. Until then, let's leave things *just* as they are."

28

BACK TO OHIO

Ourecky Homestead, Wilber, Nebraska
7:30 a.m., Monday, August 7, 1972

Ourecky sat in the kitchen, drinking strong coffee with his father, planning the repairs they would be making to the pasture fence this morning. Bea and his mother were busy making breakfast. Little Andy scampered about, chasing after a ponderously overweight calico cat.

Ourecky looked towards Bea. The past few days had been a wonderful break for them. The underlying tension between them seemed to just vanish entirely. It was as if a gloomy spell had been lifted, replaced by a glowing mantle of happiness and hope. By far, the past few days had been the happiest time in their lives together.

Bea had spent hours with his mother, poring over photo albums, soaking in the Ourecky family history, seemingly absorbed in the happy childhood she had been deprived. Ourecky watched as his mother became Bea's, not just because of the formal prescripts incumbent with welcoming a daughter-in-law into the family, but because Bea found in her the mother that she so desperately needed.

Their angelic bubble was resoundingly shattered when the phone rang in the kitchen. Before answering it, Mama Ourecky wiped her flour-covered hands across the front of her gingham apron. She listened intently, smiled, and then proclaimed, "Scott, it's for you. Some Virgil somebody wants to talk to you."

Bea was removing a pan from the oven when she heard Virgil's name. Hot biscuits flew in every direction as the aluminum pan clattered to the hardwood floor. "How *clumsy* of me," she muttered, glaring at Ourecky. She stooped down to gather the scattered biscuits, but the calico cat beat her to one. Yowling, the cat deftly batted the hot biscuit into a corner to allow it to cool.

Ourecky took the phone. The call was decidedly one-sided; he spoke little, except to occasionally utter, "Yes, sir."

He hung up the phone and turned to his mother. "Sorry, Mama, but we're going to have to skip breakfast. I have to go."

"I thought you were on furlough, son," observed Papa Ourecky.

"I was. Something came up. I have to drive to Offutt Air Force Base as quickly as possible."

"*Offutt?*" said Papa Ourecky, sticking his thumbs in the frayed bib straps of his denim overalls. "That's just south of Omaha, near Cleveland. It'll take at least two hours to drive."

"Scott, there's no sense hitting the road with an empty stomach," said Mama Ourecky. "Why don't you sit down for some sausage and scrambled eggs? I can whip them up in a jiffy. I'll also fix you some sandwiches for lunch. Spam? Peanut butter and jelly?"

"No, Mama. We really have to be going. We'll take the Plymouth, if that's all right. Bea can drive it back and stay a few days."

"Okay," replied Papa Ourecky. He pulled a ring of keys from a hook beside the kitchen door and tossed them to Ourecky. "You'll need to fill it up, though. It only has a quarter tank."

Mama Ourecky stood next to Bea and said, "I'll fix a basket while Scott packs his things."

"There's no time for me to pack," insisted Ourecky, scooping up Andy and stepping towards the door. "We have to leave *now*."

"What's this all about?" asked Bea, obviously trying to remain at least slightly composed in front of his family.

"Don't know," he replied, opening his wallet to check for cash.

"Don't know or can't say?" she asked quietly.

He jammed his wallet back in his pocket and curtly answered, "I *don't* know, Bea." He hugged his parents and said, "Bye, Mama. Bye, Papa. I love you. I'll come back as soon as I can."

Offutt Air Force Base, Nebraska
10:18 a.m.

Although they drove with the windows rolled down, the atmosphere in the Plymouth was chilly enough that they could have been cruising through the frozen wastelands of Antarctica rather than swishing past sunbaked Nebraska cornfields in late summer. They hadn't exchanged more than a dozen words since leaving his parents' house. With Andy nestled sleeping in her lap, Bea gritted her teeth and fumed silently as she watched the scenery—a virtually unchanging tableau of pastures, croplands, farm houses, barns and silos—flash by outside.

Although she knew that it was out of his control, she was furious with him for this whirlwind change of plans. Things just never seemed to change. Every time there was a glimmer of normalcy, a faint hope for a stable existence, it was snatched away as quickly as it appeared.

Since she didn't know how long they would be separated this time, she resolved herself to calm down before they said their good-byes. Closing her eyes, she realized that what she felt was really more frustration than anger.

She was frustrated with him, but she was also frustrated with the Air Force—particularly Virgil Wolcott and Mark Tew—for not making good on their many promises. And she didn't look forward to returning to Wilber this afternoon. As much as she adored his parents and enjoyed spending time with them, she knew that they would have more questions than she had answers. For some reason, they believed that he was somehow more forthcoming with her, that she had been endowed with a secret key that unlocked the mystery that was his life and livelihood.

She also felt guilty for leaving her friend Jill behind in Dayton. Jill was sick and definitely not getting any better. Her mother was there, but she was emotionally ill-equipped to care for a daughter who was slowly fading away, much less a granddaughter barely out of diapers.

They pulled in at Offutt, pausing next to a billboard that advertised *"Headquarters—Strategic Air Command, Peace is Our Profession."* The security police had obviously been notified to expect them; after a cursory check at the gate, an SP patrol vehicle escorted them to the Flight Operations building, where a vacant parking space awaited the Plymouth's arrival.

The engine sputtered to a stop. Bea stepped out of the car and glimpsed Drew Carson standing next to a gleaming white T-38 Talon. She had to admit that he struck a dashing figure; in full flight gear, confident and handsome, he looked like a gallant knight preparing to ride into battle on his trusty steed.

"Bea!" exclaimed Carson, strolling towards them and extending his arms. "I sure wasn't expecting *you!*"

"Well, Drew, we *were* on vacation," she replied, reluctantly accepting his hug. "So I sure wasn't expecting to drop everything, jump in my in-laws' station wagon, and then ride for two hours while my husband drove like a banshee to catch a jet plane to who knows where."

"What's this about?" asked Ourecky anxiously. "Mark Tew gave us an extra month. Why would they call me back so early?"

"I have no earthly idea," replied Carson. "Honestly. Virgil just told me to zip you back to Ohio immediately. Did you leave the lights on somewhere? Do you maybe have an overdue library book? Forget to put the cap on your toothpaste?"

Ourecky shook his head.

Carson nudged an aviator's kit bag with his toe. "All your zoom gear's in there," he said. "There's a locker room in Flight Ops where you can suit up. We need to be wheels up as soon as possible. I'll give these guys the nickel tour while you're getting dressed."

"Back in a minute," replied Ourecky, swinging the canvas bag up on his shoulder. He turned and walked towards the building.

"Sorry about this, Bea. You look beautiful, as always," said Carson, kneeling to hug Andy.

"Thanks, Drew, but it's not me. It's my Martha White biscuit flour," she replied, glamorously swiveling her head and placing a hand on her hip. "I use it to bring out my eyes."

"Biscuit flour, huh? Who would have guessed? Well, it certainly works wonders for you."

"So you have *no* idea of what's going on?"

"Honestly, Bea, no. Virgil mentioned that we should expect to be out of pocket for a week, possibly two, but nothing else."

"So you don't know what you'll be doing but you do know how long you're going to be gone?" asked Bea. "That makes *no* sense."

"Like I said, Bea," replied Carson, shaking his head, "I don't know. Honest."

They chatted for a few more minutes, but Bea gleaned nothing that she didn't already know. As she stood off to the side trying to maintain her composure, Carson walked Andy around the T-38. Carson held him up; as he ran his tiny hand along the gleaming fuselage, it was obvious that the child was fascinated by the plane. *Exactly what I don't need*, thought Bea.

A few minutes later, outfitted in his flight gear, Ourecky emerged from the Flight Ops building. A terrible feeling of dread gripped Bea's stomach, as if she might never see him again.

Suddenly, she realized that this was the first time she had seen her husband wearing any sort of uniform, except for the day that they initially met on the plane in Atlanta. Seeing him and Drew together in flight gear immediately reminded her of her father's squadron photograph from Korea. The two men could easily have stepped into that black-and-white image to take the place of her father and stepfather.

The scene just unnerved her. Even if he came home safe from this mysterious trip, she knew that if things continued the way they were, it was only a matter of time before he left and never came home. She reconciled herself to the notion that only something traumatic would be sufficient to wrench him from that certain trajectory.

Pulling on his flight gloves, Carson said, "I'm sorry we have to rush, Bea, but we really need to be leaving. I'm sure that we'll see you back in Ohio."

"Just a minute," she said.

Glancing at his watch, he said, "Bea, we'll miss our takeoff spot if we don't get rolling . . ."

"Drew, please give us a minute," pleaded Bea. "Please."

"Bea, I'm sorry, but Scott and I really need to *go*," he answered, donning his helmet. "*Now.*"

"Just a minute, Drew," said Ourecky calmly. "You can pre-flight. I'll be there in a flash."

"I've already *done* my pre-flight."

"You know, it never hurts to double-check things," answered Ourecky, glancing towards the T-38's angle-of-attack vane. "Safety first, right?"

"*Yeah.* Safety first. I'll take another stroll around just to be sure. That should give you two kids enough time to say your proper goodbyes."

Ourecky started to embrace Bea, but she lightly pushed him away and said, "Scott, I need to tell you something, and you need to listen. You keep insisting that you're just a run-of-the-mill engineer, basically just a nobody, but I know that the Air Force doesn't send a supersonic jet to shuttle an inconsequential nobody from Nebraska to Ohio."

"Bea, I'm sure there's a logical explanation—"

"I don't doubt that, Scott." she said, standing with her hands on her hips. "I'm certain there's a logical explanation for all of this, but right now, I don't see it. If you went to Vietnam, and were in constant mortal danger, it would be easier for me, because I see that on television every night and it's something tangible that I could come to grips with."

With Andy hugging her calves, she continued. "As it is, I have no idea what you're doing. Some days you drive off to the office like everyone else in the world, and then sometimes you just drop off the face of the earth for days and weeks and months with hardly the slightest warning, and sometimes you come back looking like hell, like you've been tortured, and you try your best to be so casual about it, but I just know in my heart that you've been in terrible danger, and it just eats me up that you won't talk about it."

"Bea, it's not that I don't want to talk about it, it's that I *can't* talk about it. I'm sorry. Now, I have to go."

"When will you be back?" demanded Bea.

"I *told* you, Bea. I'll be back when I get back. I really don't know."

She stood with her arms folded across her chest. "Scott, this roller coaster is killing me and killing us, and I need to climb off. I love you, but if you leave now, don't expect us to be there when you come home."

"*Please* don't say that, Bea."

"Scott!" yelled Carson, climbing up the ladder to the cockpit. "*Now!* We have to get moving."

Ourecky dropped to his knees and hugged Andy. "I'll see you, little man. I love you." He stood up, hugged Bea and said, "I love you, too, but I have to go now."

Aerospace Support Project
2:25 p.m., Monday, August 7, 1972

An official sedan picked up Carson and Ourecky the instant that they descended from their T-38 and whisked them directly to Blue Gemini's brick headquarters. They weren't even granted time to shed their sweaty flight suits before being escorted upstairs and directly into the generals' office.

Tew and Wolcott were there, as was Gunter Heydrich. Strangely, Admiral Tarbox was there as well; Tarbox didn't look much better than Tew; the elderly admiral looked almost frail.

The atmosphere in the office was somber, almost funereal. The table's surface was covered with engineering drawings, technical documents, legal pads, slide rules, coffee cups and ashtrays; Ourecky saw that one blueprint-sized chart depicted a cutaway diagram of something that closely resembled the MOL.

"Gents, thanks much for hightailin' it in here," declared Wolcott, gesturing for the new arrivals to sit. "Ourecky, we really appreciate you rushin' back from furlough early. We sure hated to ask you to come back so soon. I hope we didn't upset your missus too much."

"A little," replied Ourecky, taking a chair and thinking back on Bea's ultimatum at Offutt. "She's not too keen on things changing so rapidly, but I think she'll eventually come around."

"You boys hankerin' for anything?" asked Wolcott. "Coffee? Water?"

"I'd really appreciate a Coke, Virgil," replied Carson, sitting down and smoothing his damp hair with his palms. "I'm absolutely parched. It's been a long hot day."

"And it's merely startin', pard." Wolcott leaned back in his chair, punched the intercom button on his desk, and said, "Smith, hoof it downstairs, grab a couple of cold Coca-Colas and fetch them up here. Also, call the O Club and rustle up some sandwiches for these boys. I'm sure that they skipped lunch on the way."

Ourecky's stomach growled at the mention of food. Not only had he missed lunch, but he had been called away from breakfast as well, and now his metabolism was catching up.

"We're in sort of a pickle here," declared Wolcott. "Here's the lowdown. I know that we ain't informed you, and I hope you ain't too offended by our lack of candor, but the Navy has a manned platform upstairs. It's an ocean surveillance variant of the MOL."

Wolcott tapped at the diagram spread on the table. "There are two men aboard. The Navy lost communications with them several days ago. If that ain't bad enough, there was an enormous solar storm on Friday. It was a Class Three event. That's the gist of it. Admiral Tarbox will fill in the details."

Ourecky was stunned that the Navy had secretively sent up an MOL, but he was already aware of the massive solar flare. It had been on the news shortly after it occurred, primarily because it had disrupted telephone services and electrical utilities in many areas around the world. He recalled a news interview with a somber NASA spokesman, who claimed they had been very fortunate because the surge had occurred between Apollo missions; otherwise, he implied gravely, it could have been catastrophic if any astronauts had been on the moon at the time.

"To be frank, because they weren't talking, we initially assumed that the crew was dead," confessed Tarbox. "But as the situation evolved, we learned the communications outage was due to an equipment failure."

"Resulting from the solar flare, Admiral?" asked Ourecky, gradually recovering from the shock of the news. A sergeant came in and delivered a pair of Coca-Colas. Carson gulped down almost half of his immediately.

"No, their radios failed prior to that. Since we didn't have comms, we couldn't warn them about the solar event. Otherwise we could have brought them out of orbit before they were irradiated. And really, we probably would have never known what happened if your man Russo hadn't been up there. He really pulled a rabbit out of his hat."

"*Russo?*" asked Carson incredulously. "Admiral, did you say *Russo* is upstairs?"

With his eyes fixed on the table, Tarbox nodded glumly. "He is. He made contact with us yesterday. The MOL's main radios weren't working, but he salvaged enough wire to string an auxiliary cable from the MOL to their Gemini-B reentry module. He tied it directly into the circuit breakers and powered up the Gemini-B's UHF and VHF radios. It took him a few hours to establish the connection, but he persisted, and we've been talking to him ever since."

"Hah," sniffed Wolcott. "He hot-wired the danged radios. Sounds like a durned *Popular Mechanics* project. Maybe he accidentally learned something while he was here with us."

"But Russo wasn't up there by himself," observed Ourecky. He noticed a chart taped on the wall behind Tew; crudely drawn on white butcher paper, it was labeled 'Worst Case—Consumables Estimates' and showed a regression analysis of oxygen and other expendables aboard the MOL. The chart's projections didn't paint a reassuring picture. "What happened to your other man, sir?"

"He's extremely ill," disclosed Tarbox softly. His bushy left eyebrow twitched erratically. With his coarse voice at the point of breaking, he was clearly rattled by the incident. "Commander Chris Cowin. Russo told us that Cowin was outside on EVA during one of the largest fluxes on August 4, so he caught the full brunt of it. We estimate that he received roughly three hundred rems. We don't expect him to survive. He's been unconscious for the past two days, and based on what Russo has described to us, our flight surgeons don't expect him to last very long, maybe a day or two at most."

"But even three hundred rems shouldn't have been a fatal dose," claimed Ourecky. While he wasn't an authority on radiation and its physiological effects, he knew that the cumulative amount a person might receive—relative to others exposed in the same incident—was strictly luck of the draw. Simply by virtue of being outside the spacecraft's protective shielding during the solar flare, Cowin had been dealt the worst possible hand.

Tarbox's squeaky voice wavered as he replied, "You're correct. It *shouldn't* have been a fatal dose, if he could have received adequate medical treatment in a timely manner. According to Russo, Cowin had severe diarrhea and was vomiting within an hour of locking back in. He was effectively incapacitated within three hours of exposure."

"Admiral, Russo should be able to reenter and fly home by himself," said Carson. "Why didn't he just stow Cowin in the Gemini-B and bring him straight down?"

"The same reason that he's still up there by himself," answered Tarbox. "At some point during this incident, he either broke or badly sprained his wrist. Our flight surgeons are fairly certain that it's fractured."

Ourecky was sufficiently familiar with the MOL/Gemini-B configuration to immediately recognize the problem. Theoretically, Russo could probably fly the Gemini-B with one hand, but there was a more pressing problem. The Gemini-B was located at the stern end of the MOL. To board the Gemini-B, he had to transit a narrow access tunnel. To prepare for reentry, he had to close and latch three separate hatches. The first was a hatch at the end of the connecting tunnel. The second portal was a heavy circular hatch literally cut through the Gemini-B's heat shield, that opened outward from the reentry vehicle. The third was a large pressure bulkhead hatch that opened inward into the Gemini-B's restricted cabin. Ourecky knew that it was difficult enough for two men to shut and dog down these hatches, but it would be virtually impossible for one man with only one functional hand. Unless his wrist healed quickly, Russo was stuck upstairs.

Confirming Ourecky's concerns, Tarbox explained the situation with the hatches, and concluded, "Our engineers are trying to improvise a mechanical solution to assist him with the hatches, but Russo also soaked up a substantial dose of radiation. His is obviously not an acute case like Cowin's, but it's definitely taking its toll. At this point, he's very sick, disoriented and just plain weak. Even if we devise a means to ratchet the hatches shut, it's unlikely that he would make it home by himself. And closing the hatches is just one problem of many."

"There's *more?*" asked Carson. The building shook as a large cargo aircraft—probably one of the new C-5 Galaxy transports—landed on the adjacent runway.

Nodding slowly, Tarbox swallowed deeply. His hands fluttered as he replied, "Our flight surgeons also suspect that he's losing his eyesight. He reported seeing a bright dazzle in his eyes when Cowin was on EVA, immediately before the radiation alarms started sounding. His eyesight has been progressively deteriorating, so our doctors believe that his retinas were permanently damaged."

Ourecky recalled seeing vivid flickers in his eyes during a few of their missions. It was mildly amusing at first, but quickly grew to be an aggravation, particularly when he saw them as he was trying to fall asleep. The flight surgeons had not known what to make of the sporadic flashes, and were concerned about the potential for long-term damage such as cataracts. Shuddering, he imagined how terrifying it must be for Russo, stranded in orbit and slowly going blind as he also suffered from the debilitating effects of radiation sickness.

"I don't understand, Admiral," said Carson, finishing his Coke. "So if you never heard from them, were you just going to leave those men up there to rot?"

"We thought they were *dead*, Major," replied Tarbox, kneading his wizened hands together. Scowling, he looked to be at the brink of losing his composure. "We had no reason to believe otherwise."

"So you would just write them off?"

"That's *enough*, Major Carson," interjected Tew, in a voice that was barely audible. "This is an awkward situation, but you *will* show the admiral the respect that he deserves."

"Look, this whole danged situation is plenty bad enough without us second-guessin' what the Navy folks have done or ain't done," said Wolcott. "Admiral Tarbox has come here to solicit our assistance in bringin' Russo home. That's why we brought you boys back here."

"*Wait*. Let me make sure that my ears are working right, Virgil." Carson smirked, smoothing the edges of his moustache. "You did say that you wanted us to rescue *Russo*?"

"Yup," replied Wolcott, nodding solemnly. "Of course, we ain't exactly sure that it's even feasible. Gunter and his folks are looking at different options. Gunter?"

Heydrich answered, "We're still hammering out the details, but the tentative plan is for Ourecky to conduct an EVA transfer to the MOL and enter through their airlock. He will provide some initial medical treatment to Russo, check out the Gemini-B, and then reenter to Edwards."

"And just so there are no misconceptions, we don't know exactly what we're facing up there," disclosed Heydrich. "Russo absorbed a significant dose of radiation, but as the admiral indicated, we really don't know how much. We're not even sure he'll survive the trip home."

"Gents, if that ain't enough bad news," stated Wolcott. "There's a heap more. Although we know that the MOL is operational, except for its radios, we ain't entirely sure that the Gemini-B reentry vehicle is in full workin' order. Its radios are functional, which is a good sign, but there's a strong possibility that some of the other electronics got kayoed as a result of the solar storm. The onboard computer is particularly susceptible, as you two already know."

"Couldn't Russo power it up to run diagnostics?" asked Ourecky.

"*Ja.* He *could* have, but he hasn't," answered Heydrich. "Before you pass judgment on him, put yourself in his shoes. He knew from the start that he couldn't fly home by himself, so it made little sense to power up the Gemini-B and risk draining its main batteries just to see if it works. Secondly, his eyesight is so degraded that it's doubtful that he could even interpret the instruments. The fact is that the Gemini-B is either going to function or it's not. Once you're aboard, you might be able to make some minor repairs, but there's very little that can be done."

Carson scratched his head and asked, "What happens if Scott climbs into the Gemini-B and determines that it's fried? What then?"

"Honestly?" answered Heydrich, wringing his hands. "We don't have a good answer for that. Drew, we plan for you to remain on station until the Gemini-B's reentry, so theoretically, he could transfer back to your vehicle for the return."

"And leave Russo up there in the Can?" asked Carson.

"Russo might already be dead by then, hoss," grumbled Wolcott. "As a matter of fact, he might be deceased even before you two launch. If you fancy my opinion, this whole gambit's sort of a crapshoot. A danged expensive crapshoot."

"But what if he's not dead?" asked Ourecky. "I couldn't just leave him up there to die."

"We're still looking at different options for that scenario," replied Heydrich. "There are enough consumables for you to remain up there for another month, possibly even longer. Hopefully, that would give you sufficient time to resolve the problems with the Gemini-B, if there are any, or for us to find another way to bring you home."

"NASA is obviously one alternative," replied Tarbox. "And if there was no other alternative, we could approach the Soviets."

"Oh, I'm plumb *sure* the Russkies would be just danged delighted to lend a helping hand," opined Wolcott skeptically.

"Okay, assuming that the Gemini-B is on the fritz," said Ourecky. "What if we put Russo in his suit and transferred him to our vehicle for the ride home? That would return him to Earth faster for medical treatment, and I could wait until you work out something to bring me home."

"That's very chivalrous of you, Ourecky," replied Tarbox. "But that's not a viable solution. Russo *can't* don a pressure suit, even with your assistance. From his description, his upper body is badly swollen, plus he has a broken wrist. He is either returning in the Gemini-B, or he's not coming home. It's that simple. Besides, do you really think that he could exit the airlock, transit to the other Gemini, and embark in his physical condition? Again, thanks for the sentiment, but it's not a practical course of action for us."

"Agreed," noted Wolcott, slapping the table. "By the way, gents, if all this ain't complicated enough, we've got two more big chores for Ourecky to handle. Leon?"

Tarbox pointed at the diagram. "Cowin and Russo were operating intelligence systems that were gathering ocean surveillance and other data. There is over two weeks' worth of data up there, compiled on storage cassettes similar to eight-track tape cartridges. The cassettes are normally loaded into data return pods. The pods are ejected and reenter on their own. After they're under parachute, they're snatched by a cargo aircraft in mid-air. Since the crew didn't have comms with us, they couldn't eject the pods. Ideally, we want Ourecky to retrieve the cassettes before he puts Russo in the Gemini-B."

Ourecky nodded.

"Additionally," said Tarbox, gesturing at the diagram again. "The reactor needs to be shut down. It's in an independent reentry module with its own guidance system and retros. The close-out procedures are simple. Once you inactivate the reactor, you'll eject the module and if all goes well, it ends up at the bottom of the ocean, far from land, safe and sound."

Ourecky quickly studied the drawings; he was looking for solar panels or fuel cells, but saw none. "But once we shut down the reactor, sir, aren't we going to lose power in the MOL?"

"Yes," answered Tarbox. "You'll still have roughly forty-eight hours on the batteries, but once they blink out, that's it."

"Look, Scott," said Heydrich. "We're sketching out a procedure for the close-out that should minimize the risks. Our tentative plan is that you'll power up the Gemini-B, verify that the systems are operational, go back into the Can, shut down the reactor, eject it, and then reenter with the Gemini-B. That way, if you power up the Gemini-B and see that there are issues to resolve, you'll still have the reactor power."

"So what happens if Scott shuts down the reactor and the Gemini-B still craps out?" asked Carson.

"Then he has forty-eight hours to fix it, but more likely we would direct him to climb back into his suit and reenter with you," answered Heydrich.

Carson sniffed. "Well, for once, other than the *marginal* risk of riding a rocket into space, it looks like I've finally drawn the long straw. Scott's the one assuming all the risks."

Ourecky laughed nervously and said, "Gee, Drew, if you recall all the time we spent in the Tank down there in New Orleans, you always said you wish you could be doing the EVA work instead of sitting on your ass? Would you care to swap seats for this trip?"

Carson closed his eyes and was silent for several seconds. Then he said, "That sounds like a *good* idea, Scott. After seven trips upstairs, I'm confident you can handle my side of the ship. I'll do the heavy lifting on this run. After all, that would give you much better odds of . . ."

"*No*, pard," said Wolcott bluntly. "We've already pondered that option, and we've already chucked it out. Ourecky trained for the EVA work in the Tank. You ain't. You're definitely the hot hand when it comes to close-in maneuverin'. Shucks, I ain't believin' that I have to say this, Carson, but whether you like it or not, you're going to sit in the left-hand seat for this lift."

Tew cleared his throat and said, "Gentlemen, let me make something abundantly clear. This is an extremely hazardous endeavor, much more so than anything you've done up to this point. The risks are so great that we cannot and will not order you to undertake this mission."

"Virgil, I have to ask," said Carson. "What happens if we decline the mission?"

"Then we launch Jackson and Sigler, provided they accept," replied Wolcott.

"So, in other words, if Scott and I decline, you really don't have any *viable* options, do you?"

"*Yup*," answered Wolcott candidly. The grizzled former cowboy spat a stream of tobacco juice in a wastebasket and added, "That just about sums it up, hoss. Fact is, you might say that you two pretty much have us bowed over a barrel."

The room was silent for several seconds, and then Wolcott spoke. "Gents, I wish we had more time to chew the fat, but the downright fact is that this candle's burnin' and it's burnin' fast. We can give you a little while to contemplate this mess, but we need an answer pretty danged soon. You two got any thoughts on the matter?"

Carson said, "I can't speak for Scott, but I'll go if you're willing to grant me *one* wish."

"A *wish*, pard?" asked Wolcott. "You're in the wrong place. This is the Air Force. We ain't in the wish-grantin' business."

"What?" croaked Tew. "What is it that you want, Carson? What's your wish?"

Carson said, "Sir, all I ask is this: If we do this thing and make it back alive, I want a chance to fly in combat."

"Carson, I don't know if you've quite figured it out yet," said Wolcott. "But you've *already* been flyin' in combat."

"You know what I mean, Virgil. None of this will ever appear on my records, no matter what we've accomplished. All I've ever asked for was an opportunity to fly in Vietnam. In my opinion, that's more than a fair trade for what you're asking of us."

"No," replied Tew. "*Absolutely* not. It's not subject to discussion, Carson."

Wolcott chuckled and said, "Well, since we're reduced to grantin' wishes, maybe there's something that we can do for *you*, Ourecky, even though we can't make Major Carson's fondest dreams come true. Is there anything you particular yearn for? Something that might help persuade you to ride that rocket?"

Ourecky thought of his last conversation with Bea, back at Offutt in Nebraska, just a few hours ago. His life would be immensely simpler if he could just share this secret with her. "*Anything?*" he asked.

"I s'pose the sky's the limit, pardner," replied Wolcott. "Theoretically, of course."

"Honestly, there's really only one thing I want," said Ourecky. "Just for once, I would like to be able to tell Bea what I'm doing. This whole business is ripping us apart. I'm very worried that if I go up this time, it might be the end of our marriage. I know it's too much to hope for, but if there was any way . . ."

"I'm truly sorry, Scott, but that's one wish that we can't grant," replied Tew succinctly. "Surely there's something else you want."

Ourecky groped for an answer, but he found nothing. Soon, one way or another, he would be free of this burden. If he couldn't level with Bea, then there was nothing that he wanted. And then he said, "Okay. If you can't grant my wish, then at least grant Drew's. I really don't think he's being unreasonable. He's only asked you for one thing since this whole ordeal has started. Grant him that, and I'll go with him."

"*No*," vowed Tew. "And that's *final*."

"Begging your pardon, sir," interjected Carson, dropping the depleted Coke bottle in a wastebasket. "But Ourecky and I need to step outside for a chat. By your leave?"

Tew nodded.

"Make it snappy, hoss," noted Wolcott. "If we have to conjure up Jackson and Sigler, I want them here pronto, so we can get this show on the road."

"I don't think we'll be out long," said Carson, standing up and heading for the door.

As he closed the door behind him, Carson addressed the captain and tech sergeant seated in the outer office. "Shoo," he said curtly. "Disappear. Take a break. Go downstairs and grab a Coke. Bring me back one, if you don't mind. Just give me five minutes."

"Well?" asked Ourecky. "What's your perspective on this, Drew?"

"I'm going," stated Carson.

"What about flying in Vietnam?"

"I was just yanking their chain, buddy," replied Carson. "I felt I had drawn the upper hand for once, so I tried to raise the ante. Obviously, Tew isn't in the mood to play cards. So be it."

"I thought you despised Russo. You would go upstairs for him?"

"Of course. The fact that I hate his guts doesn't enter into the equation, not even in the slightest. Scott, at my core, I'm a West Point officer. I may have breezed a bunch of classes up there on the banks of the Hudson, but there are certain fundamentals that they pound into you from the first day. One is that we don't abandon folks to die on the battlefield, *regardless* of who they are. It's as simple as that. Russo needs my help, and I'm going to assist him."

"Don't you mean *our* help, Drew?"

"No, I meant *my* help. You're a smart guy, Scott, a hell of a lot smarter than I'll ever be, but I'm still going to state the obvious: there's *no* need for you to go on this jaunt. This rendezvous solution is about a thousand times simpler than anything we've *ever* flown. I could jump on the rocket five minutes from now and fly this one in my sleep, by myself. I certainly needed your big brain when we were killing Soviet satellites, but this is a simple gig, Scott. I *don't* need you."

"But the EVA?"

"Mike Sigler has spent just as much time bobbing in the Tank as you have," replied Carson. "There's absolutely no reason he couldn't go up in the right seat. Like I said, I don't need you for this junket, Scott."

"Oh, really?" asked Ourecky. He looked at Carson; not making eye contact, Carson looked at the linoleum floor. "Are you *that* comfortable about flying with Sigler?"

Carson hesitated for a few seconds, cleared his throat, and then mumbled, "Sure."

"So you're *sure*, huh? Let me give you my perspective, Drew. I'm not worried about you. Barring any significant hardware malfunctions, you'll be fine, but I'm not entirely confident that Sigler can bring Russo home in one piece, particularly if there are any complications. You may be comfortable with Sigler, but I'm not."

"Ourecky, are you *dense?*" replied Carson. "I don't want you going on this trip. Yeah, you're right; I'm going to be okay, regardless of what happens. But if you haven't figured it out, and I suspect you have, once you cross over to that MOL, there's no guarantee of coming back to our chariot. This looks too much like a one-way ticket, and I don't want you buying it. So go home to Bea. Go home to your son. That's where you need to be, Scott."

Now it was Ourecky's turn to be silent. He weighed all aspects of the situation. On one hand, he finally had an opportunity to be free of this recurring nightmare. It was the perfect out, since he was sure that Mike Sigler would be delighted to fly in his stead, if for no other reason than to vindicate himself after two failed missions.

On the other hand, the whole purpose for this mission was to bring Russo home, so it made little sense to field a second-string team. Or at least *half* of a second-string team. But Carson was right; once he crossed the gap to the MOL, there was virtually no chance of coming back. He didn't think about it too long; Russo was *dying* aloft, so time was of the essence. With the knowledge of what had to be done, he swallowed and drew in a deep breath.

"I'm in," said Ourecky softly, looking at the floor.

"No," snapped Carson. "Go *home*, Ourecky."

"I'm *in*, brother," reiterated Ourecky. "If you're flying, I'm flying. We're a team."

"*No.*"

"It's not *your* choice to make, Major Carson. Now, we can go back in there and hash out this argument in front of Virgil, Tew, and the

Ancient Mariner, or we can go back in there and present a unified front. What's it going to be, Drew?"

Carson and Ourecky walked back into the office and sat down. "We're in," announced Carson. "No strings attached."

"*Both* of you?" asked Tew, looking towards Ourecky. "Voluntarily?"

"Both of us," replied Ourecky, nodding as he nervously twisted his wedding band. "Voluntarily."

"So it's unanimous that we're going," said Carson, leaning forward. "What's the plan?"

"Gunter?" said Wolcott, deferring to Heydrich.

"Everything is en route to the PDF. We've recalled all of the launch support personnel," stated Heydrich. "But we don't anticipate having the stack at the PDF and ready to fly for another ten days."

"Ten *days*?" asked Carson, shaking his head.

Heydrich nodded. "We intend to use every minute of that time to our advantage. We're working out a training program specifically for this mission. We're still ironing out the details, but the first step is acquaint Ourecky with the MOL, so we're flying your training suits and gear to the Navy's weightlessness training facility at Buck Island in the Caribbean. You'll go there immediately after we're done here."

"Why train at this Buck Island?" asked Ourecky. "Can't we just use the Tank in New Orleans?"

"There's a full-scale MOL mock-up in the water at Buck Island," explained Tarbox. "More importantly, we have a mock-up of the MOL's airlock, as well as the Gemini-B. You'll need to rehearse the transfer procedures a few times and practice closing the hatch, since it's a reasonably safe wager that Russo will not be able to assist you with it."

Tew cleared his throat and quietly said, "Although we should be able to execute in ten days, we will *not* launch until you two are ready to go. This is an extremely risky venture, so I want you to be entirely confident with your preparations. We have a great deal of flexibility, much more so than your previous missions. Because the MOL is in a well-defined polar orbit, we have a launch window once every twenty-four hours. So let me make this abundantly clear: we will not launch until you let us know that you're ready. Agreed?"

"Yes, sir," affirmed Carson and Ourecky in unison.

"Well, hate to cut this short, but if time is at such a premium, I suppose that we need to head right down to the Caribbean," said Carson. "Virgil, I'll call Flight Ops to expedite fueling and service on our T-38. We'll patch a flight plan together and should be in the air in less than an hour."

Wolcott shook his head. "Hold your horses, buster. I know you're rarin' to saddle up and go, but from now until the moment that your hold-down bolts crack, you ain't *flying* anywhere. You and your cohort will be *flown*."

"My T-39 Sabreliner and crew will be at your disposal for the duration of your mission work-up," said Tarbox. "It's configured for VIP transport, so you'll have the entire back cabin to yourselves. There's plenty of room to stretch out and rest."

"You're going to be training around the clock until you're ready to go up," said Wolcott. "So as long as you're being chauffeured from one point to the next, I expect you to be sleeping."

"Thank you, Admiral," said Carson, nodding towards Tarbox.

"Anything else, Gunter?" asked Wolcott.

Heydrich shook his head, but quickly added, "Good luck."

"Boys, we've been through this routine enough times not to get weepy or melodramatic," said Wolcott. "You know what has to be done. Just come home when it's over."

Tarbox stood up, leaned over the table, and shook their hands. "Thank you both," he said quietly. "Good luck to you."

"Mark?" asked Wolcott. "Any parting words? We probably won't see these two stalwarts until they land back here in Ohio."

Slowly pushing himself out of his chair, Tew replied, "I'll walk them out."

"If we're breaking up here," interjected Heydrich, "I'm headed downstairs. There's still much to be done."

Tew accompanied Carson and Ourecky to the outside office. "We're indebted to you both," he said quietly. "Much more so than I can express." With tears welling in his eyes, he embraced each man in turn, which took Ourecky aback; he had never witnessed Tew express any outward emotion, except perhaps anger. As Tew hugged him, it felt almost like when his father had dropped him off at his first summer camp.

In a voice barely above a whisper, Tew said, "Scott, you know . . ."

"I know, sir," said Ourecky. "But I'm still going up."

"Well, as I said, we're indebted to you," said Tew, nodding solemnly. "Is there anything I can do for you, Scott?"

Ourecky reflected on the question, hesitated, and then replied, "I guess not, sir."

"I guess we'll see you when we get back here," said Carson.

"I hope so, gentlemen," replied Tew.

Only Wolcott and Tarbox remained after the others filtered out. Tarbox spoke. "Virgil, we've not often seen eye to eye, but I appreciate your support on this. It means a lot to me."

Wolcott chuckled, spit tobacco juice in a wastebasket, and said, "Don't delude yourself, Leon. I ain't supportin' your effort. I'm only capitulating because I know Mark Tew won't have it any other way. Personally, I think it's a fool's game to squander a danged stack to mount a rescue operation for a guy who's probably going to be dead before our men even climb up there. And even if he's not, it's mighty damned doubtful he'll endure the trip home. I'm just as danged sentimental as the next guy, but I don't much cotton to fritterin' away a vehicle that cost hundreds of millions of dollars on a high risk humanitarian gesture."

"I agree."

"*What?!*" blurted Wolcott, sputtering and spraying brown-laced saliva over the front of his starched white cowboy shirt. "*You agree?!* Then why the hell did you slink in here, hat in hand, beggin' for us to ride to the rescue?"

"I'll explain," replied Tarbox, gathering his papers. "Virgil, we don't share much, but we're both pragmatic men. Agreed?"

"Well, I reckon that you and I do have that in common," Wolcott said. "We both call a spade a spade. And as much as I don't like you, Leon, I have to admit that we're pretty much cut from the same bolt of cloth."

"Then I can be upfront with you. I could care less whether Russo comes home, but the fact is that there's a virtual treasure trove of intelligence data up there. Even though their comms gear wasn't working, they maintained their collection effort, so that platform has been sitting up there for a month, sucking up information like a vacuum cleaner. There's the synthetic aperture radar data, and all the signals intelligence, all squirreled away on those tape cartridges. *That's* what I want, Virgil."

"*Oh*," replied Wolcott. "Now things are much clearer for me. You had me pretty danged befuddled. It was sure out of character for you to act so sentimental, but I guess that's what it was: an act. Ever the opportunist, eh?"

Tarbox nodded. "Look, Virgil, if you help me, then I think I can do the same for you."

"How so?" asked Wolcott. "Maybe I ain't the brightest bulb on the Christmas tree, but I'm obviously missin' something. How could you *possibly* help me?"

"Virgil, it's a virtual certainty that my program will be scrubbed after this incident. It was hard enough to sell putting a reactor in orbit, but now that we can't guarantee that it can be safely returned to Earth, intact, without endangering anyone, I know that there's not a chance in hell that we'll send another one up."

Wolcott laughed. "No argument there, pard."

"On the other hand, you have an operational program that's proven itself over and over," said Tarbox. "I'm well aware that Tew has no desire to continue it, but there's no reason why you couldn't go forward without him. With me in your corner, it would be a virtual certainty. Plus, I can make arrangements to shift all of my funding and resources from my program to yours."

Wolcott grinned. "You're a wily varmint, Tarbox. Sure, I would appreciate your backin' and cash, but since you ain't the altruistic type,

I'll wager that there's a hitch in this bargain of yours. What do you want in return?"

"I think it's likely that Russo will stop transmitting before your men are ready to launch."

"I suspect you're right. I'm guessin' he's on his last legs as it is."

Tarbox nodded and said, "If it's apparent that he's dead, Virgil, then I'm sure that you could present a very convincing argument against this mission. All I'm asking is that when and if the time comes, you set aside your reluctance. Fair enough?"

Wolcott thought for a moment, spat a sodden lump of depleted chewing tobacco into the trash can beside his desk, and replied, "Fair enough. I think we have a deal."

"Thanks," Tarbox said. "Virgil, could you clear up something for me? I'm trying to understand why Mark Tew is so insistent against Carson flying overseas. After all, the Air Force guys assigned to the MOL were only restricted for a year after that program was shut down. Some of them have already flown in Vietnam, and even though they never went to orbit, they still had some incredibly classified information in their heads. You and I both know that if Mark Tew was willing to sign off on it, Carson could fly over there."

"I agree, pardner. I'm sure we could abate the risks."

"Then how do you explain Mark's reluctance?"

"I'll lend you my theory," answered Wolcott, furrowing his brow. "He spent time in a POW camp right at the tail end of the War. I don't think he's willin' to risk Carson being subjected to the same sort of ordeal. But before you pass judgment on Mark, you need to understand that the Air Force Chief of Staff has weighed in on this situation, so it's really out of Mark's hands."

"How so?"

"The Chief is adamant that Carson not fly overseas. Carson's records have been flagged so that even if he somehow finagles a way to Southeast Asia, it would not only be a court martial offense for him to climb into a plane for a combat mission, but also a court martial for anyone complicit in facilitating the hop. So, as much as Carson hankers to go, he ain't goin' to find too many Air Force folks inclined to aid and abet."

"Oh."

"But you have to hand it to that hardheaded sumbitch. Carson won't ever let the matter die. He's always wanglin' for a shot. He's nothin' if not persistent, but they'll be openin' a new Disneyland in Hanoi on the day that he's allowed to fly in Vietnam." Wolcott spat tobacco juice in the wastebasket. "He's pretty slick, and I'm sure that if he was left to his devices, he would sneak over there, but the reality is that there's no Air Force squadron commander dumb enough to put him in a cockpit. Not unless they want to accompany him on the next lift to Leavenworth."

Tarbox looked towards the ceiling, as if reluctant to express his thoughts, and finally said, "Virgil, if Carson's that damned insistent, then maybe there's a way . . ."

Shaking his head, Wolcott tore the filter from a cigarette. "Not even in the realm of possibilities, Leon. I don't want to discuss it."

"But I think Carson deserves *something* for his sacrifices, especially this one. Both of them do."

"Agreed." Wolcott deftly flicked open his Zippo to light his cigarette. He drew in deeply, leaned back, and puffed a smoke ring towards the ceiling. "Leon, while we're plumbing such deep and dark mysteries, how would you like to answer one for me?"

"What do you want to know?"

"I happen to know that you sit on the critical personnel assignment committee for Hugh Kittredge. And one of that committee's chores is allocating flight personnel for the different classified programs."

"That's correct."

"If that's the case," growled Wolcott, "would you care to explain why the *hell* all of our requests for additional pilots have been bounced back?"

Contemplating the question, Tarbox was silent for a moment. In his high-pitched voice, he replied, "Virgil, we both just agreed that we're very pragmatic men, right?"

"We did."

"Then I'll answer in that light. Virgil, there are *four* reasons that we did not grant you any more pilots for your first phase. *First*, Carson and Ourecky are obviously the most perfect combination to fly your missions. They're just an extremely unique fit."

"No argument there," affirmed Wolcott. "Especially with a string of seven successful missions under their belts. Hard to top that record."

Tarbox continued. "*Second*, we clearly knew that there were *no* candidates in either test pilot training pipeline, at either Edwards or Pax River, even remotely close to replacing either one of them. And that includes the current roster of all active test pilots as well."

"They are an unusual pair, like you said," noted Wolcott.

"*Third*, regardless of your past successes, it is an absolute certainty that your project would be curtailed if you had *another* fatal accident."

Wolcott took a deep draw from his cigarette, exhaled, and said, "I s'pose I can guess your *fourth* reason, Leon—you folks never really expected Carson and Ourecky to beat the odds and live this long, did you?"

Fanning away a cloud of pale gray smoke, Tarbox shrugged his narrow shoulders, nodded gravely and muttered, "Correct."

29

CAN SHOT

Aerospace Support Project
2:25 p.m., Wednesday, August 16, 1972

"Well, I have good news and bad news, buddy," observed Wolcott, ambling into the office. He took off his white Stetson, placed it on his desk, and took a seat.

"Bad news? Can it be *any* worse than George McGovern running for president?" replied Tew, folding his newspaper and taking off his reading glasses.

"Doubtful. Anyway, I'll share the good news first. The stack arrived at the PDF late last night and has already been installed on the pad. They erected it this morning and it's already topped off with fuel. Carson and Ourecky should arrive this afternoon. We should be able to launch as early as tomorrow."

"Then what's the bad news, Virgil?" asked Tew.

"I've been up in the intel shop, monitoring the weather. I have an update on Hurricane Celeste. It was a Category Four storm when it passed south of the Hawaiian Islands yesterday."

"I know. Is it something I should be concerned with?"

"Yup. It was scootin' west-southwest at about ten knots, but a few hours ago it veered onto a west-northwest track, and it's presently headed smack dab for Johnston Island."

Wolcott continued. "They're plannin' to evacuate all the permanent party off the whole dadburned island. Apparently, they hadn't considered the weather when they yanked all that mustard and nerve gas out of Okinawa and stashed it in igloos on Johnston. Someone must have realized that it might not be a safe place to be in a hurricane. Right now, they're battening down the hatches and flyin' nonessential personnel to Hawaii. That's close to five hundred folks."

Tew closed his eyes and groaned. "Then we won't be able to launch?"

"We can still shoot," answered Wolcott. "But we'll have an extremely tight window. Celeste ain't projected to hit Johnston any earlier than Saturday. Our first launch window opens tomorrow night. The second window is the followin' night, but Seibert's meteorological folks tell me that the conditions will be too dangerous by then."

"So we launch tomorrow night or not at all?" asked Tew.

Toying with the silver and turquoise slide of his bolo string tie, Wolcott nodded.

"I had told Carson and Ourecky that we wouldn't launch until they felt they were ready," stated Tew. "I guess that I'll have to break that promise."

"Not necessarily, Mark," replied Wolcott. "If they're goin' to launch tomorrow night, they need to declare their decision by tomorrow morning. If they tell us that they're ready, good. If not . . ."

"I *order* them to go."

"Right. It ain't like you have another choice."

Tew grimaced and asked, "What else needs to be done?"

"We need to move the LSTs out of harm's way, posthaste," answered Wolcott. "We should also stage a C-130 out there to snatch out our launch crew shortly after the rocket has cleared the pad."

"Is anyone from the permanent party staying behind?"

"Yup. As it stands, a skeleton crew of ten folks will remain on the island to do damage assessment after the storm passes and to run a bulldozer to clear debris off the runway. Ted Cook and four of his guys have volunteered to hunker down and stay with them."

"I guess that it goes without saying that we'll lose our security picket."

Wolcott nodded. "The Navy has already flashed them their orders. The ships are steaming out of there as we speak, but one destroyer will hold on station to monitor the island during the storm. I would hate to be sittin' on *that* tin can."

Pacific Departure Facility, Johnston Island
10:05 p.m. Thursday, August 17, 1972

Happily dreaming of his carefree childhood days in Wilber, Ourecky sputtered awake with a start. Fully expecting to glimpse Bea's face and the familiar surroundings of their bedroom in Dayton, his eyes instead came to focus on a gray instrument panel several inches before him. For a moment he thought he was hallucinating, but then reluctantly accepted the unpleasant reality that he was strapped into a Gemini-I atop a Titan II rocket, yet again, awaiting liftoff. He grimaced, flexed his fingers and yawned broadly.

The last ten days had been a virtual blur of frenzied preparations interspersed with infrequent catnaps and meals grabbed on the fly. After flying to the Caribbean, they had initially spent three days training underwater at the Navy's Buck Island facility, where he and Carson practiced EVA procedures to enter the MOL's airlock. Then it was on to Patrick Air Force Base in Florida, where he received an intensive three-day crash course in medical procedures.

Leaving Patrick, they boarded a KC-135 to endure weightlessness training parabolas as they flew across the Gulf of Mexico on their way to California. During the single longest Vomit Comet sortie in history, Ourecky practiced everything from initiating intravenous lines to donning his EVA equipment. Although they were granted an extended nap during the last hours of the flight, the grueling experience had all but wiped him out. After he was poured off the plane at San Diego,

Navy nuclear personnel spent three days schooling him on reactor fundamentals and the shut-down procedures necessary to safely send the MOL's power plant back to Earth.

Right now, he felt absolutely saturated with information. The numbing ordeal was like a hyperkinetic college cram course, except that the final exam bore life-and-death repercussions. But as much as he wanted more time to prepare, there was *no* more time; the clock was running out on Russo, if the minutes and seconds had not already ticked away. Since Tew decreed that they would not fly until *they* felt ready, their only recourse was to declare their willingness to crawl aboard a rocket bound for space. Early this morning, they did just that.

So here they were. On the plus side, compared to their previous flights, it was a whirlwind mission. *If* Ourecky successfully boarded the MOL and *if* he completed his tasks as scheduled, he could theoretically reenter to the first available recovery zone—Patrick Air Force Base—at the conclusion of their twelfth orbit, in slightly less than twenty hours. Since the Earth rotated fifteen degrees under each orbital pass, there were three other opportunities for reentry—Scott Air Force Base in Illinois, White Sands Missile Range in New Mexico and Edwards Air Force Base—at ninety-minute intervals.

"*Hey!* Are you awake yet?" asked Carson. "Come on, Scott. This is no time to be goofing off. We have places to *go!*"

"Sorry," replied Ourecky. "Wow. I was totally out of it." Not sure of how long he had been asleep, he checked the clock. Only a few minutes remained until liftoff.

"I know. I only woke you up because I was afraid I wouldn't hear the engines over your damned snoring. Not to worry, though. I've got everything covered."

"Man, I'm parched," commented Ourecky, looking longingly towards the water nozzle. His mouth was dry, like he had been gnawing on chalk. "I'd do anything for a shot of water right now."

Carson chuckled. "You know the rules. We can't crack open the bar until we reach orbit. Can you wait a few minutes?"

"I suppose I don't have much of a choice, do I?" He hoped that he could remain awake as the mission got underway. His left thigh pocket contained a glassine bag of Dexedrine capsules. It was a common practice for the "go pills" to be issued to combat pilots. Although the flight surgeons insisted that he carry them just in case, Ourecky had resolved himself not to dip into the speed unless there was no alternative.

The next few minutes passed quickly. Ourecky heard the usual sounds and felt the familiar vibrations of the former ICBM coming to life. "Launch vehicle is switching to internal power," reported the CAPCOM. "Stand by for engine gimballing."

"Standing by for gimballing," replied Carson. "Ready?"

"Ready as I'll ever be," muttered Ourecky.

"T-minus one minute," announced the CAPCOM. "Stage Two valves opening in five seconds . . . T-minus thirty seconds . . . Have a safe trip, guys . . . T-minus twenty seconds."

Carson tapped Ourecky on the shoulder, pointed at the mission clock, and said, "Yours."

The CAPCOM's litany continued: "T-minus ten, nine, eight, seven, six, five, four . . . Titan first stage ignition . . . two, one, *Zero*. Hold-down bolts fired. *Liftoff!*"

"Liftoff and the clock is started!" exclaimed Ourecky. "Scepter Eleven is headed *upstairs!*"

On Orbit
4:58 a.m., Friday, August 18, 1972 (REV 5 / GET: 6:43:00)

They overtook the MOL in four orbits, nailing the rendezvous precisely as planned. Forgoing their normal station-keeping and approach procedures, Carson quickly coaxed the Gemini-I into a parking position roughly eighteen feet away from the MOL, directly above the airlock hatch, and then carefully nulled out any remnants of relative motion. By this time, he had the process down pat, like parking a car on a quiet suburban street.

"Time to gear up for your little jaunt across the void, Scott," said Carson, unwrapping the foil from a stick of Juicy Fruit chewing gum. "I'll prep your umbilical while you get dressed."

"Man, Drew, I'm worn out and I haven't even done anything yet," replied Ourecky, yawning and stretching. He snapped some frames to finish a roll of film, changed it, and stowed the camera. He groped in his side storage pocket, fished out two pre-printed checklist cards labeled 'ELSS Donning Procedures' and 'Pre-EVA Suit Integrity Check' and secured them to his instrument panel with an alligator clip. "I would give *anything* to crawl into bed for a week."

"Scott, are you sure you're ready for this?" asked Carson, pulling the white umbilical from its storage pouch. Pinning it between his knees, he carefully "stacked" it in S-folds so that it would feed out cleanly. The twenty-five foot lifeline would convey oxygen, electrical power and communications to Ourecky during his abbreviated excursion. It was relatively easy to manipulate right now, but once charged with oxygen, it would take on a life of its own, behaving like a sluggish and inflexible anaconda.

"I'm just *really* tired," replied Ourecky, rubbing his eyes. "But that can't be fixed."

"Look, no one's heard a peep from Russo since we've been upstairs. We don't know if he's even alive. We can stall for another rev so you can grab a nap," said Carson. "No one's going to fault you for that, Scott. You've been burning the candle from both ends for the past ten days. I would feel a lot more confident if you logged some decent rest before you went over."

"Me too, but I don't think we have that luxury, particularly since this clock is ticking down from the other side. Anyway, I appreciate the offer, but there's no sense delaying the inevitable."

"Okay, but it's your funeral." Carson removed dust covers from the umbilical fittings and stowed the plastic caps in his thigh pocket.

"Thanks for reminding me, brother." Ourecky reached back over his left shoulder to unstow the ELSS—EVA Life Support System—chest pack from its storage cabinet between their heads. The bulky ELSS contained a thirty-minute emergency oxygen supply, heat exchanger, back-up batteries and a carbon dioxide scrubber.

Connected to the umbilical, the ELSS functioned as an interface, not unlike a placenta between an expectant mother and fetus, to filter and condition the raw ingredients necessary to sustain life in the incredibly inhospitable environment of space. To prepare for this unique mission, Ourecky's ELSS had been modified with quick-release fittings for the umbilical's oxygen and electrical connections, so that he could undo them swiftly with his hands encased in the awkward pressure suit gloves. As an additional safety precaution, a supplemental thirty-minute oxygen tank was also attached to the bottom of the ELSS.

The ELSS chest pack was only one piece of a two-part EVA Support Package. The second component was a large MMU—Modular Maneuvering Unit—backpack, which would be carried aloft in the Gemini's adapter section. Looking almost like a "Buck Rogers" contraption, the MMU contained a substantial supply of oxygen, more batteries, a UHF radio, and an integral set of small thrusters—powered by hydrogen peroxide—which would allow an untethered spacewalker to operate independently from the Gemini.

Originally designed for the Air Force MOL effort, the MMU was supposed to have been tested during NASA Gemini missions but was not. Since this mission's EVA was of extremely short duration—hopefully—the MMU was not a viable option; besides, Ourecky had only trained with the ELSS, so the chest pack would have to suffice for their requirements.

As simple as this excursion seemed, it was not. After he entered the MOL's airlock, Ourecky would disconnect the ELSS from the umbilical, temporarily becoming a self-contained one-man spacecraft. He dreaded that phase of the mission, mainly because the ELSS lacked a radio. There hadn't been adequate time for the Blue Gemini engineers to jury-rig one, so he would be incommunicado until safely inside the MOL. Once aboard, he could talk to Carson and the ground over the Gemini-B's radios, but in the interim, if he encountered a problem, he was entirely on his own, since he would no longer have the hard-wired intercom link in the umbilical.

As he fastened the ELSS's connections to the matching fittings on his suit, he recalled that he had a back-up radio, a hand-held UHF transceiver that he could use to communicate with Carson, inside a transfer case that would accompany him. The hermetically sealed case, an aluminum-shelled container roughly the size of an overnight suitcase, contained medical supplies and other items. Ourecky would have preferred to bring a wider assortment of supplies and tools, but the amount of materials was constrained by the cubic dimensions of the airlock.

6:20 a.m. (REV 6 / GET: 8:05:00)

On orbit, *nothing* was simple. Even a seemingly mundane task like opening the door and venturing outside was a time-consuming and intensive endeavor. After Ourecky donned and activated his ELSS, the two men spent over an hour crawling through pre-EVA checklists, inflating their suits, performing suit integrity checks, and depressurizing the cabin. Finally, they were ready to unlatch and swing open Ourecky's hatch.

After struggling with it in the Tank, he found that the hatch was much easier to manipulate than he anticipated. Slowly standing up in his seat, he marveled at the partially obstructed view of the Earth passing by below. As he swiveled to look towards the rear of the Gemini-I, he saw long strips of expended primacord fluttering in the vacuum, flailing like a squid's tentacles from the gleaming white rim of the adapter section.

He took in a deep breath and steeled himself for his big leap. Carson had maneuvered the Gemini-I to within five feet of the MOL, so Ourecky had to cross the gap between the vehicles, aiming for a handrail next to the MOL's airlock hatch. The handrail was about four feet long, with approximately the same diameter as a large broomstick.

He knew that if he missed the handrail on the first attempt, Carson could reel him in by the umbilical for another try. That safety net was available on his short trip between the Gemini-I and the MOL, but he was painfully aware that if he experienced any setbacks in the airlock, after he had disconnected the umbilical, Carson could provide with him only very limited assistance.

A return trip would be a dicey undertaking, since if he misjudged the distances or inadvertently bounced without securing a handhold on the Gemini-I, he could readily find himself adrift with a limited supply of oxygen. Carson might successfully maneuver to catch him in time, but it was highly unlikely.

With just a gentle push, he launched himself and caught the MOL's handrail on the first attempt. Gripping the bar tightly with his left hand, he popped the exterior hatch inwards with his right hand. The hatch was spring-loaded, with a catch to hold it in the open position. "Exterior hatch is moving smoothly," he reported. "It's holding just as advertised."

"Excellent," replied Carson over the intercom. "You're looking good out there."

"I'm poking my head in to eyeball the hatch seal," stated Ourecky, edging headfirst into the airlock. He visually inspected the composite rubber gasket to ensure that it was still intact after prolonged exposure to temperature extremes and vacuum. He saw no obvious defects, but the moment of truth would come later when he attempted to pressurize the airlock. "The seal looks good. Ready with the transfer case?"

"Ready," replied Carson.

With his upper body protruding into the airlock, Ourecky tugged on the case's Dacron lanyard. It was another potentially dangerous move in this complex ballet. Slowly pulling hand over hand, like taking in a fishing net, he had to draw in the cord carefully to avoid snarling the case's tether with his umbilical. The lanyard was routed through a D-ring attached to his parachute harness, so once he reeled it in, he had to transition it from the front of his body to the rear.

Encumbered by the stiff EVA suit and the unwieldy ELSS chest pack, he found that moving the case was much more awkward than he had previously anticipated. The effort expended at least fifteen minutes longer than planned. Finally getting it in position, he wriggled into the airlock. Tugging on its lanyard, he wedged the case snugly behind his thighs.

Since it was the only way to enter the MOL, the rescue effort was entirely dependent on the airlock functioning properly. The airlock was similar to the lock-out rescue chamber on a submarine. Its exterior hatch opened inwards. Once Ourecky entered and pressurized the airlock, the exterior hatch would seal itself. As more oxygen flowed in, the airlock pressure would equalize with the MOL main cabin, then he would be able to open the interior hatch. In simplest terms, the whole thing functioned somewhat like the flapper valve at the bottom of a toilet tank.

The fact that the exterior hatch was open was a tremendously good sign. According to Tarbox, Russo had been instructed to prepare the airlock for just such a contingency. If the airlock had been left pressurized, as it normally was, then Ourecky would have been unable to crack open the exterior hatch, and all of this effort would have been in vain.

To allow him adequate room to close and open the hatches, the airlock's interior was three feet in diameter and roughly seven feet long, so he could shift back and forth as necessary. He could see a sliver of blue Earth below, and saw that it was quickly growing dim. Having lost track of time, he checked his watch and realized that they were going into their forty-five minute interlude of orbital darkness. That wasn't good, since they had planned to complete the transfer and lock-in during the light phase. His wrestling contest with the transfer case hadn't helped the situation.

"I know we're dragging behind schedule, but I'm going to press on." Situating himself in the airlock, Ourecky was the verge of becoming the loneliest man in the universe, if but only briefly. He opened the supply valve on a supplemental oxygen tank strapped to the bottom of the ELSS; the thirty-minute reservoir should offer him plenty of time to board the MOL. "Drew, I'm ready to chop the umbilical," he said calmly. "Are you ready over there?"

"I am," answered Carson. "Are you *absolutely* sure?"

Ourecky swallowed, blinked, and replied, "Yeah. Russo is waiting for me in there."

"Okay, buddy. Good luck. Give me a five count when you're ready for me to cut the cord."

"Okay. Five . . . four . . . three . . . two . . . one . . . *mark*." As he counted down, with his gloved hand poised on the umbilical's quick disconnect, Ourecky stared apprehensively at the small instrument panel on the top of the chest pack, focusing his attention on an indicator marked "ELSS BATT." The rectangular light blinked on, indicating that Carson had switched off power and oxygen flow. Ourecky disconnected the umbilical's oxygen and electrical connections and then unsnapped the restraint tether attachment on his parachute harness.

Except for the sound of his breathing, it was eerily silent. With his link severed, Ourecky was alone. Moments later, tugged by Carson, the umbilical slowly receded through the exterior hatch opening like an albino eel disappearing into an underwater cave. Once he was confident that there was no danger of the umbilical fouling the hatch, Ourecky shoved himself down in the airlock and used his booted feet to push the spring-loaded exterior hatch closed against its seal.

He adjusted his position to prepare for the next phase, pressurizing the airlock. If all went well, that would be just a simple matter of opening the flood valve to allow oxygen to flow freely into the space. Recessed into a pocket, the oversized valve was made to operate with unwieldy gloves. Ourecky twisted it open.

He watched the airlock internal pressure gauge next to the valve; its needle should fluctuate as oxygen surged into the small space, but there was nothing. Then he realized that the flood valve had two stages, with a safety shut-off inside the MOL. Russo should have left that safety valve open, but apparently had neglected to do so.

Ourecky groaned; it was an extremely disheartening turn of events. The consequences of the simple oversight would quickly cascade. If the airlock was not filled with oxygen to equalize the pressure, the exterior hatch would not seal, the interior hatch could not be opened, and this rescue mission would be all for naught. Ironically, Russo might be only a few feet away from the airlock, but without a radio or other means of communications, short of frantically banging on the interior hatch in the hope that he would realize his dilemma, Ourecky had no way to urge him to simply twist open the safety valve.

Striving to remain calm, he wiggled his fingers and then tried the flood valve again. It swiveled freely; he twisted it completely shut and then rotated back to the full open position, but there was still no outflow of air. In anticipation of this very contingency, he and Carson had sketched out a "crawfish" emergency withdrawal plan.

To signal that he was in trouble, he would open the exterior hatch and then push down so that his legs protruded from the airlock. Carson would then maneuver as close as he could and blink the floodlight twice, indicating that he was ready to receive Ourecky back into the Gemini-I. Ourecky would then launch himself, climb back into his seat as swiftly as possible, and wait for Carson to reattach the life-giving umbilical. Afterwards, they would wrestle the hatch closed, re-pressurize the cabin and prepare for reentry. It was an extremely risky maneuver, and the potential danger would be greatly compounded since they would be forced to execute it in orbital darkness.

Meditating on his circumstances, he closed his eyes and did some quick volumetric calculations on his mental blackboard. If he was *right*, his supplemental tank would fill the airlock to yield sufficient pressure to open the interior hatch. If he was *wrong*, he could be trapped in the airlock with no means of communication, with less than thirty minutes of air to breathe, doomed to die in soundless solitude. The only sane recourse was to cut his losses and return to the Gemini-I to head home, damning Russo to oblivion.

The only sane recourse? If only it was that simple. He flew up here to help Russo, and the only thing that separated them was a mere eighth-inch thickness of titanium. Of course, it was an eighth-inch of titanium drawn snugly in place by a brutally perfect vacuum.

Ourecky also suspected that Tew had reluctantly dispatched him to the MOL with the knowledge that once he was here, he would not turn back. And for whatever reason, he thought of Haiti; he knew little of what had transpired there, except that he was painfully conscious that men had risked *their* lives to save *his*, and now he felt compelled to settle the balance.

Even with the oxygen flow shut off, his suit still contained a couple of minutes of breathable air. With no time to waste, he decided to go for broke. He closed the supply valve on the supplemental tank and disconnected its short hose from the ELSS's MMU oxygen fitting. He mashed in the Oxygen Flow selector knob on the ELSS's instrument panel and rotated it to the 'Off' position. Then he twisted the supplemental tank's valve open again and triggered the emergency purge device to more swiftly vent the cylinder.

Now, the oxygen cylinder intended to buy him roughly thirty minutes—more than adequate time to transit the airlock—was spilling its precious contents into the vacuum. It was a momentous wager, the biggest bet that he would ever place. *Would the exterior hatch seal? Would there be sufficient pressure to equalize the chamber and allow the interior hatch to open? Will I get to Russo in time?*

He unexpectedly found himself jammed against the wall of the airlock and then realized that the gas swiftly jetting from the purge port acted like a runaway maneuvering thruster. As the residual oxygen in his suit ran out, he gasped and then held his breath. In moments, he felt lightheaded, his fingers tingled, his vision started to dim, and his heart pounded in his chest. Anxiously watching the pressure gauge needle creep up the scale, he waited until it indicated five pounds per square inch before unlocking and opening his helmet visor. Still holding his breath, he gratefully listened to the weak whooshing sound of oxygen flowing from the supplemental tank. It was *unbelievably* cold inside the small chamber, at least a hundred degrees below zero, probably much colder.

He exhaled through his nose; he heard a faint crackling sound as his warm breath instantaneously exploded into a cloud of minute ice crystals. He hesitantly sucked in a frigid mouthful of air. Even though his lungs ached immediately, he savored the breath, promising himself that he would never again take precious oxygen for granted. Seconds later, he heard a hissing sound and faint pop as the airlock's pressure equalized with the MOL cabin's, and the interior hatch spontaneously opened.

He scrambled though the hatch and swiftly shed his helmet, gloves, and the ELSS chest pack. His face burned and his eyes watered. It was uncomfortably hot inside the MOL's main cabin; the chilly air flowing in from the airlock quickly dissipated. He was taken aback by the transition; it was like being transported from the arctic to the tropics in an instant.

He opened the transfer case, tore away an insulation layer, switched off a battery-powered heating element, and yanked out the handheld UHF radio. He turned it on, adjusted the volume, and reported, "Drew, I'm inside." He was amused at the sound of his voice; in the MOL's mixed atmosphere of helium and oxygen, it bore a slightly squeaky and high-pitched tone.

A moment passed before Carson replied, "Excellent. Now I can breathe again. Any problems with the airlock?"

"Nothing significant," replied Ourecky, almost nonchalantly. He rubbed his tingling nose and smiled to himself. "I still have to locate Russo, so I'll be off-air for a while." As he bled off suit pressure, he took a moment to orient himself to his new surroundings. Compared to the cramped Gemini-I, the MOL was lavishly spacious. He was filled with a sense of euphoria; after seven trips into orbit, crammed into the Gemini-I cabin, he was finally able to float free. And the freedom was not just a brief interlude, like the half-minute spells of weightlessness he experienced on the Vomit Comet parabolas. It was the stuff of his childhood dreams, and despite the urgent circumstances, he reveled in it.

Simultaneously, he was aghast. The station was in shambles, like the aftermath of a raucous frat party in orbit. A revolting stench permeated the stiflingly warm air. A grubby constellation of waste and debris— crumbs, litter, discarded food wrappers, and clumps of hair—floated in the cabin. It was disgusting and eerie at the same time.

Strange orbs, most about the diameter of a finger, hovered in the midst of the clutter. Curious, he looked closely at a few and realized that they were globules of vomit; some had already dried, while others—glistening, undulating yellow-green spheres tinged with blood—were very new. Overcome by spasms of nausea, he retched; he tasted a surge of sour bile, but managed to retain his last meal. His every movement set the air into motion, and the blobs drifted to and fro, many splattering into the walls and yielding yet many more projectiles.

In addition to the airlock, the MOL's forward area contained the occupants' closet-like sleeping compartments. One enclosure was open and vacant; the curtain was tightly drawn on the other. Ourecky yanked open the flimsy curtain and gasped as he encountered Cowin's lifeless body. Despite his medical briefings on what to expect, what he saw appalled him. If he didn't know that Cowin was human, he might have guessed that he was some sort of alien.

He had never met Cowin, so he was curious what the Navy astronaut looked like when he was still alive. Cowin's chest and abdomen were grossly distended with edema, and his hairless head was swollen like a child's balloon. Set deep in his grotesquely bloated face, his eyes were wide open, locked in a perpetual stare of bewilderment. His swollen arms were extended in front of him, as if he was grasping for an answer to his untimely demise.

Cowin wore long underwear. Stretched to the breaking point, the cotton fabric was drawn absolutely taut over his inflated body; it looked as if his carcass was squeezed into a white sausage casing. Ourecky briefly contemplated zipping Cowin into his cocoon-like sleeping restraint. He squeamishly reached out and nudged one of Cowin's outstretched arms, and realized that stuffing the corpse into the bag would be a futile waste of time and energy. Rigor mortis had long since set in, and the stiff body would not be moved.

He didn't linger to pay his respects. He could do nothing for Cowin and had yet to find Russo. He remembered that one of his more ghoulish tasks was to photograph Cowin's corpse—and Russo also, if he found him dead—but he decided that the macabre chore could wait. He pulled the compartment's curtain closed. Dodging odious blobs, he gradually worked his way aft, traversing the filthy galley space, and found Russo in the operations module. Although he was still alive, if only marginally, Russo didn't look much different than Cowin.

Russo still retained most of his hair, although there were several bald patches on his scalp. Purple blotches marked his swollen face. Gradually emerging from a stupor, he tried to speak, but voiced nothing but incoherent babble. Ourecky examined his eyes; the pupils were slightly opaque, like cataracts. Ourecky now understood why no one had heard from Russo lately.

Turning away from Russo, Ourecky made his way to the reactor control station located in the next compartment aft. He pulled out a reference card prepared by the Navy's reactor specialists. Comparing the numbers to the reactor's controls, he verified that the nuclear furnace was functioning within acceptable parameters.

According to his cursory assessment, everything was in order; the reactor was perfectly healthy, and could probably continue operating for months. Then he worked his way forward to the stern compartment, retrieved the medical bag from the transfer case, and returned to Russo.

He keyed the handheld radio and spoke. "Drew, I've checked the reactor and it's secure. I'm with Russo, and I'm commencing the treatment protocol."

"I'll relay the reactor status to the ground," replied Carson. "How does Russo look?"

"Much worse than I expected. I doubt he'll survive reentry. He's pretty damned frail."

"Roger," answered Carson. "I'll pass the news on."

Using a strap, Ourecky anchored Russo to a bulkhead. After four attempts, he successfully threaded a catheter into a vein in the crook of Russo's elbow and connected it to a Viaflex infusion bag of whole blood. Lacking the assistance of gravity, Ourecky patiently kneaded the bag to force the blood through the intravenous line and into Russo's circulatory system.

Once he drained the first bag, he replaced it with another, and then another until he administered four units altogether. Next, he injected Russo with pre-loaded Tubex syringes that contained antibiotics, potassium iodide, Valium, and morphine.

As he monitored his patient's vital signs, Ourecky laughed to himself; although he was not a physician, he was fairly certain that he was performing the first house call in orbit. The watershed moment would be yet another unseen entry in the annals of human spaceflight.

After receiving the serum and drugs, Russo was slightly more lucid. "Can't . . . see . . . well," he mumbled. "You . . . you . . . *Ourecky?*"

"Yeah. It's me, Ourecky. I'm here to bring you home, Russo."

"But . . . you're . . . not a pilot."

"True, but this is hardly the time to quibble over technicalities," answered Ourecky, stowing away the depleted IV bags and expended Tubex cartridges. "But if it makes you feel any better, you'll be sitting in the left-hand seat on the ride down. Besides, I'm more comfortable flying from the right side, anyway."

Smiling feebly, Russo faded back to sleep. Ourecky paused long enough to climb out of his cumbersome space suit. The action wasn't listed on the flight plan, but with the pervasive heat in the cabin, he knew he wouldn't last long otherwise. Stripped down to his long underwear, which were already soaked with perspiration, he went to work.

On Orbit
1:43 p.m. Friday, August 18, 1972 (Rev 11 / GET: 15:28:03)

For an instant, Ourecky found himself suspended in the ethereal twilight between sleep and wakefulness. For the second time today, he was a child again, soaring over Wilber, and he had finally succeeded in grasping an elusive object of his dream world, a souvenir he could take with him into consciousness, a secret talisman that would allow him to fly while he was awake.

Suddenly, he realized that he wasn't *dreaming* that he could fly; he *was* flying. It was a painfully disconcerting sensation, something like what a dozing driver feels when waking at the wheel of a speeding car. He looked at his hand and saw that the object he had snatched over from the dream realm was in fact one of the many errant tape cartridges he had been chasing.

He had been aboard the MOL for over six hours, toiling against a deadline, agonizing over everything that had yet to be accomplished. He was operating at the ragged fringes of his endurance; the sleep deprivation was definitely taking its toll. For some inexplicable reason, his chest was terribly sore and he was developing a persistent cough; he attributed both to the putrid air that was likely heavily laden with bacteria and other forms of unhealthy crud.

There were hundreds of tape cartridges to collect and account for. He had to hand it to Russo; even as he was losing his sight, he still managed to change out the surveillance data cartridges on schedule. Unfortunately, most of them—especially those bearing the data collected in the past few days—weren't labeled, so they were woefully disorganized.

To make matters worse, Russo had crammed the loaded cartridges into mesh storage bags, but apparently had not realized that one of the bags had ruptured a seam. At least fifty or more tape cartridges floated free, jumbled from one end of the station to the other.

As he alternated between frantically packing tape cartridges into storage cases and keeping watch on Russo, he heard Carson's voice on the radio.

"Hey, Scott, can you bump to Channel Three?" asked Carson. "We need to chat."

"Bump to Three? Yeah. Wait."

Still groggy, Ourecky dialed in the new channel, paused, and then transmitted. "Drew?"

"Busy over there?"

"Busy? I'm busier than a one-armed paperhanger," replied Ourecky. Surely Carson knew better than to call him up just to have a conversation.

"Yeah. Hey, something just came up that might ease your workload considerably. If nothing else, you can at least slow down, maybe even grab some sleep."

"If that's the case, I'm just *dying* to hear anything you might have to say," replied Ourecky.

"Virgil sent us a message on my last contact. He, Mark Tew, and Tarbox apparently had a long powwow after you sent down your medical assessment of Russo. Are you sticking with your conclusion that he's not going to make it down alive?"

Ourecky looked towards Russo, who was still apparently unconscious, and then curtly replied, "Yes. Drew, I don't think he's going to make it home."

"Okay. You should be aware that the flight surgeons at Wright-Patt concur with your diagnosis, based on the vitals you've sent. They don't think he'll survive the G-load on reentry."

So what's the point? thought Ourecky. "I guess I should have gone to medical school instead of becoming an engineer." He paused to cough repeatedly. Wheezing to catch his breath, he keyed the radio and said, "I'm missing something here, Drew. Can you clarify?"

"Yeah," replied Carson. "Based on what the docs say, Virgil and Tarbox are recommending that we shift the priorities of the mission. They want you to stay up for another twenty-four hours, to ensure that you can recover all the data from the SAR and signals intercept gear. Right now, they want you to cease what you're doing, snatch two or three hours of decent sleep, and then resume work. Once you've salvaged all the tape cartridges, you should still have time to grab some more rest before reentry."

As he contemplated the prospects of staying up another day, Ourecky gazed at the chaos. As horrible as he felt right now, and as disgusting as the conditions were, he *liked* it up here. While he would have preferred a more pristine environment than a metal-walled petri dish filled with floating trash and exploding globs of puke, he found the experience of prolonged weightlessness exhilarating.

And a chance to *sleep?* The daylong extension was almost more than he could hope for, to be able to enjoy this unfettered world rested and without having to rush. Then he looked towards Russo. Reflecting on his bloated and discolored face, things came more clearly into focus. Ourecky had come up here on a *mission*, not for a field trip or pleasure cruise. He resolved himself not to yield to temptation.

He coughed several times, cleared his throat, blinked his tired eyes, keyed the radio and said, "Hey, Drew, since Cowin is dead and Russo is incapacitated, that leaves me as the only functional officer aboard this Can. I suppose that formally makes me the MOL mission commander, right?"

"Yup," replied Carson. "That's correct. Congratulations on your command of a naval vessel. That's quite an accomplishment for an Air Force guy."

"Good. If that's the case, I'm making my first command decision." He paused to cough several times and then caught his breath before continuing. "Please convey to Virgil that we are reentering for White Sands on Rev 15, as currently planned. I'll do my best to salvage all the tape cartridges that I can grab, but my first priority is to carry Russo home."

"Aye, aye," replied Carson. "*Good* decision. I'll relay to Sheriff Wolcott, and maybe he can talk some sense to the Ancient Mariner. Hey, Scott, your coughing is starting to sound really bad. Are you *sure* you're okay to continue?"

"Do I have a choice?" asked Ourecky, wheezing as he struggled to breathe. "We came up here to give Russo a lift home. That's what . . . I'm . . . going to . . . do."

"Okay. Be aware that I am going to let the flight docs know that you're under the weather. Maybe they can come up with some magic potion to make you feel better."

"Sure. Can't hurt." Ourecky smiled weakly, coughed deeply, and keyed the microphone twice. As he stuffed a rubber-banded batch of cartridges into a storage case, he heard a weak groan. Floating effortlessly in the air, he turned his head to see that Russo was awake, although just barely so.

"*Scott*," muttered Russo weakly. "Thank . . . you."

Realizing that it was the first time Russo had ever addressed him by his first name, Ourecky grinned and replied, "You're welcome, Ed."

3:31 p.m. (REV 12 / GET: 17:16:20)

Ourecky was in the cramped Gemini-B, working to stow a storage case that he had filled with tape cartridges. On his third attempt, he successfully slid the awkward case into its tight slot and clicked shut the latches that locked it securely in place for reentry. As he began to worm his way back through the circular hatch opening to return to the MOL cabin, the radio interrupted him. Coughing, he turned up the volume and pressed the handheld transceiver to his ear.

"How are you holding up, Scott?" asked Carson. His voice carried a very concerned tone. "Feeling any better?"

"Uh . . . no. The coughing is just getting worse . . . and . . . now I'm coughing up crap. My chest is really sore, I feel like I'm running a high fever, and I'm really . . . weak."

"Okay. I'm relaying all this down to the docs. Wait." The radio was silent for several seconds, and then Carson asked, "You said you had a *productive* cough?"

"Huh?"

"You're coughing up crud?"

"Yeah. It's pretty damned ugly."

"The flight surgeons want to know what it looks like."

Ourecky coughed deeply, spit out the results, and then examined the resultant lump hovering before him. "Uh, it's greenish-yellow mucus, really thick and lumpy, tinged with spots of blood. And there's other stuff in there . . . like little particles."

"Are you nauseous?"

"No. Not . . . nauseous . . . at all."

"Okay. I'll get back to you shortly."

Aerospace Support Project
5:35 p.m.

With his Tony Lama cowboy boots on the edge of his desk, Wolcott leaned back in his chair and reviewed the remaining details of the flight plan. He heard a faint tapping and looked up to see a young captain poking his head into the glassed-in office space. He immediately recognized him as a flight surgeon newly assigned to Blue Gemini. At present, he was the only physician in the mission control facility; the more experienced doctors had been dispatched to the different recovery sites in anticipation of Russo's return to Earth.

"Is General Tew here?" asked the bespectacled flight surgeon.

"He's indisposed, Sawbones," replied Wolcott. "Something on your mind? I'll pass it on to him when he gets back."

"Yes, sir," said the flight surgeon. Bearing a clipboard, he stepped inside the enclosure and closed the door. "May I be candid, sir?"

"Tarnations! Of course, son. If you've got something to say, then just spit it out. Savvy?"

The flight surgeon nodded gravely and said, "I'm really concerned about the prospects of your man making it safely to Earth."

"Russo?" replied Wolcott. Striving to keep a serious visage, he stifled an urge to chuckle. "Really, twixt you and me, pard, I don't much suspect that he'll make it down alive. We had to lend it the ol' college try, though."

The flight surgeon looked at him as if perplexed, and clarified, "I wasn't talking about Russo, sir. I meant Major Ourecky. I just came from communications, where I talked with the other flight medical personnel by radio. We're very concerned that Ourecky isn't going to make it back."

"How so?"

"Based on what Major Carson is reporting, Ourecky has apparently acquired a very significant respiratory infection. It could be just a really bad cold, but it also might be as serious as pneumonia or severe bronchitis."

"I'm aware that he's ailing," Wolcott said. "I thought he was just a little under the weather, but is it really *that* bad?"

The flight surgeon adjusted his wire-framed glasses. "It is. We suspect that it's some sort of bug that he picked up before flight. Uh, sir, was the crew not placed in quarantine before the flight? You know, NASA has some established protocols for—"

"This ain't NASA and besides, we ain't flyin' to the moon," said Wolcott impatiently. "Most of our flights are of such short duration that the boys don't even have time to get sick. Besides, before they fly, they rarely spend too much time around very many people, so we ain't yet seen the need to further isolate them. Besides, it doesn't really matter at this point."

"Sir?"

"If Ourecky's that sick, the genie's already way out of the bottle. Don't waste my time or General Tew's by tellin' us what *should* have been done. It's too danged late for that. Now, tell me, right now, you're assumin' that he brought the bug up with him, but is there any chance that he caught something up there?"

The young flight surgeon nodded. "It's possible, but if that's the case, it would have to be a very virulent strain of bacteria. But from what he has described, the MOL would be an exceptionally fertile environment for bacterial growth. To be honest, it's a very confusing situation. If it's an infection, the onset was extremely rapid, much faster than we would have ever anticipated."

Wolcott scratched his head and asked, "How about the danged radiation? Could he have been exposed to radiation like Cowin and Russo?"

"It's doubtful, sir. He has checked the monitoring instruments every hour since he's been up there, and they are no greater than normal background."

"Well, what if it's some sort of super-bacteria?" asked Wolcott, conjuring up images from recent science fiction movies about deadly germs from outer space. "Radiation killed Cowin and has almost killed Russo. If there was already bacteria up there, could it have accelerated its growth? Could it have caused some kind of mutation?"

The flight surgeon shook his head and answered, "No, sir, we would expect exactly the opposite effect from the radiation. After all, as I'm sure that you know, irradiation is a widely accepted process to destroy bacteria in packaged food."

"Well then, if it might only be a bad cold, Sawbones, why are you danged concerned about Ourecky not making it home?"

"He's already very debilitated, sir," answered the flight surgeon. "Both of them are. The heavy G-loading during reentry is traumatic enough for healthy subjects, but it could be even more detrimental for him. If he keeps coughing like he is now, and can't clear his airway during reentry, he could aspirate sputum and choke to death. And quite simply, sir, he could be so weak that he just passes out. I guess I don't have to spell out the consequences of him losing consciousness during or immediately after reentry."

Wolcott shook his head. "Okay, doc. I'm just a mite perplexed here. What you're describing seems to be entirely out of our hands. What is it that you would have us do?"

"We are almost positive that Russo won't make it," asserted the flight surgeon in a curt, matter-of-fact tone. "And considering his current medical condition, it's a tremendous risk for Major Ourecky to attempt to reenter on his own. All things being equal, we think the most sensible option is for Ourecky to discontinue the rescue mission, suit up, and execute an EVA transfer back to the Gemini-I while he is still physically able."

"And just leave Russo up there to die?"

Nodding his head, the young flight surgeon calmly uttered, "Speaking on behalf of all the physicians, we think it's for the best. Better to conclude this mission with one fatality rather than two."

30

THE LONG WAY HOME

On Orbit
6:44 p.m. Friday, August 18, 1972 (REV 14 / GET: 20:29:00)

As it turned out, the extra twenty-four hours weren't even necessary. Toiling relentlessly like a man possessed, shuttling back and forth through the station while snatching tape cartridges out of the reeking air, Ourecky had succeeded in recovering all the data, and had secured all four storage cases in the Gemini-B.

His next endeavor was to load Russo into the Gemini-B. None of the training at Buck Island or on the Vomit Comet had adequately prepared him for what he thought would have been the simplest chore of all. It took over an hour to negotiate the short tunnel between the MOL's stern compartment and the Gemini-B reentry vehicle.

It took him three attempts to realize that the two of them could not transit the tunnel at the same time. After repeatedly banging his head—literally—he fastened a strap around Russo's ankles, went up into the Gemini-B, and then dragged him upwards through the tunnel. Once aboard the Gemini-B, it took another thirty minutes to lash Russo's swollen body into the left seat, using a collection of special webbing and friction buckles sent up in the transfer case.

Finally, he was faced with the moment of truth. Referring to a reference card detailing the procedures to restore power to the Gemini-B, he sucked in a deep breath, switched on the main batteries, and then depressed a series of circuit breakers. Without hesitation, the long dormant spacecraft quietly flickered to life. He slowly let out his breath, coughed several times, and then began the first phase of setting the controls for their reentry. And even though things had been hectic to this juncture, the next two hours would be a relentless race against the clock.

Satisfied that the Gemini-B was ready for reentry, he keyed the radio to speak to Carson. "Drew . . . I've loaded Russo and . . . powered up. Going back to . . . scram the reactor."

"Wait," replied Carson.

"Wait?"

"There's a new development, straight from Tew and Wolcott. They want you to scuttle the rescue, suit up, lock out, and come back over to reenter with me."

"*What?*" blurted Ourecky. His tormented body was overcome with spasms as he fell into another prolonged fit of coughing. He examined the resultant clump of thick mucus; alarmed, he saw that his phlegm was still marbled with specks of blood and bits of dark matter that he couldn't identify. Obviously, something was seriously wrong with him, not just a simple cold or passing case of the flu.

"They're sure Russo won't make it, and they're concerned that you'll probably lose consciousness. How long will it take you to kit up? It will take me at least an hour to move into position, depressurize, pop the hatch, and be ready to receive you."

He shook his head. After all the agonizing pre-mission preparations and all the work that he had already accomplished, they wanted him to *abandon* the mission? It was insane. He was *this* close. In just slightly more than an hour, the deed would be done.

Besides his unwillingness to throw in the towel, two very practical matters precluded locking out of the MOL and returning with Carson. If he expedited the process and cleared the airlock in short order, he could be back aboard the Gemini-I in a matter of minutes. Despite that, even after his umbilical hoses were reconnected, it would take at least thirty to forty-five minutes to close his hatch and repressurize the cabin. In his current state, that was an extremely *long* time to remain suited up; at the rate he was coughing up fluid, he could literally drown in his hermetically sealed helmet.

More significantly, because he had already exhausted his ELSS's supplemental oxygen tank to pressurize the airlock on his way in, he would be entirely reliant on the primary tank—rated at thirty minutes— to return to the Gemini-I. And he had already used some of the primary tank; for rough planning purposes, he estimated that he had no more than fifteen minutes remaining. He *might* make the transit with fifteen minutes of air, but that meant that there was absolutely no margin for error. It would be a huge gamble, since even the slightest fumble or misstep meant almost certain death.

Ourecky had been reluctant to disclose the incredible risk that he had taken in the airlock, even to Carson. He plainly knew that Tew and Wolcott would be furious with him if they even suspected it. If he stuck with the plan and returned with the Gemini-B, no one would ever be the wiser, since the only tangible evidence—the depleted oxygen tank on the ELSS—would remain in orbit.

Even though he had avoided it so far, he had to tell Carson. He coughed up another lump of bloody mucus, keyed the mike and stammered, "Drew . . . I can't . . . lock out."

"Why?"

"I . . . uh . . . accidently . . . vented my supplemental tank."

"You *what?* How?"

"I triggered the emergency purge valve," explained Ourecky. At least that part was factual, if not entirely revealing of the whole truth. Oddly, as he tried to paint the picture for Carson, he felt almost exactly as he did when he attempted to justify his strange conduct and prolonged absences to Bea. "Once the purge starts . . . it can't be . . . stopped."

"Oh," answered Carson. "You still have some residual in your primary tank, don't you?"

"Yeah . . . maybe . . . fifteen minutes."

Carson was silent for over a minute, then stated, "Fifteen minutes? You can make it over, Scott. It will be tight, but you can do it. I'll take you back."

"No. I'm going to . . . stick it out," vowed Ourecky, in a hoarse and raspy voice. "I'm bringing . . . Russo . . . home."

Carson was silent again, but finally said, "I wish you would reconsider, Scott, but I understand. I don't agree with you, but it's your decision and I respect it. I'll let them know downstairs. I'm sure they won't be happy with you."

Ourecky coughed several times and said, "I'll . . . live with that. See . . . you . . . in a bit?"

"Yeah, you will. Be careful, brother."

7:05 p.m. (REV 14 / GET: 20:50:12)

The reactor reentry vessel was endowed with its own computer. It was similar to the Gemini's onboard computer, except that it bore a singular purpose: once the reactor was jettisoned from the MOL, the computer would guide its reentry to ensure a controlled descent into an unpopulated remote location on Earth. Like the Gemini's computer, it was designed to automatically receive an updated navigational fix uploaded from a ground station. But since the MOL's data uplinks were on the fritz, along with the rest of the station's finicky communications gear, the positional data had to be updated manually before the reactor could be ejected.

Straining to remain conscious, Ourecky frantically punched in the last few sets of numbers. With two minutes to spare, he double-checked each entry against the data relayed by Carson. After suffering a protracted fit of coughing, he keyed the radio and said, "Drew, I've checked everything . . . let's verify the touchdown point entries just to be triple-sure . . . uh, Address 10 is Neg 16. Address 11 is 98 . . . Correct?" If the numbers were entered correctly, the reactor vessel—which was packed with radioactive isotopes that would remain hazardous for thousands of years to come—would descend under parachute roughly 1700 miles west of Australia and then sink into the murky depths of the Indian Ocean.

"That's affirm, Scott. Address 10 should read Neg 16 and Address 11 is 98. You're set."

Ourecky breathed a sigh of relief. The final step was to key the computer to accept a three-dimensional position in space, which he had previously entered, once Carson verified it with a star shot.

"Okay, Scott, I have my sextant up and I'm tracking my star now. Ready to lock in the fix and dump the reactor?"

"Ready." Holding his breath, Ourecky threw the SCRAM switch; a light flickered, confirming that the control rods had slid into position to halt the fission process.

"Stand by . . . stand by . . . stand by . . . On my mark . . . Four . . . Three . . . Two . . . One . . . Mark."

Ourecky methodically tapped the key sequence that obligated the computer to correlate the navigation fix. As a light flashed green, indicating that the computer had digested the data, he shoved the red plunger to jettison the reactor vessel.

The JETT REACT confirmation light blinked on; hopefully, the pod's explosive bolts had fired and the reactor would soon be on its long descent into the atmosphere. In any event, as he hustled forward towards his own ride home, he realized that it was futile to continue contemplating the reactor's fate, since that matter was now well out of his grasp.

Arriving at the transfer tunnel, he took a last look around and glimpsed Cowin's swollen right arm poking through a gap in his sleeping compartment's privacy curtain. Ourecky squeezed through the narrow tunnel for what seemed like the hundredth time today. Inside the Gemini-B, the accumulated data cartridges were safely stowed away in four boxes, and Russo was still blissfully comatose in the left seat. Ourecky wrestled the heavy heat shield hatch into place, locked it, and then swung the pressure hatch closed. It sealed with a satisfying click. He dogged the hatch secure, then squirmed into his own seat and fastened his restraint harness.

There were still many niggling details to attend to, but Ourecky was too tired and too busy to be apprehensive. He was now effectively on his own; Carson no longer flew in formation alongside the MOL. After the reactor was jettisoned, he had executed a minor burn to boost the Gemini-I into a slightly higher orbit. As a result of the maneuver, which was intended to create a safety margin as Ourecky made his final attitude adjustments for reentry, the Gemini-I now travelled marginally slower than the MOL. If everything went according to plan, Ourecky and Russo would touch down at White Sands less than an hour from now, and Carson would land at Edwards Air Force Base—approximately seven hundred miles to the west—roughly ninety minutes later.

Wheezing to catch his breath, Ourecky updated the Gemini-B's computer with reentry data before talking himself through the sequence that fired explosive charges to physically separate the Gemini-B reentry vehicle from the MOL. A few minutes later, after orienting the spacecraft, he watched the clock, counted down, and jabbed the button to light the retros.

The familiar thump of the first retro reverberated through his back. Within seconds, all six of the Gemini-B's retrorockets were burning exactly as designed. At this point, only one thing was absolutely certain: they might not make it home, but they were definitely leaving orbit.

After seven flights into space, Ourecky had never grown very fond of reentry, even though the fiery ordeal signified that the perilous journey was swiftly drawing to a conclusion. For the next few minutes, there was little for him to do except monitor the instruments, since the computer would control the different rolling maneuvers to generate lift. He thought of Bea and prayed that she would not make good on her ultimatum. He also prayed that he would complete his mission to ferry

Russo home alive. For good measure, he prayed for Carson's safe return as well.

Through his window, he watched the pinkish-red radiance steadily build as the heat shield absorbed friction. Soon, the rosy glimmer was replaced by a brilliant orange glow and finally by a dense rippling maelstrom of flame. A pervasive vibration set in; the shaking was so violent that it literally rattled his teeth. As the cabin heated up and the clamor rose, he continued to hack up crud. As the G's piled on, he could scarcely draw a breath; it felt like a circus elephant was perched squarely on his chest. His head throbbed, his guts churned, and his heart pounded in his chest. Rivulets of sweat trickled from his forehead, stinging his eyes and clouding his vision.

He fought to summon all the tricks he had learned at the Wheel centrifuge in Johnstown; grunting loudly, he forced himself to maintain his lungs partially inflated so that they wouldn't collapse altogether. As globs of mucus accumulated in his throat, he tilted his head to the side and choked them up. Barely lucid, he struggled to remain conscious. He knew that falling unconscious meant almost certain death, since if he failed to initiate the sequence to deploy the paraglider, the spacecraft would plummet to Earth and eventually bury itself in the desert.

Wheezing, compelling himself over and over to breathe, he focused on the instruments as he waited for the appropriate moment to dump the drogue chute. If he could remain conscious until then, just a little while longer, their chances for survival would be greatly enhanced. As a safety measure, the paraglider was intentionally rigged so that it would initially assume the "half-brakes" configuration after opening; at half-brakes, the paraglider had minimal forward airspeed, so it behaved almost like a conventional parachute, descending more or less straight down. So even if he passed out and wasn't able to steer the paraglider to the landing strip, the fabric wing should deliver them to Earth intact, and with any luck, they would be quickly found and rescued.

Outside the cabin, the undulating flames gradually diminished and were finally snuffed out altogether as the Gemini-B plunged through the upper reaches of the atmosphere. Finally, the 60K telelight blinked on. Reaching out, with his arm as heavy as lead, Ourecky verified that the landing squib bus circuit was armed. Seconds later, he pushed the button to manually deploy the drogue. Through the window, he saw the parachute stream out and inflate. Without prompting, he completed the other busywork that was part of the frenzied post-reentry sequence. He switched off the Reaction Control System—RCS—thrusters, installed the D-rings for their ejection seats, switched the computer to the Rate Command mode, and verified that his restraint harness straps were locked. He was beset by another spell of violent coughing, but kept his eyes on the instruments.

Watching the altimeter, he heard a loud *clunk* as the fiberglass cylinder that contained the paraglider automatically ejected from the blunt nose of the spacecraft. Like clockwork, the paraglider smoothly deployed. Finally, dreading this step in the process, he threw the switch to transition the spacecraft to the two-point suspension mode. Compared to the equivalent experience in the Gemini-I, he was extremely surprised—and pleased—at the relative softness of the transition in the stripped-down Gemini-B.

Since they were over dry land, with no ocean or open water within hundreds of miles, he went ahead and lowered the three skids for landing; a green light flickered on, indicating that they were fully down and locked in place. He breathed a sigh of relief; they were almost home.

As he activated the gas-powered control motors for the paraglider, he heard a faint but familiar voice in his earphones: "Trident One, this is Charger Two on Channel Three. Do you read?"

"I . . . read . . . you . . . on Channel Three," gasped Ourecky.

"Trident One, Charger Two, roger. This is Mike Sigler. Parch and I are flying chase. Ground radar reports that we are four miles north of you, approximately five thousand feet below. We are maneuvering to close, will pick up your six, and will follow you down. Have you acquired TACAN?"

Ourecky checked the TACAN board on his instrument panel; the acquisition light was green and the DME—Distance Measuring

Equipment—display was clicking numbers into place to indicate his distance from the touchdown point. "I . . . have . . . TACAN . . . on Two," he declared. Forgoing the controllability and stall checks, he steered the paraglider to pick up the heading to the TACAN beacon.

"Trident One, this is Charger Two. We know you're pretty smoked, so we're going to relay all instructions to you and help you with the landing. Okay?"

"Okay," croaked Ourecky.

"Trident One, you're on heading, but you need to burn off about a thousand feet of altitude to acquire proper glide slope," offered Sigler. "I recommend that you perform S-turns for approximately sixty seconds."

Following Sigler's advice, Ourecky gently rocked the hand controller back and forth, causing the paraglider to execute a series of long, graceful turns that maintained his general heading but depleted his excess altitude.

"Altitude looks good now," stated Sigler. "You're cleared for a straight-in approach to Runway 17. Winds are four knots out of One Nine Zero. Altimeter is 28.50. Field elevation is 3913 feet. Surface is dry lakebed. I see your gear are down. Can you verify that they are locked?"

Ourecky checked the light again. "Gear are . . . locked," he replied. "I copy . . . straight-in to Three-Five . . . winds four out of One Nine Zero . . . altimeter two-eight-five-zero . . . uh . . . field is three-nine-one-three feet, dry lakebed."

"Trident One, good copy," replied Sigler.

"Field in sight . . ." announced Ourecky a few seconds later. "I have . . . the lights . . . gear down . . . and locked. On final . . ."

"Just fly the ship, Scott," declared Sigler. "Don't talk. Save your breath. You're lined up on the strip. Just hold what you have, and gravity will do the rest."

Gravity? Groggy, eyes burning, gasping for breath, Ourecky struggled to remain alert. Watching the marker lights march towards him, he verified his glide slope, confirmed his alignment on the runway and center-indexed the hand controller. With his task all but completed, his consciousness seemed to gradually fade away, and then he was gone.

Now, gravity *did* have the controls, at least for the few remaining seconds of this mission.

Northrop Strip, White Sands Missile Range, New Mexico

Gazing at the northern sky through binoculars, watching the Gemini-B's approach on short final, Admiral Tarbox waited on an elevated wooden platform that overlooked the landing area. A senior enlisted Air Force firefighter, responsible for coordinating the emergency response actions at the site, stood alongside him.

Alarmed, Tarbox abruptly realized that the landing was not going exactly to plan. For whatever reason, possibly because he was unconscious, Ourecky didn't flare or jettison the paraglider as he touched down, so the reentry vehicle still had a considerable amount of forward speed as it made contact. Accompanied by a terribly shrill screech, the three skids threw off a shower of sparks as they scraped the packed gypsum of the runway. The spacecraft plowed fairly straight for a hundred yards and then veered at the last moment, finally coming to rest at the right edge of the marked runway. Apparently ignited by the glowing heat shield, the nylon paraglider slowly caught fire as it fluttered behind the vehicle. Overhead, a pair of T-38 chase planes buzzed by before swooping skyward.

A small fleet of emergency vehicles converged on the returned capsule. Normally, their procedures called for the emergency personnel to wait at least thirty minutes to allow the spacecraft to cool down before making an approach, but this situation was far from normal. A pair of boxy red O-11 crash-rescue firefighting trucks rumbled in close; one drenched the scorching spacecraft with a dense blanket of "A-Triple F" firefighting foam and the other sprayed water to extinguish the smoldering paraglider. Dripping with white froth, billowing clouds of scalding steam, the spacecraft's still-sizzling metal shingles popped, crackled and squealed.

A crash-rescue truck's water cannon doused the Gemini-B with a cooling stream as a thick-skinned M113 armored personnel carrier joined the armada. The rear ramp of the hulking troop carrier hissed down, disgorging a squad of firefighters. Encased in heat-reflective "proximity" protective suits, the rescuers bore an arsenal of crash axes and specialized tools. Resembling faceless aliens from a science fiction movie, the silver-garbed men cautiously approached the spacecraft from the front. One removed an emergency hatch-opening tool from a bracket in the blunt nose. He inserted the tip of the bar-shaped tool into a small receptacle in the right hatch and vigorously twisted it to actuate the unlocking mechanism. In seconds, the firefighters had popped open both hatches and were reaching inside.

Staring through his binoculars, Tarbox impatiently monitored the proceedings. He suddenly realized that the Air Force firefighters were risking their lives to gain access to the crew. Since Ourecky had not flared and released the paraglider upon touchdown, probably because he had lost consciousness, it was a virtual certainty that he had also not safed the ejection seats and various pyrotechnics in the cabin. A minor blunder was more than adequate to detonate any of a multitude of explosive charges. If the ejection seats spontaneously fired or the hatches inadvertently blew open, the resultant force could immediately kill the rescuers.

The firefighter standing alongside Tarbox listened to a terse message on his handheld radio, acknowledged it, and then reported, "Admiral, they're both alive but unconscious."

Lowering the binoculars from his eyes, Tarbox relaxed for the first time in several hours. "Thank you," he replied quietly, letting the binoculars dangle from the lanyard around his neck.

Even as the emergency workers carefully extracted Ourecky and Russo from the Gemini-B, a UH-1 "Huey" medevac helicopter was arriving to ferry them to the hospital at nearby Holloman Air Force Base. The transfer was quick; only moments later, the helicopter clattered into the air and departed towards the east. In the west, the sun was falling under the horizon. Night would come soon; in anticipation, a crew of workers set up portable generators and light sets to illuminate the spacecraft for the detailed recovery operations that would go on until morning.

Anxious to take possession of the cases that contained the intelligence data accumulated during the MOL's short-lived mission, Tarbox clambered down the ladder and strode towards the spacecraft. Midway there, he paused to reflect on what had just occurred. With the knowledge that the two men were alive, he breathed a sigh of relief.

Thinking of how Ourecky had risked his life to bring Russo home, he smiled to himself. The indefatigable whiz kid had delivered the mail more times than the Pony Express, but had never sought any recognition for himself. Tarbox wished that there was a way that Ourecky and Carson could be publicly recognized for their bravery. It could *never* happen, but then again . . . *maybe* it could. He grinned as a tiny seed of an idea took root in the back of his thoughts. Farfetched, the oddball idea would have to wait, but it was one that might answer the fondest wishes of both men. Moreover, thought Tarbox, looking towards the Gemini-B, it might advance *his* cause as well.

Aerospace Support Project
6:15 p.m., Sunday, August 20, 1972

With his hands on his hips, Tew stood behind his desk and gazed out his small window at a tranquil blue Ohio sky. The past couple of weeks had been momentous, although fraught with some very trying moments. There was much to be thankful for, but still plenty to mourn over. But although Tarbox had lost a man in orbit, *his* men were safely on the ground.

Outside, the weather could not be any more beautiful or inviting. It seemed ironic, considering that they were awaiting news of the damages wrought by Hurricane Celeste when it passed just twenty-five miles north of Johnston Island yesterday. It was now morning there, but a preliminary report wouldn't be available until a recon flight, flying out of Hawaii, passed over the island.

Even if the damages appeared to be minor, reoccupying the evacuated atoll would still be a laborious process. Besides the PDF launch facility, the island was home to a chemical weapons stockpile relocated from Okinawa. Before most of the island's five hundred workers could return, rubber-suited chemical warfare specialists would sample the air and physically check the casks inside the storage igloos to ensure that they weren't breached or weakened by the storm.

The door creaked open. Looking as if he had just witnessed a massive train wreck, Wolcott trudged slowly into the office. He plopped his Stetson on his desk, sat down, opened a drawer, and pulled out a bottle of Jack Daniel's finest. Opening another drawer, he extracted two shot glasses and filled them with whiskey. He looked towards Tew and said, "Tragic news, pard. Fancy some anesthesia? It'll sure make this go down easier."

"Thanks, but no," replied Tew. "And isn't it a bit early for you to be drinking, Virgil?"

Wolcott threw back one shot and replied, "Not today, it ain't. And if you ain't partakin', pardner, it would be a shame to let this other one go to waste." He gulped down the other shot and wiped his mouth with the back of his hand.

"Bad news?"

"Well, we have one more stack, but we ain't got a place to launch it from," replied Wolcott somberly, handing Tew a folder. "Read it and weep, pard. That bitch Celeste slammed Johnston Island much harder than anyone anticipated. On a good note, the chemical weapons stockpile looks like it weathered the storm, but our end of the Island got hammered. Our pad was leveled and the dock facilities were washed away. The Project 437 Thor gantry was flattened, too."

Tew nodded solemnly as he scanned the report and examined an electronically transmitted photo taken by a reconnaissance jet hours earlier. The stark image corroborated Wolcott's bleak assertions. The coral island's hardpan surface had been scoured by gusts exceeding a hundred miles an hour. High winds and heavy tide had scraped most of the PDF complex—including the gantry, the massive concrete launch pad and the dock facility—into the surf. The sandblasted blockhouse appeared to be still intact, or relatively so, but the sensitive electronics that it housed had assuredly been ruined by floodwaters and drifting sand. He expected it to be bad, but not quite *this* bad.

"I guess you know that it's going to take us months to rebuild," noted Wolcott. "Of course, that's assumin' that we secure funding for the second phase. And that's contingent on . . ."

"Virgil, we're *not* rebuilding," declared Tew. "We're *done*."

"Done? You're just going to surrender the fight and scuttle this thing?"

"I am. If we're ordered to rebuild the PDF and continue Blue Gemini, then I'll do my utmost to make that happen. Otherwise, I am not pursuing any effort to prolong this, and I'm asking you to do the same."

"But . . ."

"No," interjected Tew emphatically, not displaying even the slightest inclination to budge. "We've run these men through the wringer, Virgil. Not just Carson and Ourecky, but all of them, every single one. It's an absolute miracle that we've only killed two men in this process. It's time to shut it down, and if this damned hurricane isn't the omen that causes you to see that, I don't know what will."

Tew slowly walked over and put his trembling hand on Wolcott's shoulder. In a faltering voice, he said, "*Please* accept that it's over, Virgil. My doctors tell me that I have less than a year to live, but they also said I have less than a month if I don't abandon this pace. After the dust settles, I want to go home and live out that year. Can you grant me that? We made it to orbit, Virgil, which is what we always wanted. Isn't that enough?"

"Yeah, Mark," muttered Wolcott, pivoting his head to look up at his friend. "I s'pose you're right."

Lackland Air Force Base, Texas
12:15 p.m., Wednesday, August 23, 1972

Carson strolled down the hospital corridor and nodded at a security guard posted outside the hospital room occupied by Ourecky and Russo.

He quietly entered the room, eased into a chair by his friend's bedside, leaned back against the wall, and opened the daily San Antonio newspaper. He had just returned from a brief visit to the base headquarters, where he had a brief conversation with Virgil Wolcott via a secure telephone.

Shortly after Ourecky and Russo touched down at White Sands, after they had received preliminary treatment at Holloman, they were flown to the Air Force's flagship hospital—Wilford Hall Medical Center at Lackland Air Force Base, Texas—for more definitive care.

After his solo landing of the Gemini-I in California, Carson took possession of a T-38 and immediately zoomed to Lackland. He pledged to remain at Wilford Hall for the duration, until Ourecky was sufficiently healthy to return to Wright-Patt. During their phone call, amongst other issues, Wolcott had authorized Carson to remain in Texas for as long as he saw fit.

Lifting his arm, Carson quickly took a whiff of his left armpit. He was getting a bit ripe. Since arriving here in a borrowed flight suit five days ago, he had slipped over to the base exchange to purchase a pair of khaki chinos, a knit sport shirt and a cheap pair of canvas deck shoes.

He read the front section of the paper, set it aside, and looked up to watch Ourecky. Attired in hospital issue blue pajamas, with his mouth and nose covered by a translucent oxygen mask, the sleeping engineer really looked none the worse for wear. Of course, that was merely an illusion, since Ourecky still was very much on the mend.

Shortly after arriving in Texas, Carson finally learned what had happened when his friend entered the MOL. It was his first inkling of what the engineer had faced after his umbilical was severed, as well as the life-and-death decision that he had made to continue with the mission. Carson had been shocked to learn that Ourecky had willingly sacrificed his supplemental oxygen to repressurize the MOL's airlock, but was happy that there was finally a logical explanation for his seemingly inexplicable sudden decline in health.

In the first breath he drew before leaving the airlock, Ourecky had unknowingly exposed his respiratory system to a frigid atmosphere that was nearly two hundred degrees below zero. As tests would later show, the momentary exposure to super-cooled air had flash-frozen a substantial number of bronchioles and alveoli. That explained Ourecky's persistent cough for the first few hours aboard the MOL, an annoyance that swiftly deteriorated into a hacking cough and inflamed lungs. When describing his phlegm in orbit, the "little particles" he saw were in fact granules of dead tissue than had begun to slough off from the interior of his lungs.

Shortly after he had arrived at Lackland, after sheepishly recounting his ordeal in the airlock, Ourecky had been subjected to an extensive battery of chest X-rays and other tests. The doctors diagnosed him with severe bronchitis and confined him to a hospital bed where he received a daily regimen of painful respiratory treatments and a course of broad-spectrum antibiotics to stave off further infection. Although Ourecky's coughing grew progressively worse as he expelled still more dead tissue, the attending pulmonologist assured Carson that the lingering inflammation would clear up in a few weeks.

Carson looked towards the opposite side of the room, towards Russo. Except for an occasional twitch or faint sputter, Russo could readily pass for dead, but even though he had only briefly regained consciousness twice since returning to Earth, he was also healing. Although his situation was still dire, he was expected to eventually recover. According to the doctors treating him, although Russo would probably never regain his full eyesight and would suffer lasting health effects as a result of his radiation exposure, his overall prognosis was fairly positive.

Slowly stirring, Ourecky pushed his oxygen mask to the side and weakly spoke. "Hey. Where did you go?"

"Hey yourself, buddy," replied Carson, folding the newspaper before sticking it under the chair. "I had to go over to the commo shop at the base headquarters to hear the latest news and gossip from our friend Virgil."

"Anything . . . new?"

Nodding, Carson leaned over the bed and quietly said, "They received the recon report from Hawaii."

"And?" asked Ourecky. "Is the PDF okay?"

"Nope. It's kaput. *Defunct*. No more," answered Carson. "Almost totally destroyed."

"You're kidding," muttered Ourecky. His expression seemed to suddenly brighten at the news. "It's *gone*? That means . . . that means . . . we're *done*?"

"Well, not quite. There's still another stack, and Virgil isn't too inclined to let it go to waste. He said that they're looking at alternatives for contingency missions."

"Oh. For a moment, I thought we were off the hook."

"Well, *you* certainly are, at least for the time being, until you heal up," replied Carson. "But you know, Scott, your overall stats really aren't looking too great. You've flown eight missions, and landed in a hospital bed twice. Maybe you should just fly in pajamas from here on."

"Funny. You think that the suit techs could sew me some PJs out of Nomex?"

Carson heard a noise in the hall, nudged Ourecky's shoulder and quietly announced, "Your lunch is on its way."

Ourecky coughed several times and hoarsely muttered, "Oh, great. More Jell-O and cottage cheese. Yum."

Glancing towards Russo, Carson said, "At least you're awake and not getting your chow through a tube."

Ourecky nodded weakly. "I *am* thankful for that."

"I believe that Ed's thankful for it also. I think that he's also probably thankful that you weren't willing to throw in the towel and abandon him upstairs. I would be, anyway."

Carson looked towards a wheelchair parked against the wall. "Hey, Scott, do you want me to roll you downstairs to that payphone again today? You can try to call Bea again. Maybe you won't miss her *this* time."

"Maybe," replied Ourecky glumly, looking towards the wall. "Thanks, but I'm wondering if Delta hasn't changed her flight schedule again. Besides, I'm really not looking forward to explaining where

I've been or why I'm coming home with another mystery medical condition."

"Sorry." Carson studied Ourecky's suddenly circumspect face. They had worked together long enough that he knew when his friend was hiding something from him, and Ourecky's abrupt change in expression telegraphed that something was worrying him, but he was obviously not yet ready to divulge his concern.

Carson decided to change the subject, in hopes that it might brighten the somber mood. "Hey, Scott, you know that nurse on the early morning shift?" he asked, grinning as he twisted the right end of his moustache. "The one that comes in around seven?"

"The really cute little redhead?" replied Ourecky. "Jeanne?"

"Yeah. That's the one. Turns out that she shares a house with a couple of the other nurses, and they have their own washer and dryer. She said I could come over and wash my clothes, so I was thinking . . . If you don't mind . . ."

Ourecky rolled his eyes, coughed, and said, "Some things just never change . . ."

31

NO GOOD DEED . . .

Simulator Facility
Aerospace Support Project, Wright-Patterson Air Force Base, Ohio
11:25 p.m., Saturday, August 26, 1972

Even though Ourecky and Carson weren't scheduled to arrive from Texas until nearly midnight, and even though it was the weekend, the entire Blue Gemini staff jammed into the simulator hangar to wait for their return. Like most of those present, Gunter Heydrich was very tired, almost to the extent of sheer exhaustion; the past three weeks had been an emotional roller coaster and had worn deeply on him.

Certainly, as the two men walked in, the air burst into echoing applause and heartfelt hurrahs, but compared to similar such gatherings in the past, this homecoming was a rather somber affair. The hastily conceived rescue mission was a resounding success, a momentous triumph, but most present were painfully aware that they had almost lost Ourecky—the quietly tenacious engineer beloved by all—in the effort.

More so than anything else, an underlying air of uncertainty drew a damper on their collective enthusiasm. Although the Project still had one more Titan II/Gemini-I combination waiting at the HAF, the dedicated PDF launch complex had been obliterated. Adding to the insecurity were the presidential elections looming in November. The nation was weary of the slogging war in Vietnam, and it looked as this collective discontent might yield the Democratic candidate— Senator George McGovern—a viable shot at occupying the White House. And that possibility didn't bode well for Blue Gemini, since there was virtually no chance of a second phase if the Democrats won the Presidency. Depending on the timing, there was a very strong possibility that the twelfth and final mission of the first phase might not even fly.

Consequently, most of the people present—at least those civilians who worked under some form of contract—had little idea of what the future held. The only thing that *was* certain was that after achieving the goal of successfully landing men on the moon, NASA's manned space program was being significantly curtailed. The last mission to the lunar surface—Apollo 17—was scheduled to fly in December. The remaining three missions had been cancelled, and except for a makeshift space station—Skylab, cobbled together of recycled Apollo parts, scheduled to launch next year—the future of spaceflight was vague and uncertain. Aerospace engineers, who had spent the past decade as commodities in high demand, were suddenly finding themselves part of an ever-expanding surplus of talent. For the first time in their professional lives, most of the brilliant men packed into this hangar were facing the specter of prolonged unemployment.

After the initial applause, the remainder of the reception ceremony was quiet and understated. Apparently in deference to Ourecky's frail condition, Mark Tew decreed that there would be no elaborate celebration or lengthy speeches. Heydrich escorted the two returning heroes to the center of the hangar, to the base of the short platform that supported the Box, where they were seated to receive those waiting to greet their homecoming.

As the staff queued up, Heydrich glanced down and noticed some faint blotches on the painted concrete, and realized that the stains were Tim Agnew's blood, marking the spot where he had collapsed to the floor during a break over three years ago.

As the crowd diminished and the well-wishers faded away to make their way home, Tew and Wolcott were among the very last to offer their congratulations.

"It looks like you took the long way home," said Wolcott. "How are you feeling, hoss?"

Ourecky coughed, nodded his head, and replied, "Much better, sir."

"Well, that's good. I can't say that I approve of all the risks you took, but there ain't any arguing with the results. We're beholden to you, son, as always."

"We'll catch up later," said Tew, bending forward to weakly embrace Ourecky's shoulders. "Go home now. Go see your wife and son."

Heydrich noticed that tears were welling in Ourecky's eyes and assumed that he was anxious to be home. Watching as Tew and Wolcott walked away, he said, "I'm so glad that you two gentlemen are home, safe and sound."

"Thanks, Gunter," replied Carson, pushing himself out of his chair and stretching. "We couldn't have done it without you and your guys."

"We owe you," added Ourecky. "As always."

"Ready to hit the road, Scott?" asked Carson. "Ready to put this trip behind us?"

"I am," sighed Ourecky, allowing Heydrich to help him to his feet. "Very much so."

Dayton, Ohio
1:07 a.m., Sunday, August 27, 1972

After being feted for their momentous accomplishment, riding home in Carson's Corvette, Ourecky was absolutely wrung out. It was long past midnight as they pulled up to Ourecky's house. His heart sank when he saw that the house was dark, definitely an ominous sign, but he didn't want to alarm Carson.

"Here we are, home sweet home," noted Carson, looking towards the house as the Corvette idled. "It's late. I guess Bea's already gone to bed."

"Yeah. Probably," replied Ourecky hoarsely. "Anyway, I appreciate the ride, Drew. Man, I feel like crap."

"Get some sleep, Scott," said Carson.

Dog-tired, Ourecky stiffly climbed out of the car and waved good-bye. Turning towards the front porch, he saw that the yard was long in need of mowing. As he heard Carson pull away, he hacked up a clump of mucus and spat it out in the shaggy grass. On wobbly legs that felt like old rubber bands, he slowly staggered up the concrete steps. He felt horrible; his throat was sore and his chest still ached.

He groped for the extra key under the welcome mat, found it, opened the door, and went in. After turning on the lights and walking around the house, it didn't take long for him to painfully realize that Bea had made good on her promise.

Her pictures were gone from the walls and fireplace mantel. Her clothes were gone from the bedroom closet and her dresser. Andy's room was cleaned out as well: no bed, no clothes, no toys, no son.

She left a tersely worded letter on the coffee table. It contained no anger or animosity, but was just a clear statement of facts. In it, she told him that she had gone to stay with her friend Jill, who had been tentatively diagnosed with ovarian cancer. She said that he was welcome to visit whenever he wanted, but that she would not return to him until his circumstances changed. Except for the short note, every vestige of his wife and his son was gone.

Famished, he went to the kitchen to forage in the cabinets. The cupboard wasn't exactly bare, but the pickings were mighty slim. He slathered two slices of stale bread with peanut butter and honey, sat at the table, and slowly ate his sandwich.

He was almost surprised that he didn't feel angry. He expected this, although he had hoped against hope that it wouldn't come to this, but he also understood why she felt like she did. If anything, he was angry with himself and frustrated with his impossible circumstances.

He resolved himself to respect her wishes and grant her some time to think, but at this point, he was just too numb to contemplate the situation. He knew that it might not have been like this, but he made his choices and now he would have to live with the consequences. Slightly more than a week ago, he knew what it was like to be the loneliest man in the universe; now, he had to come to grips with being the loneliest man on Earth.